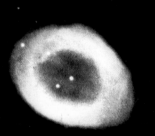

The Ring Nebula of Lyra is a shell of gas expanding about a very hot star. (*Hale Observatories.*)

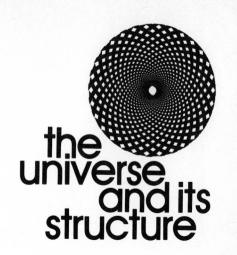

the universe and its structure

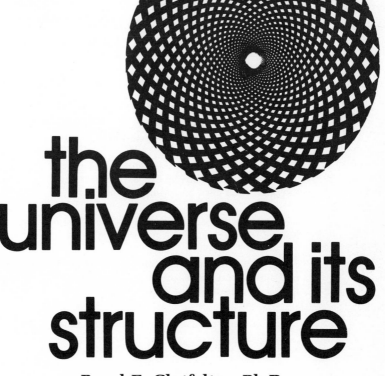

the universe and its structure

Beryl E. Clotfelter, Ph.D.

Williston Professor of Physics
Grinnell College

McGraw-Hill Book Company

New York   St. Louis   San Francisco   Auckland   Düsseldorf   Johannesburg
Kuala Lumpur   London   Mexico   Montreal   New Delhi   Panama   Paris   São Paulo
Singapore   Sydney   Tokyo   Toronto

THE UNIVERSE AND ITS STRUCTURE

234567890 KPKP 79876

This book was set in Melior by Textbook Services, Inc.
The editors were Robert A. Fry, Janis M. Yates, and Michael LaBarbera;
the designer was Jo Jones;
the production supervisor was Leroy A. Young.
The drawings were done by Eric G. Hieber Associates Inc.
Kingsport Press, Inc., was printer and binder.

Cover art reproduced with permission from "The Spiral Way" by Gerald Oster, *Natural History* Magazine, August-September, 1974.

The photographs used on the endpapers of this text are used with permission of the Hale Observatories. Copyright by the California Institute of Technology and the Carnegie Institution of Washington.

Library of Congress Cataloging in Publication Data

Clotfelter, Beryl E   date
    The universe and its structure.

    Bibliography: p.
    1. Astronomy. I. Title.
QB43.2.C57   523   75-6714
ISBN 0-07-011385-8

contents

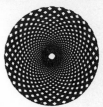

Preface   vii

SECTION I   MEN, PLANETS, AND IDEAS   1

  1. Introduction   6
  2. The Earth in Space   29
  3. The Solar System   46
  4. Mechanics   82
  5. Optics   102
  6. Nuclear Energy   137

SECTION II   STARS AND GALAXIES   146

  7. The Sun   153
  8. The Stars—Appearances   171
  9. The Stars—Distances and Intrinsic Properties   183
 10. Double Stars   198
 11. Variable Stars   211
 12. Our Galaxy   224
 13. Stellar Evolution   242
 14. The Universe of Galaxies   262
 15. Life in the Universe   282

SECTION III   THE OLD, THE NEW, AND THE UNKNOWN   298

 16. The Cosmological Problem—A First Look   304
 17. Quasi-stellar Objects   316
 18. Pulsars   330
 19. Primordial Blackbody Radiation   343
 20. Geometry of Space and Space-Time   353
 21. Black Holes   375
 22. Matter-Antimatter   387
 23. Gravitational Radiation   394
 24. Cosmology—A Second Look   401

Appendix 1   Physical and Astronomical Constants   415
Appendix 2   Nomenclature of Astronomical Objects   416
Appendix 3   The Constellations   418
Appendix 4   Twenty Brightest Stars   420
Appendix 5   Stars within 5 Parsecs   422
Bibliography   426

Index   433

v

preface

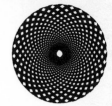

This book is based on a one-semester course designed for students who are not majoring in science, who have slight science and mathematics backgrounds, and who take the course by choice rather than to fulfill a requirement. The mathematical competence assumed is that provided by high school algebra, and no knowledge of physics is assumed. Because this course is possibly the only science course the students will take in college it attempts to show science in its cultural setting by presenting some of the history of astronomy, particularly the Copernican revolution, and to present science as an ongoing process rather than as a catalog of facts. The emphasis throughout this book is less on facts about astronomical objects than on the methods by which those facts are obtained and the uncertainties in our knowledge of them.

The most important difference between this book and other introductory astronomy texts is the emphasis given here to very modern research topics, particularly those with a bearing on cosmology. Nine of the twenty-four chapters deal with subjects which have developed entirely or been substantially invigorated since 1960. Several reasons may be suggested for giving more emphasis to these modern topics than has been common in astronomy textbooks. One reason is that some of the ideas are familiar to many students. Quasars, pulsars, and black holes, in particular, have been the subjects of so many articles in newspapers, newsmagazines, and Sunday supplements that many students know the names. This book attempts to present the problems associated with those concepts and to review recent work on them in such a way that the reader is prepared to understand the significance of further research that may be reported in the future. Surely a legitimate goal of such a book is preparing its readers to read with comprehension the science articles in the popular and semitechnical press.

A second reason for giving great emphasis to topics which are "hot" research areas is that study of incomplete and evolving areas can convey a better feel for the nature of science and scientific progress than focus on more settled areas can. No explicit discussions are given of the scientific method, but the ways in which scientists formulate and attempt to answer questions are presented in many specific examples.

Finally, a third reason for the relatively large amount of space devoted to current research topics is that students find this material interesting. One must be blasé indeed not to be **vii** fascinated by black holes, curved space, and radiation which originated almost at the beginning of the universe. Careful attempts have been made in discussions of all such topics, however, to emphasize experimental results, and questions which can be answered experimentally, rather than speculations.

All the standard topics of astronomy texts and courses are included here, but some are compressed to make space for the more modern material. On the other hand, current research related to some of the traditional topics is included in unusual detail, as in the unit on the sun, which includes a discussion of the solar neutrino problem. The book can be used for a traditional course, with the later sections used as "frosting on the cake," or an instructor may emphasize the modern parts and give less attention to some earlier sections. In my own course I spend approximately equal amounts of time on each of the three sections of the book.

I wish to thank my wife, Mary Lou, and my son, David, both of whom have read parts of the manuscript and offered helpful comments. It is also a pleasure to acknowledge the valuable criticisms and suggestions made by Professor Michael Shurman of the University of Wisconsin at Milwaukee, and by Professor Frank Bash of the University of Texas.

Beryl E. Clotfelter

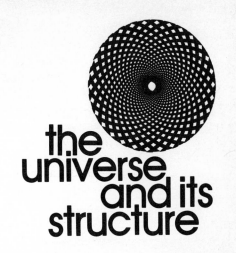

the
universe
and its
structure

men
planets
and
ideas

one

In a time long forgotten, in the childhood of mankind, a man recently become a man looked at the sun and wondered about its movement. That observer's descendant points a telescope at the heavens and photographs objects which were ancient when the sun formed. That early human looked at the moon and wondered what it is; his great great grandchild walks on the moon. And some unknown early astronomers noticed that certain stars in the sky wander; their descendants have sent television cameras to those planets.

The way has been long from those unspoken questions about the sun and moon, and we have passed many landmarks on the road from that starting point to our position. Along the road were the ziggurats of Babylon, where priest-astrologers observed the heavens. Along the road were the astronomers of Egypt who watched the rising of Sirius in order to predict the flooding of the Nile. And perhaps somewhere along the road were the unknown builders of Stonehenge. The road was long and progress was slow.

The road still stretches ahead of us endlessly, but within the past few hundred years the rate of passage has quickened, so

2

1. The moon. A composite of photographs taken at first and third quarters. (*Lick Observatory photograph.*)
2. Polar star trails, formed by pointing a camera at the North Star and leaving the shutter open for several hours. (*Lick Observatory photograph.*)
Spiral reproduced with permission from "The Spiral Way" by Gerald Oster, *Natural History* Magazine, August-September, 1974.

that now we seem to be hurry-
ing forward. We perhaps acquire more information about
the heavens in a day than ancient astronomers did in a mil-
lennium, but we press eagerly
forward, driven by the same curiosity and the same kinds of
questions which inspired our
ancient forebears.

The most primitive observation of the heavens reveals patterns that repeat cyclically: The
sun rises and sets daily; the
moon goes through its phases in
about 29 days; the position of
the rising sun on the horizon
changes in the course of a year.
Substantially more sophisticated observation reveals patterns in the repetition of
eclipses of the sun and the
moon. Observing, recording,
and recognizing the cycles undoubtedly was the first kind of
astronomy, and it required few
and simple concepts.

Human beings are not long
content with descriptions. They
want to know *why* things move

3

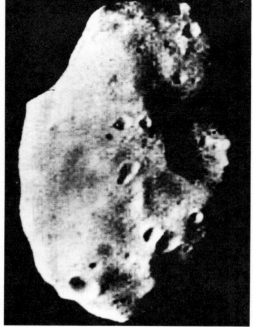

3. Venus, seen by Mariner 10 television
cameras on February 6, 1974, from a distance of 720,000 km. The photograph was
taken in invisible ultraviolet light. (*Jet
Propulsion Laboratory*.)
4. Phobos, the inner of Mars's two
moons, taken by Mariner 9 from a distance of 5,540 km.. (*Jet Propulsion Laboratory*.)

4

as they do, and so they construct models—fanciful descriptions of objects in the heavens that cannot be directly examined. The sun is a chariot of fire, drawn across the vault of the heavens each day. Or the earth is a sphere, and the sun is a ball of fire, circling the earth. The planets are stars moving on circular paths about the earth. Trying to explain the apparent motions of the planets, Greek astronomers first applied mathematics to science in a way we can recognize as similar to what we now think of as "modern" science, and from the time mathematics was applied to astronomy to make it the first science worthy of the name, the rate of progress along the road quickened. The concepts the Greek astronomers introduced into the model of the heavens were mathematical concepts: circles, angles, lines, and moving figures.

In time, however, observers wanted to make their models

6

5

7

5. The comet Arend-Roland, photographed April 30, 1957. (*Lick Observatory photograph.*)

6. Jupiter, photographed in blue light with the 200-in Hale telescope. The great red spot, atmospheric bands, the satellite Ganymede, and Ganymede's shadow on the planet are visible. (*Hale Observatories.*)

7. Saturn, photographed in blue light with the 100-in telescope. (*Hale Observatories.*)

8

more real, and they hypothesized that the planets and the
moon are bodies similar to the
earth. They introduced into the
description of the heavens new
concepts, those appropriate for
mechanics: mass, gravitational
force, acceleration, inertia.
Again, the introduction of a new
set of concepts increased the
rate of progress along the road.

For uncounted millennia
people had looked at the heavens and observed the stars and
the sun and moon and planets,
but what they could see depended upon the quality of their
eyesight. When the telescope
was invented, and after that the
spectrograph and other vision-
extending equipment, the rate
of progress changed from a
stroll to a brisk walk. Now familiar sights could be seen with
new detail, and things never
before seen could be observed.
A planet was no longer merely a
point of light or a massive object, but a world with surface

8. The 120-in reflector at Lick Observatory. (*Lick Observatory photograph.*)
9. The tower telescope at Sacramento
Peak Observatory. This telescope is used
only to study the sun. Sunlight enters the
tower through a 30-in window 136 ft
above ground level, passes downward
through a 350-ft tube from which the air
has been evacuated, strikes a 64-in concave mirror, and returns to the observing
level to form an image of the sun 20 in in
diameter. The telescope is designed to
show extremely small features in the
sun's photosphere and chromosphere.
(*Sacramento Peak Observatory, Air
Force Cambridge Research Laboratories.*)

9

detail and an atmosphere, and the composition of the atmosphere and the temperature of the surface could be determined.

Observers who would truly see the heavens must take with them when they look up not merely a chart of the constellations, and not merely a telescope, but also an assortment of concepts with which to interpret what is seen. A truly knowledgable observer must understand mass and doppler shifts, centripetal force and blackbody radiation, parallax and absorption spectra. In this part of the book we supply the reader with that assortment of concepts needed to look at the heavens with understanding. We also review the changes in the understanding of our solar system home, where planets have changed from lights in the sky to worlds with weather and erosion, and where the earth has changed from the focal point of creation to one of the planets.

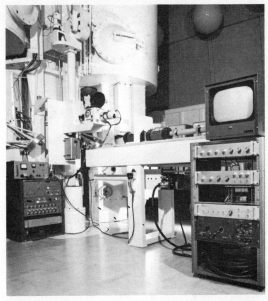

10

10. Interior of the tower telescope at Sacramento Peak Observatory. (*Sacramento Peak Observatory, Air Force Cambridge Research Laboratories.*)

11. The Arecibo Observatory. The reflecting surface of the radio telescope is 1,000 ft in diameter and is immovable, but the direction in which the telescope "looks" can be varied about 15° from the vertical by movement of the detector. (*Office of Public Information, Cornell University.*)

11

1
introduction

Science and the humanities are often presented as antithetical, but in reality nothing is more genuinely humane than science. The desire to understand and explain the world around them has been one of the most characteristic traits of homo sapiens, and as the method has changed from creation of myths to analysis of starlight and creation of models the demand upon imagination and creative human thought has increased rather than diminished. Surely the concepts used to explain the universe now—black holes, white holes, curved space, an expanding universe, etc.—are more bizarre and make more demands upon the imagination than the products of the mythmakers of ancient times. The structure of the universe does not present itself to be perceived by the casual observer, nor does the evolution of the universe reveal itself at a glance. What we know about the structure and evolution of the universe is learned by imagining models, studying the models, and then comparing their properties with what can be seen of the real world. In the end we can never know whether we are correct in our description; we can only know sometimes that we are wrong.

Since ancient times the view of the universe held by educated people has undergone a radical revision. The universe is no longer a realm of caprice, moved by the whim of gods, but a world governed by laws which can be discovered. From a closed box of small size, it has become an almost limitless expanse of space. The elements in it have changed from the few planets and few thousand stars which can be seen by the naked eye to the billions of billions of stars which make up uncounted galaxies. This knowledge of the extent and nature of the universe provides a substrate for the thinking of literate human beings, and almost certainly it is an important and perhaps an essential ingredient in the world view which is characteristic of the technological cultures of the twentieth century.

Fundamental changes in the prevailing view of the nature of the universe have occurred several times in the past. In the seventeenth century the fact was discovered that the same laws of mechanics which operate on the surface of the earth also govern the motion of planets. In the nineteenth century the distances to stars first began to be measured and the compositions

6

of stars were determined, with the result that for the first time
we knew that the stars are made of the same elements as the
earth. In the early twentieth century the fact that the sun is not
at the center of the galaxy was first recognized, the expansion of
the universe was discovered, and the theory of general relativity was revealed as a better explanation of gravitational phenomena than Newton's theory. Now in the last half of the twentieth century discoveries have been made which may force further significant changes in our view of reality. Probably nothing
so dramatic as displacing the earth and our solar system from
the center of the universe will occur, but tentative explanations
are being offered for the behavior of the nuclei of some galaxies
and of quasars which are fully as dramatic (if less traumatic) as
that.

The history of astronomy is a significant part of the intellec-
tual history of humanity, and no book on the current view of
the structure of the universe would be complete without some
mention of the most significant changes of the past. The discussion here must be so short, however, that the most which can be
hoped is that it will pique the curiosity of the reader and send
him or her to more complete accounts.

ASTRONOMY BEFORE COPERNICUS

The earliest astronomical observations may have been inspired
by curiosity about the regular changes visible in the heavens or
by the needs of navigation, but the first systematic study of astronomy over periods of time probably was produced by the
need for a calendar. The most obvious change in the sky which
is useful for marking time—other than the succession of days
and nights—is the phases of the moon. As soon as human
beings began to engage in agriculture, however, it became
desirable to be able to predict the passage of the seasons. In
other words, the year, marked by the passage of the earth
around the sun, became more important than the month,
marked by changes of the moon. Unfortunately for the early astronomers the moon does not go through its phases an integral
number of times in a year. Therefore if the length of the month
is determined by the passage of the moon through all its phases,
the number of months in a year will be 12 and a fraction more.
If one tries to combine a measure of time based upon lunar
changes and one based upon the year in some sort of calendar,
it is necessary to make the months of unequal length, or let
some years have 12 months and others have 13, or in some
other way take into account the fact that there are not an integral number of lunar periods in a year. Devising such systems
required careful observations over a number of years, and the

need for a calendar was one of the most important reasons that such observations were made. Later, as calendars became more accurate, the fact that the year does not contain an integral number of days also caused problems, and our present calendar with its use of leap years is a reasonably satisfactory solution.

No matter how beleaguered the early calendar makers may have thought themselves because of their problems fitting months and days into years, the results of their work were very fortunate. Sufficiently detailed observations were carried out and recorded that several hundred years before the beginning of the Christian Era, astronomers in Mesopotamia, Egypt, and China had determined the length of the year with fair accuracy. At the same time that information leading to the calendar was being gathered, observations were made of eclipses, the motion of the planets, and, at least in the case of the Chinese, of novae, which they called "Guest Stars."

Although a considerable body of astronomical knowledge was developed in the Far East, it contributed little if any to the development of similar knowledge in Europe. The knowledge of Egypt and Mesopotamia provided the foundation for work in Greece, however, and through the Greeks it had an influence on later European thought.

Greek Contributions

Without attempting to indicate a line of development or to be at all complete, we shall mention a few high points in the development of early Greek astronomical thought. It should be noted that much of Greek astronomy was lost in the destruction of the library at Alexandria about A.D. 390 and important works may have existed of which we are totally unaware.

Pythagoras of Samos, who lived from about 580 to 500 B.C., was one of the first to recognize that the earth is round. Parmenides, in about the middle of the fifth century B.C., conceived of the universe as a continuous series of concentric spheres with earth as the center. Philolaos of Croton, a contemporary of Socrates, described the universe as having at its center the central fire (also referred to as the hearth of the universe), with 10 bodies rotating about it—the counterearth, the earth, the moon, the sun, the five planets, and finally the fixed stars. His description implied that the earth must rotate on its axis, for the earth always turns its back side toward the central fire. The counterearth lies between the central fire and the earth.

Eudoxus Eudoxus (fourth century B.C.) is called by George Sarton "the founder of scientific astronomy and one of the greatest astronomers of all ages" because of his theory of homocen-

FIGURE 1.1 A planet shows retrograde motion when it
stops its forward motion as viewed against the background stars, reverses direction for a time, and then
resumes its original motion.

9

INTRODUCTION

tric spheres, which attempted to explain the motion of the
planets by giving a mathematical description of the positions of
the heavenly bodies at all times. As the ancients, without telescopes, looked at the heavens, they could see only "stars" in
addition to the sun and moon, but those stars were of two
kinds: fixed and wandering. The fixed stars are those which
maintain their positions year after year, for example, the stars
comprising the Big Dipper. The wanderers are the planets, and
from week to week and year to year their positions among the
fixed stars change. Their paths are complicated, and their
speeds of movement among the fixed stars are variable. At
times a planet may even show retrograde motion, which means
that a planet which has been moving in one direction through
the fixed stars for some time will stop, reverse its direction, and
move backward for a while before it resumes its original motion.

Since the only motion of the fixed stars that can be seen by
the unaided eye is a 24-h rotation about the earth, one might
plausibly describe them as celestial bodies attached to a giant
sphere whose center is at the center of the earth, rotating once
every day. A similar explanation of the planets is not easy,
however, and it is the problem of devising such an explanation
which Eudoxus solved. Because Eudoxus accepted the notion
which had been advanced by the Pythagoreans that all the motions in the heavens must be circular and uniform, he tried to
explain all the peculiarities of motion of the objects in the heavens in terms of the circular motion of a body carried on a
sphere. For example, as a first approximation, a planet was assumed attached to the equator of a sphere which rotated about
one of its diameters; the center of the sphere was the center of
the earth, which was considered to be stationary. That rotation
caused the planet to move about the earth in a circular path. As

FIGURE 1.2 Eudoxus described the motion of the sun, moon, and
the planets by a system of spheres within spheres. The earth was at
the center of the system. Each of the other bodies (sun, moon,
planet) was attached to the equator of a sphere which rotated about
an axis passing through the earth. But the axes of that sphere were
attached to a second sphere, which was in turn mounted on a third
sphere. Three spheres were required for the sun, three for the moon,
and four for each of the planets.

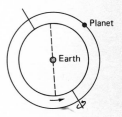

a second approximation, the axis on which the sphere rotated was assumed not to be fixed but to be attached to another sphere, also centered on the earth, which rotated on its own axis. The resultant motion of the planet depended upon the tilt of one of the spheres relative to the other and upon their relative angular speeds. If these two parameters could not be adjusted to give satisfactory agreement between the prediction and the known motion of the planet, the second sphere could be placed inside a third with its own axis of rotation, and so on until the agreement was adequate. To explain the apparent trajectories of the fixed stars, Eudoxus needed 1 sphere, to explain the sun and moon he needed 3 spheres for each, and for each of the planets he knew (Mercury, Venus, Mars, Jupiter, Saturn) he needed 4 spheres, for a total of 27 spheres. Eudoxus did not speculate on the reality of the spheres or on the force which makes them move; it was sufficient for him that the theory explained the positions of the planets and other bodies. It is worth noting that besides being a cosmologist, Eudoxus was an observer, and one of the observatories he used was still being pointed out 300 years later.

Aristotle Apparently because of a desire to believe that the spheres bearing the planets are real, Aristotle introduced buffering, or "reacting," spheres between the sets which Eudoxus had proposed, so that the sets would not be independent of one another. Aristotle proposed a continuous series, each nested inside the next. The purpose of the reacting spheres was to negate the effect of part of the others; thus inside the four spheres which explained the motion of Saturn he placed three to negate the effect of the last three outside and to provide a base for the spheres carrying Jupiter. When Aristotle finished the modification of the system, the total number of spheres had risen to 55, but the descriptive value of the system had not been improved.

Aristarchus In the third century B.C. Aristarchus of Samos made two contributions to astronomy which are worthy of note. The first was to measure the radii of the sun and moon and their distances from the earth in terms of the diameter of the earth. His measurements were very inaccurate, but the mere fact that he made such measurements at that time is significant. His other contribution was to propose a model of the universe in which the fixed stars and the sun remain unmoved, and the earth revolves in a circular path about the sun. He further said that the sphere of the fixed stars is so large that the circle made by the earth going around the sun is like a point by comparison. The assumption about the large size of the sphere carrying the fixed stars was probably made to explain the absence of annual parallax. Parallax is the apparent motion of a distant object because of the motion of the observer, and if the observer is riding on a moving earth, the direction in which the observer must look in order to see a particular fixed star should change in the course of the year. But no such apparent shifting in position of the stars had ever been observed, and the failure to see such an effect can be explained only by conceptualizing the earth as stationary, or the stars as very far away.

Aristarchus thus developed a heliocentric hypothesis 18 centuries before Copernicus. He was accused of impiety for putting into motion the "hearth of the universe," and his work seems to have had little effect on his contemporaries. Copernicus referred to Aristarchus's hypotheses as a justification for his own temerity in proposing that the earth moves around the sun rather than the reverse, but for 500 years after Aristarchus Greek cosmologists continued to refine the geocentric system.

Eratosthenes One of the striking measurements carried out by ancient astronomers was the determination of the size of the earth, a calculation made by Eratosthenes of Cyrene, who was born about 273 B.C. and died about 192 B.C. That the earth is spherical had been long recognized, and that it is not huge had also been recognized, but Eratosthenes measured the circumference with amazing accuracy for the first time. Aristotle had estimated the circumference of the earth as 400,000 stadia and Archimedes had estimated 300,000 stadia, but Eratosthenes measured it as 250,000 stadia, a figure which probably is very nearly correct.

Eratosthenes's method was to measure the difference in latitude between two points whose separation was known and which he believed to lie on the same meridian. At Syene (modern Aswan) a vertical pole casts no shadow at the time of the summer solstice (the time the sun reaches its most northerly position). It is said that he recognized that the sun was directly above Syene at that time because at noon the sunlight illuminated the water in a deep well without casting a shadow on the walls. By observing the length of the shadow cast by a vertical rod at Alexandria at the same time, Eratosthenes determined that the difference in latitude between Syene and Alexandria is one-fiftieth of a circle. Since the distance between the two cities is 5,000 stadia, the circumference of the earth must be about 250,000 stadia. From repeated measurements he arrived at the figure 252,000 stadia.

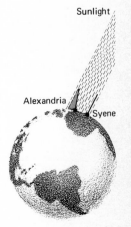

FIGURE 1.3 Eratosthenes's measurement of the circumference of the earth. When the sun was directly above Syene, Eratosthenes erected a vertical pole at Alexandria, 5,000 stadia to the north, and observed the length of the shadow cast by the pole. From the measurement of the shadow he deduced that the distance from Syene to Alexandria is $\frac{1}{50}$ of the circumference of a circle.

The value of the stadium differed from time to time and place to place, so that we cannot be certain of the value used by Eratosthenes. Sarton considers the best value to be one which converts 252,000 stadia to 39,690 km.* Since the correct value for the earth is 40,120 km, Eratosthene's measurement is off the true value by only about 1 percent. If other estimates of the stadium are used, the error in his measurement rises to 4 or 6 percent—still spectacularly good. The accuracy is partially fortuitous, for the measurements cannot have been good to 1 percent; and in fact, Alexandria and Syene are not actually on the same meridian, a fact which should introduce a slight error into the calculation.

If readers recall having been told that the sailors on Columbus's ships believed that the earth was flat and that they might fall off the edge, and perhaps even that the scientists of Columbus's day believed that the earth was flat, they may wonder how the knowledge of the Hellenistic world about the shape and size of the earth came to be lost in the fifteenth century. The fact seems to be that the nonsense about fears of falling off a flat earth was an invention of writers after Columbus, perhaps of Washington Irving, who wrote a dramatic account of the voyage. Neither Columbus's journal of the trip nor his son's account of it mentions any such notion. Columbus undertook the trip to India because he had a grossly incorrect idea of the size of the earth, believing it to be much smaller than it is. The skepticism about the plan expressed by geographers of his day was based upon their more accurate knowledge of the size of the earth.

Ptolemaic System

We now choose to neglect other astronomers of the later Hellenistic period and move to Claudius Ptolemy of Alexandria, an astronomer and geographer of the second century A.D. Ptolemy wrote a book which we know by the Arabic name *Almagest* (a contraction of the Arabic phrase meaning "the greatest of books") in which he used contributions of many predecessors together with his own original work to construct a model of the universe which dominated astronomical thought in the Western world until the time of Copernicus.

Assumptions   Ptolemy accepted five preliminary assumptions:

1 The heaven is spherical in form and rotates as a sphere.

---

*George Sarton, "A History of Science," p. 105, Harvard University Press, Cambridge, 1959.

The Ptolemaic system as shown by Gassendi in 1658. (*Yerkes Observatory photograph.*)

2 The earth, too, viewed as a complete whole, is spherical in form.

3 The earth is situated in the middle of the whole heaven, like a center.

4 By reason of its size and its distance from the sphere of fixed stars the earth bears to this sphere the relation of a point.

5 The earth does not participate in any locomotion.

Using these assumptions and three devices—the eccentric, the epicycle, and the equant—Ptolemy constructed a geocentric model of the observable universe. The devices were made necessary by the acceptance of the old pythagorean assumption that the only path suitable for a heavenly body was the circle. Since the moon, the sun, and the five known planets obviously did not travel on circles concentric with the earth, modifications were required, but they all worked on the basis of combinations of circles. Although ancient tradition also assumed that the motions were uniform, Ptolemy used the equant to "fudge" on that requirement.

Eccentric   The eccentric is a circle whose center is not at the center of the earth. If the sun moves on an eccentric circle, one can understand why the noontime sun appears larger in the winter than in the summer—it is closer to us. This device also explained in part the variations in brightness of the planets.

Epicycle   Epicyclic motion was a device to explain retrograde motion of the planets. If one observes the positions of the planets in the stars over a period of time, they can be seen to move forward for a time, then stop and move backward (retrograde) for a shorter time, then reverse again and move forward. In the 12 years Jupiter takes to make a circuit around the background stars, it makes 11 such hesitations and backward steps. The explanation given by astronomers before Ptolemy and adopted by him is that the planet moves about a circular path (the epicycle) whose center moves around a larger circle (the deferent) whose center is at or near the earth. To explain the motion of Jupiter, it was hypothesized that the planet makes 11 circuits around the epicycle while the epicycle's center is making one trip around the deferent. Four variables were at the disposal of the model maker to permit matching the prediction to observation: the radius of the deferent, the radius of the epicycle, the speed of motion around the deferent, and the speed of motion around the epicycle.

Equant   The device which we call the equant (from *punctum aequans*, or equalizing point) distorted badly the old belief that speeds about the center should be constant, for it replaced

FIGURE 1.4 The eccentric, one of the devices used by Ptolemy, is a circular path of a heavenly body which is not centered on the earth.

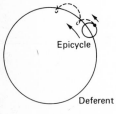

FIGURE 1.5 The epicycle is a small circle, about which a planet was supposed to move while the center of the epicycle moved around the deferent, whose center was at the earth. Epicyclic motion could explain the retrograde movements of planets.

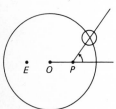

FIGURE 1.6 The equant was a device used to explain why planets do not always move about their orbits at constant speed. The center of the orbit of the planet (or the center of the deferent) is represented by *O*, and the earth is at *E*. Point *P* is located as far from *O* as *E*, but on the other side, and a line pivoted at *P* rotates with constant angular speed. The planet is at the point of intersection of the line and the circle.

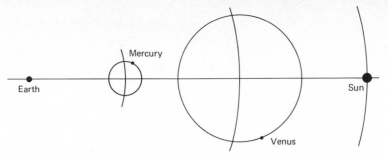

FIGURE 1.7 Mercury and Venus are never seen far from the sun, and Ptolemy explained that fact by hypothesizing that Mercury and Venus each move on an epicycle and that the centers of their epicycles always lie on the line joining the earth and the sun.

constant rate along the deferent by constant angular speed about another point called the equant. In the diagram (Fig. 1.6), the center of the epicycle moves about the deferent whose center is O; the earth E is away from that center, and the equant is point P, as far from O as E but in the opposite direction. The center of the epicycle moves so that the line joining it to P rotates at constant angular speed; the result is that the epicycle center moves with variable speed on the deferent.

The fact that the interior planets, Mercury and Venus, are always near the sun was explained by assuming that the centers of their epicycles always lay in a straight line with the sun; thus their maximum angular distances from the sun were determined by the radii of their epicycles.

Until the time of Copernicus, in the early 1500s, the system of the world proposed by Ptolemy was virtually unchallenged in Europe. It was cumbersome, requiring more than 70 separate motions to describe the behavior of the sun, the moon, the stars, and the five planets, but describe that behavior it did. To the accuracy with which measurements could be made with the instruments in use, the predictions possible with the Ptolemaic system were satisfactory. Further, at the time it was developed it was in agreement with Greek philosophical ideas, and when it was reintroduced into Europe from the Arab world near the end of the "dark ages" it was incorporated into Christian theology by the scholastics. Both common sense and man's sense of his own importance were satisfied by a theory which placed the earth at the center of the universe and made the sun and other bodies move about it.

COPERNICAN REVOLUTION

A radical break with ancient thought occurred in the sixteenth century, a break referred to as the *Copernican revolution*. Four names are prominent in the story of that revolution: Copernicus, Brahe, Kepler, and Galileo.

Nicholas Copernicus. (*Yerkes Observatory photograph.*)

Copernicus

Nicolaus Copernicus was born February 19, 1473, in Torun, Poland. In 1491, when he was eighteen years old, he went to Cracow to attend the university, and he was in his second year at the university when Columbus discovered America. In 1496 he went to the university at Bologna, Italy, to study canon law, but at that time he seems to have had some interest in astronomy. Eventually he entered the service of the church and remained a loyal member of an order until his death. Although we think of him primarily as an astronomer, he made a mark in many other areas—calendar reform, monetary reform, etc.—but the one for which he was best known during his life was medicine. His astronomical masterpiece was the final work of his life, and he saw a printed copy of it for the first time when he was on his deathbed, in 1543.

Description of Copernican system* Copernicus's major work was entitled *Six Books Concerning the Revolutions of the Heavenly Spheres.* In it he set forth a heliocentric system of the

---

*The term *Copernican system* is sometimes used to refer to the system of the world developed by Copernicus, with circular orbits, epicycles, and other details, and at other times it is used to refer to any heliocentric system. In this chapter the term is used only to mean the specific detailed system of Copernicus, which Kepler later proved incapable of describing the world.

world, but he still held to the view that the paths of the planets must be circles. One of his motivations was the elimination of the equant, a device which he found particularly offensive because when it was used the body moved with uniform velocity neither on the deferent nor the epicycle. In searching for a better system to describe the motion of the planets, he discovered that some of the ancients, including Aristarchus, had suggested that the earth moves and that the sun is the stationary center of the world. He began to investigate that idea to determine whether it might yield a more satisfactory description of the known motions than the Ptolemaic system. Finally, by lengthy calculations, Copernicus demonstrated that with circular orbits centered (or almost centered) on the sun, with a few epicycles and eccentrics, and without any equants, the motions of the planets could all be described by a heliocentric system as well as by the older system. His own description of the system is this:

And though all these things are difficult, almost inconceivable, and

The Copernican system as shown in Copernicus's book, *De Revolutionibus Orbium Coelestium*, printed in 1566. (*Yerkes Observatory photograph.*)

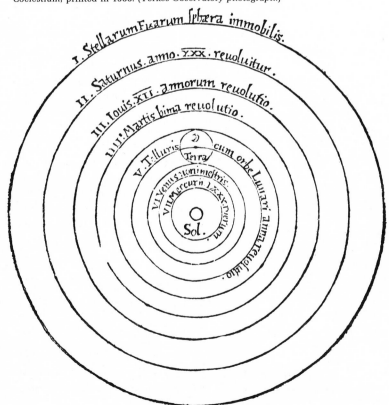

quite contrary to the opinion of the multitude, nevertheless in what follows we will with God's help make them clearer than day—at least for those who are not ignorant of the art of mathematics.... then the order of the spheres will follow in this way—beginning with the highest: the first and highest of all is the sphere of the fixed stars, which comprehends itself and all things, and is accordingly immovable. In fact it is the place of the universe, i.e., it is that to which the movement and position of all the other stars are referred. For in the deduction of terrestrial movement, we will however give the cause why there are appearances such as to make people believe that even the sphere of the fixed stars somehow moves. Saturn, the first of the wandering stars follows; it completes its circuit in 30 years. After it comes Jupiter moving in a 12-year period of revolution. Then Mars, which completes a revolution every 2 years. The place fourth in order is occupied by the annual revolution in which we said the Earth together with the orbital circle of the moon as an epicycle is comprehended. In the fifth place, Venus, which completes its revolution in $7\frac{1}{2}$ months. The sixth and final place is occupied by Mercury, which completes its revolution in a period of 88 days. In the center of all rests the sun. For who would place this lamp of a very beautiful temple in another or better place than this wherefrom it can illuminate everything at the same time? As a matter of fact, not unhappily do some call it the lantern; others, the mind, and still others, the pilot of the world.... Therefore in this ordering we find that the world has a wonderful commensurability and that there is a sure bond of harmony for the movement and magnitude of the orbital circles such as cannot be found in any other way.*

Objections to Copernican system It is apparent from the last sentence that a significant part of the attraction of the heliocentric system for Copernicus was esthetic, but he recognized that that argument would not convince everyone. Consequently he tried to anticipate some of the objections which might be raised and to answer them. In particular he tried to show that his system was as compatible with church dogma as the Ptolemaic system, and he carried out computations to show that his system was at least as capable as the Ptolemaic of predicting future positions of the planets.

The question of parallax was an old one, and although Copernicus gave the correct answer, it proved to be unconvincing to his critics. Copernicus's explanation was that the sphere of the fixed stars must be so large that the effect of parallax cannot be seen, but his contemporaries were unable to conceive of a universe so large. (Not until 1838 was the annual parallax of

---

*Nicolaus Copernicus, "On the Revolutions of the Heavenly Spheres," in *Great Books of the Western World*, vol. 16, p. 526, Encyclopedia Britannica, Inc., Chicago, 1952.

stars observed, for the effect is much too small to be seen with the instruments available to Copernicus.)

Retrograde motion was very simply explained by the Copernican system. Copernican theory did not require that each planet move on an epicycle in order to explain such motion; the natural combination of motions of the earth and the planet sufficed (Fig. 1.8).

From our vantage point, it may seem that the advantages of Copernicus's system should have been obvious, but to his contemporaries such was not the case. The notion of the moving earth was attacked on religious grounds by some Catholics, Protestants, and Jews. To practicing scientists the slight computational simplifications of the new system probably were not sufficiently attractive to justify learning a new system and rearranging all one's thinking, and in fact the accuracy was not better than the old system afforded. Perhaps one of the most serious objections to the Copernican system, however, was that it made no sense in terms of the physics of the day, and Copernicus did not offer a new physics. This point requires some elaboration.

Aristotle had given an explanation for the motions of bodies which was still in vogue in the sixteenth century, although it had been criticized and parts of it had been known to be wrong for at least three centuries. In simplified form, the explanation was this: Matter in the terrestrial region consists of four elements: earth, water, air, and fire. Each has its natural place, and they arrange themselves in the order named, earth taking the lowest position and fire the highest. Each of these has a natural motion, the natural motions of earth and water being downward and those of air and fire being upward. Every object we handle is made up of a combination of these four elements, and its natural motion is that of the element present in greatest abundance. Thus when water is heated, fire is added to it until sufficient quantity has been added to make the combination (steam) rise. If part of the fire later separates from the water, the natural motion of the water becomes dominant and it falls again. In addition to these natural motions, which are straight up or down, there are violent motions, such as those imparted to arrows or projectiles. Aristotle held that constant motion

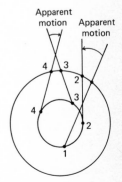

FIGURE 1.8 Retrograde motion as explained by the Copernican system. The earth is the planet moving on the inner orbit, and one of the outer planets is on the outer orbit. The corresponding positions of the two planets are numbered the same, and the lines of sight from the earth to the other planet are indicated. The planet appears to move one direction between positions 1 and 2 but to move backward between positions 3 and 4.

requires the application of a constant force; when the force ceases to operate, motion stops. To explain the motion of an arrow after it leaves the bowstring, persons believing Aristotle's theories had to invent elaborate pictures of air, first pushed aside by the point, then swirling to the back and pushing on the back end of the arrow to maintain its forward motion. There was one exception to the rule that a force act so long as motion persisted—Heavenly bodies were held to be made up of a fifth element, the *quintessence*, also called the *aether*, and its nature was to move in circular paths. Therefore no force was required to make the heavenly bodies move in circles; it was their natural motion. Furthermore, they were light, unlike the massive earth. It was the nature of the earth to remain stationary, just as it was the nature of the heavens to revolve, and all this fit together with Aristotle's explanations for falling objects, for projectile motion, and for other terrestrial phenomena.

All that science of motion had to be rejected by the person who took Copernicus seriously, for he was proposing that the massive earth, whose nature was to remain immobile, was in reality moving in a circular path, whereas the sun, whose nature was to move in a circle, was stationary. Copernicus offered no explanation of the force which must be causing the earth to move. Thus although the geometrical aspects of the new system might be attractive, its implications for physics made its acceptance very difficult, and not until Galileo and Newton replaced Aristotelian physics with a new theory of motion was this objection eliminated.

Copernicus had made observations of the stars and planets, but they were relatively crude. His positions had an uncertainty of several minutes of arc. More accurate measurements were needed if the theory of the structure of the world was to develop further, and the man who provided those was Tycho Brahe.

Brahe and Kepler

Tycho Brahe was a Dane, born in 1546. He became interested in astronomy at the age of fourteen when an eclipse was accurately foretold, and in spite of the efforts of the uncle who was rearing him to direct his attention to more practical matters like the law, he persisted in his study of the stars. He was an observer, not a theorizer, and he never saw any sense in the theory of Copernicus. Frederick II gave Brahe money to build an observatory on the island of Hven, between Denmark and Sweden, and there he built *Uraniborg*, the Castle of the Heavens. For 20 years Brahe worked with students and assistants measuring positions of stars and planets with an accuracy previously

Tycho Brahe. (*Yerkes Observatory photograph.*)

unattained—often to 2 minutes of arc. (A minute of arc is one-sixtieth of a degree.) Brahe designed and had built in his own shops elaborate measuring instruments, but the telescope had not been invented, and his accuracy of measurement was limited by properties of the human eye. His goal had been a catalog of 1,000 stars, but he reached only 777. Late in his life Brahe went to Prague; there he met Kepler, to whom he gave his astronomical observations. Brahe died in Prague in 1601. It is worth noting that Brahe never accepted the Copernican system. He devised his own, which is called the Tychonian. It proposed that the earth is stationary, the sun moves around it, and all the other planets revolve around the sun. From the standpoint of observation, that is equivalent to the Copernican system, for the choice of reference system (sun or earth) is arbitrary, but from the standpoint of the new physics which was to be developed the difference is profound.

Kepler's first law Tycho Brahe and Johann Kepler exemplify the division between experimental and theoretical science which has become formalized in our own time. Brahe was an observer and experimenter. Kepler (1571–1630) was a German who took Brahe's observations ostensibly for the purpose of constructing better tables of planetary motion than were available; his real purpose was to perfect the heliocentric theory. He was a mystic, who spent much time and effort trying to fit the spheres upon which the planets might move into the five regular solids, and he published the "music of the spheres" in standard musical notation. Finally, however, he set about

Johann Kepler. (*Yerkes Observatory photograph.*)

the task of improving the Copernican system using Brahe's observations, and it was then he discovered that the Copernican
system is impossible. Copernicus had proposed a system of
circular orbits, and when Kepler began serious work on that
scheme, trying to find the orbit of the planet Mars, he came to
the conclusion that the best possible circular orbit still
predicted the position of Mars 8 minutes of arc away from the
observed position according to Brahe. Eight minutes of arc is
not much—the angle subtended by a penny at 27 ft—and if
Kepler had had measurements only of the quality made by
Copernicus he could have attributed the discrepancy to observational errors. But he was confident that Brahe's measurements were better than that. When he found that the circular
orbit could not be made to fit the facts, Kepler took the step
which perhaps more than any other marks the difference between astronomy up until that time and astronomy since. He
dropped the preconception that the orbits of the planets must
be circles just because the circle is the perfect figure and God
could not have made the paths anything else. He asked simply:
What is the path of Mars? The answer, when he finally worked
it out, was simple: The path of Mars is an ellipse, and the sun is
at one focus. This is the genesis of what we know as Kepler's
first law of planetary motion: The path of each planet is an
ellipse with the sun at one focus.

The ellipse is a geometrical figure which had been known for
centuries, one of the "conic sections." It is a flattened circle; in
fact, a circle is a special case of the ellipse. The ellipse has two
foci, and the distance from one focus to a point on the curve
and back to the other focus is a constant for all points on the

curve. As a consequence of that property, an ellipse can be

drawn by placing two pins (foci) in a sheet of paper, tying a string longer than the distance between the pins to them, and then moving a pencil in the loop of string, with the string kept taut, marking all points that can be reached (Fig. 1.9). A little thought, or the experience of drawing an ellipse by this method, will convince one that an ellipse can be made into a circle, a shape so thin that it is almost a line, or anything between. The amount of flattening of the elipse is described by a number called the *eccentricity*, lying between 0 and 1 and defined as the ratio of the separation between the pins (placed at the foci) to the length of the string. If the pins are placed together, their separation is 0 and the eccentricity is 0; the figure is a circle. That is one limiting form of the ellipse. If the pins are moved apart, the eccentricity becomes larger, but it can never be larger than 1; an ellipse with an eccentricity equal to 1 has degenerated into a straight line—the other limiting figure. The circles used in older descriptions of planetary orbits worked as well as they did because the orbits of the planets have very small eccentricities. The eccentricity of the earth's orbit is 0.0167, which means that it is the shape produced by a string 1 m long tied to two pins 1.67 cm apart.

If one is to have any appreciation for the work of Kepler, it is important to realize that he had no general mathematical method to apply, but rather worked with numerical values, trying and discarding one idea after another until he found one which worked. The amount of labor involved was prodigious.

Kepler's second law   Kepler's first law lets one know the path of a planet if enough is known about its orbit to characterize the ellipse, but it does not tell how fast it moves on the orbit. That is done by Kepler's second law, which is often referred to

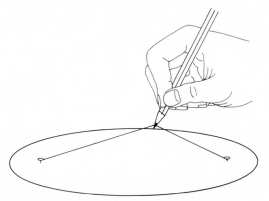

FIGURE 1.9 An ellipse may be drawn with two pins, a piece of string, and a pencil. The positions of the pins are the foci, and the line is to be drawn everywhere the pencil can move with the string kept taut.

FIGURE 1.10 Kepler's second law. The areas swept out by the line joining a planet to the sun are the same in any two equal time intervals. If the time required for the planet to go from 1 to 2 is the same as the time to go from 3 to 4, then the two sectors have the same areas.

as the *law of constant areal velocity* (Fig. 1.10): A line drawn from the sun to the planet sweeps out area at a constant rate. This law predicts differing speeds at different parts of the orbit; for the earth, the speed varies from 18.2 mi/s at the far point from the sun (*aphelion*) to 18.8 mi/s at the nearest point (*perihelion*). Each planet has a different rate at which the line sweeps out area, so that there is one value for Mercury, another for Venus, etc. Each of the first two laws applies to each planet individually. Therefore each has an elliptical path with certain characteristics; each has a particular areal velocity. Neither of these laws relates one planet to another.

Kepler's third law   Ten years after the publication of the first two laws, Kepler found and published a third: The squares of the sidereal periods of the planets are proportional to the cubes of the semimajor axes of their orbits. The *sidereal period* is the time required for the planet to make one complete revolution around its orbit, as viewed from the stars. All effects on our observations of the motion of the earth have been eliminated from consideration when the sidereal period is determined. The square of the period is simply that number multiplied by itself. If a line is drawn the long way in an ellipse, passing through both foci, it is called the *major axis*, and half that is called the *semimajor axis*. (The line which bisects the major axis is the minor axis.) Let us call the period of a planet P and the semimajor axis a; then Kepler's third law says that $P^2 \propto a^3$. One can always convert a proportionality into an equation by introducing a constant of proportionality, and if we call such a constant K, the third law is

$$P^2 = Ka^3$$

The important thing about this relation is that the same K applies to all the planets. The value of K depends upon the units used to measure the period and the semimajor axis. If we use seconds for the period and meters for the axis we will find

FIGURE 1.11 The semimajor axis of an ellipse.

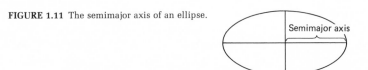

Semimajor axis

one numerical value for K, but if we use days for the period and
miles for the distance we will find another value. There is a
particular choice of units that makes the value of K easy to work
with, and consequently it is often used. The equation applies to
the earth as well as to the other planets, and so we can write

$$P_E{}^2 = Ka_E{}^3$$

where the subscript E indicates earth. Now if we take the
period of the earth to be 1 year, and define the length of the
semimajor axis of the earth's orbit to be 1 astronomical unit
(AU), then $P^2 = 1$ and $a^3 = 1$, so that K must also be 1.
Therefore, if the periods are measured in years (earth time) and
semimajor axes are measured in astronomical units, the equation reduces to

$$P^2 = a^3$$

One way of testing Kepler's third law is to determine
whether the same K actually does apply for all the planets.
Table 1.1 gives the semimajor axes in astronomical units and
the periods in years (earth time). Three-figure accuracy has
been used, and it is apparent that the constant is truly the same
for all the planets. When we discuss gravitation we will see
why that is so.

Because Kepler's three laws of planetary motion are so important, we repeat them all together:

First law The path of each of the planets is an ellipse with
the sun at one focus.

Second law A straight line joining the sun to a planet
sweeps out area at a constant rate.

Third law The square of the period of a planet is proportional to the cube of its semimajor axis.

TABLE 1-1

Planet	Semimajor axis, AU	Period, yr	P^2, yr^2	a^3, AU^3	P^2/a^3 $(= K)$
Mercury	0.387	0.241	0.058	0.058	1.00
Venus	0.723	0.615	0.378	0.378	1.00
Earth	1.00	1.00	1.00	1.00	1.00
Mars	1.52	1.88	3.53	3.51	1.01
Jupiter	5.20	11.9	142.0	141.0	1.01
Saturn	9.54	29.5	870.0	868.0	1.00
Uranus	19.2	84.0	7060.0	7080.0	0.997
Neptune	30.1	165.0	27200.0	27300.0	0.996
Pluto	39.4	248.0	61500.0	61200.0	1.00

26 Galileo

The fourth person who won acceptance for the new point of view regarding the structure of the universe was Galileo Galilei (1564–1642). Galileo not only made important new discoveries and propagandized for the new ideas in astronomy, but he also helped found the "new mechanics," to be developed more fully by Newton, which provided a logical explanation for the motion of heavenly *and* terrestrial bodies.

Galileo learned that Lippershey in Holland had combined two lenses to make distant objects appear closer, and in 1609 he made a telescope for himself. He immediately trained his telescope on the heavens and made a number of discoveries which contradicted predictions of the old Aristotelian science. He observed sunspots and the mountains on the moon, thus proving that the sun and the moon are not "perfect" as had been supposed. He observed bulges on Saturn, although he was unable to recognize the ring system. He saw four of Jupiter's moons thus demonstrating motion occurring about a center other than the earth, and he observed that the Milky Way is made up of a huge number of stars. He also saw the phases of Venus, which was crucial in refuting the geocentric theory. According to Ptolemaic theory Venus moves on an epicycle whose center always lies between the earth and the sun (to explain the fact that it is never more than about 47° from the sun), so that the planet can never be seen from the earth illuminated fully from the front. According to heliocentric theory, Venus should exhibit all the phases which the moon shows. Until the telescope was used it was impossible to determine which predic-

Galileo Galilei. (*Yerkes Observatory photograph.*)

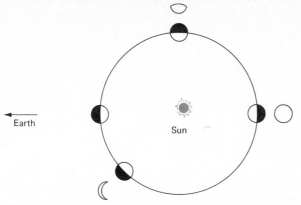

FIGURE 1.12 As seen from the earth (with a telescope) Venus goes through the same cycle of phases as the moon. This fact is predicted by the Copernican theory but cannot be explained by the Ptolemaic system.

tion was correct, but Galileo was able to see that Venus does indeed go through all the phases (Fig. 1.12).

After some earlier writings had brought him into conflict with the Church, Galileo wrote *Dialogue Concerning the Two Chief Systems of the World,* in which both the Ptolemaic and the heliocentric theories are presented in dialogue form, but it is clear that the heliocentric view has the better of the argument. Church authorities took more offense at this work than Galileo apparently had expected, and he was forced to recant the heretical notion that the earth moves around the sun.

The type of scholastic argument which Galileo struggled against is exemplified by this quotation from the Florentine astronomer Francesco Sizi, explaining why there could not be any satellites around Jupiter:

> There are seven windows in the head, two nostrils, two ears, two eyes, and a mouth; so in the heavens there are two favorable stars, two unpropitious, two luminaries, and Mercury alone undecided and indifferent. From which and many other similar phenomena of nature such as the seven metals, etc., which it were tedious to enumerate, we gather that the number of planets is necessarily seven.... Besides, the Jews and other ancient nations, as well as modern Europeans, have adopted the division of the week into seven days, and have named them from the seven planets: now if we increase the number of planets, this whole system falls to the ground.... Moreover, the satellites are invisible to the naked eye and therefore can have no influence on the earth and therefore would be useless and therefore do not exist.*

---

*Oliver Lodge, ''Pioneers of Science,'' p. 106, Macmillan and Company, London, 1893.

Many of Galileo's contemporaries to whom he tried to show the things he had discovered refused even to look through the telescope.

Galileo died early in 1642. Isaac Newton was born on Christmas Day of that same year. By the time Newton published his great work on mechanics in 1687, the issue of a geocentric system was all but dead, at least in England. Newton accepted the heliocentric system and worked out its details with an elegance and power unimagined before.

QUESTIONS

1 Suppose that on the day that the sun was directly above Cyene at noon, Eratosthenes had found that the shadow of a pole at Alexandria indicated that Alexandria is $\frac{1}{36}$ of a circle north of Cyene. What circumference would he have computed for the earth?   *Ans* 180,000 stadia

2 Suppose that the pole at Alexandria, set vertical, was 2 m high. How long was the shadow observed by Eratosthenes? (This requires the use of trigonometry.)   *Ans* 0.25 m

3 Contrast the underlying assumptions of the Ptolemaic, Tychonian, and Copernican systems.

4 Draw the Tychonian system. Why is it equivalent, observationally, to the Copernican system?

5 What predictions do the Ptolemaic and Copernican systems make for the appearances of phases of Mercury and of Mars?

6 Which of Kepler's laws implies that the speed of a planet varies continuously?

7 A comet is observed to have a period of 31.7 years. Approximately what is the length of the semimajor axis of its orbit measured in astronomical units?   *Ans* ≈ 10 AU

8 Near the extreme points of an elliptical orbit the rate at which the line from the sun to the moving point sweeps out area is approximately $\frac{1}{2}rv$ where r is the distance to the sun from the orbital position and v is the orbital velocity at that point. Consider a comet which at its closest approach to the sun (perihelion) comes $\frac{1}{2}$ AU from the sun and at that point is moving 80 mi/s. If at the most distant point (aphelion) it is 20 AU from the sun, what is its speed at that point?   *Ans* 2 mi/s

9 Compute how far from your eye a penny must be placed so that it will subtend an angle of 1°, 1′, 1″. (A penny has a diameter of $\frac{3}{4}$ in or 19 mm.)   *Ans* 1.1 m; 65 m; 3,900 m

10 State the objections to the Copernican hypothesis that might have been voiced by a contemporary of Copernicus trained in Aristotelian physics.

the earth in space

Before we continue with a detailed study of either the solar system or of the stars beyond, perhaps we should take time to consider the earth in relation to the space around it. We shall step off the earth and look at it spinning and whirling about the sun, and we shall also take a brief look at the moon, and its effects on the earth. By considering the stars as they look when viewed from earth, we shall also begin to develop a nodding acquaintance with the appearance of the nighttime sky.

THE EARTH

In this section we shall discuss the most obvious motions of the earth—its *rotation* on its axis, producing night and day, and its *revolution* about the sun, producing the succession of years—as well as the complicated effects of the earth's *tilt*.

Motions of the Earth: Revolution and Rotation

Let us think of the sky as it looks with the stars apparently pasted on the inside of a huge globe concentric with the earth. That globe is called the *celestial sphere,* and we can describe the motions of the sun, moon, and planets in terms of positions on it. One way of describing the motion of the earth about the sun is to say that during the course of a year the sun appears to move completely around the celestial sphere. Of course when the sun is in the sky one cannot see the stars near it and thus observe its motion from month to month, but one can notice the stars which appear in the Western sky immediately afer sunset and then compute or estimate the changing position of the sun relative to them. The path the sun appears to take among the stars is called the *ecliptic*.

The time the earth requires to make a trip around the sun is a year, but because the motion of the earth involves more than a simple rotation on its axis and motion in an ellipse about the sun, we can define two different kinds of year; these will be discussed in more detail later.

As we well know, the earth rotates once on its axis every 24 h, producing the alternation of day and night. The day, then, is the time required for one rotation. But as inspection of 29

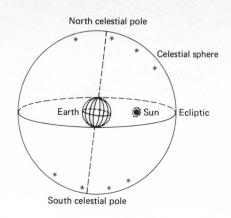

FIGURE 2.1 Stars, as viewed from the earth, appear to be fastened on the inside of a huge sphere, called the celestial sphere. The earth's annual motion around the sun occurs in a plane called the plane of the ecliptic, and the intersection of that plane with the celestial sphere is the ecliptic. From the standpoint of the observer on the earth, the sun moves around the ecliptic in the course of a year, and the plane of the ecliptic is the plane defined by the sun's motion.

Fig. 2.2 will show, if we were standing out among the stars, we would see the earth take one time to complete a rotation, but if we were standing at the sun, we would see it take a different time. The discrepancy arises because during the time the earth is rotating once it is also moving about 1° on its annual trip around the sun. We can therefore define the day in two equally important and legitimate ways—in terms of motion measured by reference to the stars or in terms of motion measured by reference to the sun. The first is called the *sidereal day*, and it is about 4 min shorter than the *solar day*. But even the term solar day is ambiguous, for because of the ellipticity of the earth's orbit and the variation in speed of the earth moving about that orbit, the length of time it takes for

FIGURE 2.2 The sidereal day is the time in which the earth rotates 360° on its axis, so that as viewed from the stars, it has returned to its original position. The solar day is the time in which the earth returns the same face to the sun, and because of its movement about the sun, the earth must rotate approximately 361° to complete a solar day. Consequently the solar day is about 4 min longer than the sidereal day.

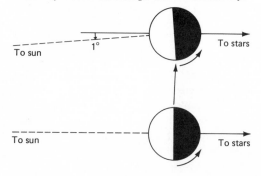

the earth to turn relative to the sun varies. Thus we use the term *mean solar day*, the point of reference for which is an imaginary sun which always moves through the heavens at the same rate. Ordinary time, then, is based on the mean solar day, which we divide into 24 h. Astronomers trying to locate stars must use sidereal time, and observatories usually have clocks which run on sidereal time, so that the positions of stars in the sky can be easily correlated with the reading of the clock. For most purposes, however, the position of the sun is more important than the positions of the stars, and we customarily follow mean solar time, or to be exact, we use a system of time zones based on mean solar time.

Tilt of the Earth

The seasons, the variation in length of days from winter to summer, and the varying positions in the sky of the noontime sun all result from the fact that the axis of the earth is tilted with respect to the plane of the ecliptic. Think of the earth as a spinning top, turning about the polar axis. The line around the earth halfway between the poles is the equator, and the plane it defines is called the equatorial plane. The plane in space along which the earth makes its annual circuit of the sun is the plane of the ecliptic. If the polar axis of the earth were perpendicular to the ecliptic plane, then the equatorial plane and the ecliptic plane would be the same. In that case day and night would always be the same length, we would have no seasons, and the sun would always appear at the same position in the sky at the same time of day.

In reality, however, the axis of the earth is tilted $23\frac{1}{2}°$ from that perpendicular position; hence the planes of the equator

FIGURE 2.3 The axis of the earth is tilted 23.5° from the perpendicular to the plane of the ecliptic. The northern hemisphere has summer when it is tilted toward the sun, as in the left half of the diagram; when it is tilted away from the sun, the northern hemisphere has winter.

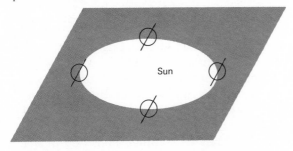

Sun

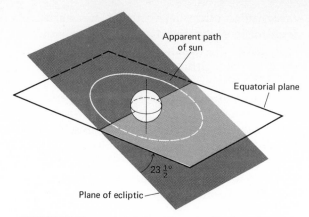

Apparent path
of sun

Equatorial plane

$23\frac{1}{2}°$

Plane of ecliptic

FIGURE 2.4 The equatorial plane (determined by the earth's equator) and the plane of the ecliptic intersect, making an angle of 23.5°. The sun appears to move on a path lying in the ecliptic plane, and hence for 6 months of the year it is above the equatorial plane and for the other 6 months it is below that plane. The line of intersection of the planes marks the position of the sun at the equinoxes.

and ecliptic are also tilted $23\frac{1}{2}°$ relative to one another. Consequently, at one time of the year the northern hemisphere of the earth receives more direct sunlight than does the southern hemisphere, and 6 months later the situation is reversed. When the sun is above the northern hemisphere that part of the earth has summer; when the sun moves south, summer moves with it and the northern regions have winter. The earth is nearer the sun when the northern hemisphere is having winter, but the variation in distance from nearest to farthest is only about 3 percent, and that effect is much less important than the $23\frac{1}{2}°$ tilt.

People living on the earth's equator see the noonday sun north of them during half the year and south the other half. Its maximum variation is $23\frac{1}{2}°$ to the north and then $23\frac{1}{2}°$ to the south. It reaches its maximum northward position on June 22 and its maximum southward position on December 22. On March 21 and September 23 it passes directly overhead. March 21 is called the *vernal equinox*; September 23 is the *autumnal equinox*; June 22 is the *summer solstice*; and December 22 is the *winter solstice*. On the summer solstice, a person living at $23\frac{1}{2}°$ N latitude sees the sun directly overhead at noon.

The astronomer visualizes two intersecting planes (equatorial plane and plane of ecliptic) and sees the sun apparently moving along the ecliptic plane. When the sun reaches the position where the two planes intersect, it is directly above the earth's equator, and days and nights are of the same length. For that reason those dates are called the equinoxes. But that line of intersection of the two planes also defines a position in space,

and its intersections with the celestial sphere serve as reference
points to locate positions on that sphere.

Precession of the Equinoxes

The line of intersection between the equatorial and ecliptic
planes which the sun crosses on March 21 is the vernal
equinox, and the solar year is the time from one crossing to the
next. If that line remained fixed in space, as viewed from the
stars, the solar year and the sidereal year would be the same.
But the line actually moves slowly, and as a result the ordinary
year (also called the *tropical* year) based on the sun is about 20
min shorter than the sidereal year. The sidereal year is
365.2422 . . . mean solar days.

In order to understand the motion of the earth that causes the
difference between the sidereal and the tropical year, think of a
child's top, spinning on a table. If the axis were precisely vertical, the top would spin without other motion, but since the
axis is tilted, the pull of the earth's gravity causes the top to
precess, which means that its axis traces out a cone. Very likely
the top also shows *nutation* as it precesses, which is a slight
up-and-down variation.

Exactly the same thing happens with the earth. The moon,
attracting the equatorial bulge of the earth, provides the force
that causes precession, and the axis of the earth moves on a
cone, taking about 26,000 years to make one revolution. The
earth also shows a very slight nutation, with a period of about
19 years. The precession of the earth causes the line of intersection between the equatorial and ecliptic planes to move
westward about 50 seconds of arc each year, and that movement causes the difference between the tropical and the
sidereal years.

At the present time the axis of the earth points very nearly
toward a star which we call Polaris. That has not always been
true, however, nor will it remain true. As the axis of the earth
moves along its cone, it will leave Polaris and point more or
less at other stars. About 12,000 years hence the bright star
Vega will be the North Star, as it was some 14,000 years ago. It
is interesting to note that the movement of the earth's axes of
rotation was discovered by Hipparchus approximately 150 B.C.

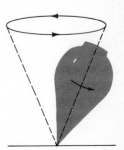

FIGURE 2.5 A spinning top which does not stand precisely vertical
precesses, which means that its axis traces out a cone. The same
force (its weight) which would cause the top to fall over if it were
not spinning causes it to precess if it is spinning.

The moon accompanies the earth in its annual trip around the sun, but because both the earth and the sun pull the moon strongly, its motion is quite complicated. We shall not discuss most of the details of lunar theory or motion.

Synodic and Sidereal Months

The plane of the moon's motion is tilted about 5.2° to the plane of the ecliptic, and therefore we can see the moon both fully illuminated and totally dark. The condition in which the moon appears to us to be dark (because it is almost precisely between us and the sun and therefore the side away from us is illuminated) is called the new moon. The condition in which we see the fully illuminated side is called the full moon. At other positions we see the moon partially illuminated. The moon goes through a complete cycle of positions with respect to the sun in approximately 29.5 days, or one *synodic* month.

From the viewpoint of the stars, however, an observer would see the moon go through its phases in only about 27.3 days, or in one sidereal month. The difference arises because during the sidereal month the earth has moved approximately one-twelfth of a revolution around the sun, and hence the moon must move farther to place itself in the same position relative to the earth and sun than to reproduce its position relative to the stars.

Motions of the Moon

The moon rotates on its axis once per revolution about the earth, so that it always presents the same face toward the earth. It is said to be *tidally locked* to the earth, which means that the earth produces a tidal bulge on the moon which is attracted to the earth with sufficient force to keep that face always turned toward us.

The sun exerts an even greater pull on the moon than does the earth, and hence the moon's orbit is always concave toward the sun. The distance from the earth to the moon is approximately 240,000 mi; the distance from the earth to the sun is 93,000,000 mi. The sun-earth distance is therefore approximately 400 times the earth-moon distance. If the orbit of the earth about the sun is represented by a circle of radius 10 in, the distance from the moon to the earth will be $\frac{1}{40}$ in, and it is easy to see that the curve representing the moon's orbit will always be concave toward the sun. What is not apparent from the geometrical configuration alone is that the concavity is a result of the fact that the sun has a greater pull on the moon than the earth has.

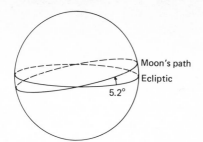
Moon's path

Ecliptic

5.2°

FIGURE 2.6 The plane in which the moon moves about the earth is tilted 5.2° with respect to the plane of the ecliptic.

Eclipses

If the moon moved in the plane of the ecliptic, we could expect an eclipse of the moon and an eclipse of the sun each month. An eclipse of the sun occurs when the moon comes between the earth and the sun so that the shadow of the moon strikes the earth. An eclipse of the moon occurs when the shadow of the earth strikes the moon. But because the plane in which the moon moves is tilted with respect to the plane of the ecliptic, eclipses can occur only when the sun, earth, and moon are in line at the time the moon is passing through the plane of the ecliptic. The line of intersection of the plane of the moon's orbit and the plane of the ecliptic is called the *line of nodes*, and eclipses can occur only when the moon is at or near that line. A little reflection reveals that eclipses of the moon can occur only when the moon is full, and eclipses of the sun can occur only when the moon is new.

Because the sun is an extended source of light rather than a point source, the shadows cast by the earth and the moon have two parts, an *umbra* and a *penumbra*. When the shadow of the moon passes across the earth, the eclipse of the sun is total for persons within the umbra, and partial for those in the penumbra.

Solar eclipses can be spectacular, with the disk of the sun

FIGURE 2.7 The parts of a shadow cast by an object illuminated by an extended source such as the sun. Part A is the umbra, where all view of the source is blocked; B is the penumbra, where part of the source is blocked. An observer at C would see a ring of the source around the blocking object. If the blocking object is the moon, a total eclipse of the sun is seen within region A, a partial eclipse within region B, and if the earth is far enough from the moon that it is in part C rather than A, an annular eclipse can be seen within region C.

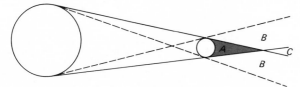

covered and only a rim showing all around, because of a strange coincidence. Although the sun is 400 times as large as the moon, it is also 400 times as far from the earth. Both bodies subtend angles of about 0.5° in the sky, so that when they are in line, the moon can just cover the sun. (Variations in the moon's orbit can bring it nearer or take it farther from us so that the match in apparent size is not always precise.)

Tides

Gravitational attraction decreases with distance, in a way to be described later, and it is the variation with distance which produces tides. The moon is relatively near the earth, and it attracts the earth, but it attracts the part of the earth turned toward it more strongly than the center of the earth or the far side of the earth. The difference in attraction causes bulges on the earth on the faces turned toward and away from the moon. A simple way to think of the two bulges is this: The side of the earth turned toward the moon rises because the moon is attracting it more than the other parts of the earth; the side of the earth turned away from the moon rises because the moon attracts the main body of the earth more than that far side. The "solid" earth has tides up to about 9 in, but when the bulge occurs in the deep ocean, it rises about 2 ft. When the tidal bulge being carried along by the spinning earth turning below the moon reaches a coast, the tide may become very large.

The sun actually attracts the earth with a force 180 times as great as the force with which the moon attracts the earth, but the tides produced by the sun are only 45 percent as great as those produced by the moon. The reason for that difference is that the sun, being so much farther away than the moon, attracts all parts of the earth almost equally, whereas the moon, even though all its attractions are much smaller, is close enough to attract the near and far sides of the earth with a different force. The sun does produce a tide, however, and the ac-

FIGURE 2.8 The moon's greater pull on those parts of the earth nearer to it produces two tidal bulges on the earth, one turned toward the moon and the other on the opposite side of the earth.

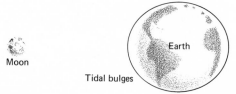

Moon

Earth

Tidal bulges

tual tides on the earth are a combination of the effects of the
moon and the sun. Near shores, the interaction of the sea with the ocean bottom has a great effect on the tidal effects observed.

We know that the day is now about 0.016 s longer than it was 1,000 years ago, and for at least the past 2,000 years the length of the day has been increasing about 0.0016 s/century (0.0016 s/day/century). Evidence for this are ancient records of solar eclipses. In addition, fossil records indicate that 300 million years ago the year had about 30 more days than it has now. This lengthening of the day—a slowing of the rotation rate of the earth—is produced by the tides.

We think of the tidal effect in two parts. The sloshing seas, drawn by the moon and rushing across the sea bottoms, suffer friction, and energy is taken from the spinning earth and converted into heat by that friction. At the same time, the bulge on the earth produced by the moon is always a little ahead of the moon because the earth and moon turn in the same direction, but the earth turns faster. Therefore the moon is always pulling back on the earth's bulge, tending to slow the rotation, and at the same time the bulge is pulling forward on the moon, tending to increase its speed. The result is that the earth slows slightly, and the moon, as its speed increases, drifts away from the earth. It is drifting away at about 3 cm/year. Eventually the moon will reach such a distance, and the earth will have slowed enough, that the length of the day and the month will be the same, about 50 of our present days.

THE STARS

For the purpose of locating stars, we visualize them as placed on the inside of the celestial sphere. The earth is at the center of the sphere. Of course we know that some stars are near to us and some are very far away, but for the purpose of describing how the heavens look we ignore that fact. Astronomers must have a way of identifying positions on the celestial sphere so that they can find the same object repeatedly and so that they can communicate with one another about objects they are studying with confidence that people in different observatories are viewing the same object. For this purpose angular positions are sufficient.

Declination and Right Ascension

Imagine the axis of the earth extended outward until it intersects the celestial sphere. The points at which it strikes the celestial sphere are called the celestial poles (north and south). Now imagine the plane of the earth's equator extended outward

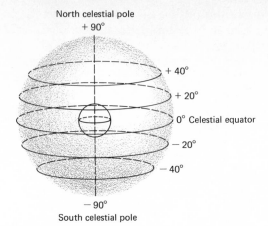

FIGURE 2.9 The earth and the celestial sphere. Declination is measured on the celestial sphere in degrees, 0° to +90° north from the celestial equator and 0° to −90° south from the celestial equator. The parallels of declination are the projections onto the celestial sphere of the parallels of latitude.

to intersect the celestial sphere; this line is the celestial equator. Positions of stars north or south of the celestial equator are measured in degrees, just as latitude is measured on the earth. Now imagine the celestial sphere with lines drawn on it parallel to the celestial equator and spaced 10° apart. The equator is 0°, the north celestial pole is +90°, and the south celestial pole is −90°. This coordinate, measured north or south from the celestial equator, is called *declination*. Declination is related to latitude on the earth in a simple way; a star whose declination is +40° moves directly above the circle defining a latitude of 40° north, and if a person is situated on that latitude parallel, the star passes over the person's head once each 24 h.

The latitude lines of the earth can be projected onto the celestial sphere without any problem, thus providing one coordinate for objects in space, but the longitude lines cannot be projected so easily. As the earth rotates, the longitude lines seem to pass around the celestial sphere. Somehow those lines must be projected onto the sphere and stuck in position there. That implies that we must choose some reference point on the celestial sphere from which to measure positions east and west. We need a point like the prime meridian through Greenwich, England, which provides a starting point for longitude measurements on the surface of the earth. The reference point chosen is the intersection of the vernal equinox with the celestial sphere. In other words, the line from the earth to the sun at the moment about March 21 when the sun crosses the equato-

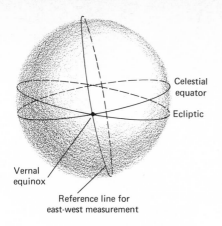

FIGURE 2.10 The reference line for east-west measurement (corresponding to longitude on the earth) is the great circle passing through the celestial poles and through the vernal equinox. The vernal equinox is the point on the celestial sphere where the sun appears to cross the equatorial plane in the spring, about March 21.

rial plane is extended to the celestial sphere, and the meridian line from pole to pole passing through that point of intersection is the starting point to measure positions east and west.

Such a point, where the line of intersection of two imaginary planes meets an imaginary sphere, may seem hopelessly hard to find. There is no mark in the sky to tell us where the vernal equinox is, but it can be located with great precision. It is located in the constellation Pisces, and it is visible in the evening sky during the autumn. One end of the constellation Cassiopeia is almost on the reference meridian, and it can be used to orient oneself in the sky.

Distances east and west from the reference meridian can be measured in degrees, just as longitude is measured on the earth. Degrees are measured to the west from the reference point. More commonly, however, another system is used, in which positions are described not in terms of degrees but of hours, minutes, and seconds. During 24 h the earth carries each point around a circle, 360°. In 1 h, then, a point moves 15°, and posi-

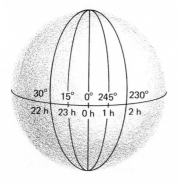

FIGURE 2.11 The celestial sphere, viewed from outside, showing the east-west marking in degrees and in hours.

tions on the celestial sphere can be measured in terms of the time required for an observer to be carried from the reference position to the position desired. The hour markings go eastward from the reference position, and positions indicated by hour, minute, and second are said to be *right ascensions* (RA). If a star is located at declination $+ 30°$, RA 8 h, it is 30° north of the celestial equator and 8 h (= 120°) east of the vernal equinox.

Because of the precession of the equinoxes the location of the vernal equinox constantly shifts; it will rotate around the sky in about 26,000 years. For this reason the RA coordinates constantly shift slowly, and if a star is to be located precisely, the date for which the coordinates were calculated must be indicated. Most star positions now are referred to the year 1950.

Constellations and Stars

The sky has been divided into 88 regions called *constellations*, and each is named. Some of these have been recognized for centuries, such as the Pleiades, Ursa Major, Cassiopeia, and Hercules, and some are of recent origin, chosen to complete the coverage of the celestial sphere. The fact that a group of stars provides a pattern in the sky which is recognizable in some way does not indicate that the stars are actually related, however, and in some cases the constellations are made up of stars moving in different directions, so that thousands of years hence the shape will have been entirely destroyed.

We shall describe a few constellations and stars to help the

FIGURE 2.12 The Big Dipper, Polaris, and Ursa Minor as they look in the evening in September. The pointer stars of the Big Dipper can be used to find Polaris. Alcor can be seen very near Mizar, in the handle of the Big Dipper.

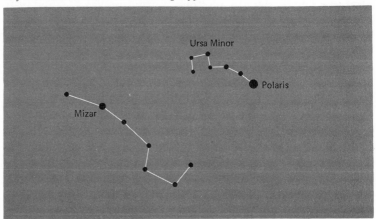

The stars of the Big Dipper in Ursa Major. (*Yerkes Observatory photograph.*)

beginner find examples of some of the things which will be described later in the book.

Probably the easiest grouping of stars to recognize, and the most commonly known, is the Big Dipper. The Big Dipper is not a constellation itself, but is part of the larger group called Ursa Major, the Big Bear. Early in the evening in the fall the Big Dipper is low in the northwest. It has two noteworthy features: The two stars which form the outer edge of the bowl (called the pointer stars) point to the North Star, and the second star from the end of the handle is a complex double. Move your eyes from the bottom of the bowl, along the pointers, and they arrive at Polaris. Although Polaris is rather faint, it stands almost alone in a large area so that it is easily recognized. It is so nearly on the line of the earth's axis (at the north celestial pole) that it appears to hardly move at all as the earth rotates. In addition to being the North Star, it is a Cepheid variable, one of a class of stars which fluctuate in brightness. These will be discussed in detail later. The star next to the end of the handle of the Big Dipper is Mizar, and very near it is a fainter star, Alcor. The ability to see Alcor used to be a test for acuity of vision. These two stars actually are not related; they are what is called a *visual double* because they appear to be almost in line. Mizar, however, is an optical double, and in a small telescope both components of it can be seen. They are separated by 14 seconds of arc, and they revolve about one another.

Polaris is the end of the handle of the Little Dipper (correctly called Ursa Minor, the Little Bear). This constellation is faint, but the Little Dipper is turned as if it were pouring into the Big Dipper.

Directly across the Pole Star from the Big Dipper, and about the same distance from Polaris, is Cassiopeia, a constellation

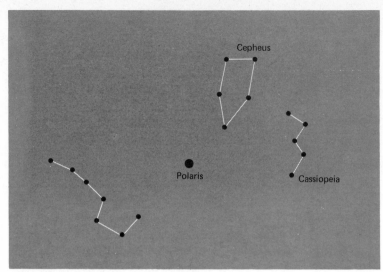

FIGURE 2.13 The Big Dipper, Polaris, Cepheus, and Cassiopeia as they appear in the evening in September.

which looks like a W. Very close to Cassiopeia is Cepheus (the King), looking like a square with a triangle on the top. Cepheus is notable because one of its stars, Delta Cephei, was the first variable of its type found, and it gave its name to the entire class, members of which are known as *Cepheids*.

Again starting with the Big Dipper, we can follow the curve of the handle about 30° beyond the handle to the bright star Arcturus, which is in the constellation Boötes (the Herdsman). (Remember this phrase: Follow the arc to Arcturus.) Boötes looks more like a kite than a man, although it can be seen as a man seated and smoking a pipe. Arcturus is a red giant, about 25 times the diameter of the sun, with a luminosity about 100 times that of the sun.

About 9 in the evening, early in September, Cygnus (the Swan), also known as the Northern Cross, is directly overhead. At the head of the cross is Deneb, a bright white star. Deneb is about 2,600 light years from the sun. At the foot of the cross is Albireo, an unimposing star to the naked eye but a beautiful double when seen through even a small telescope. The Milky Way passes through Cygnus, and if one is away from city lights on a dark night, this faint band of luminosity which is the combined effect of some 100 billion stars can be seen.

Directly east of Cygnus can be seen four stars forming an approximate square, the Great Square of Pegasus. The square is useful as a landmark to assist in finding the only spiral galaxy visible to the naked eye, the Great Galaxy of Andromeda. As the

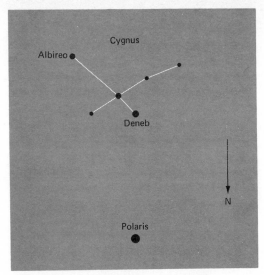

FIGURE 2.14 Early in the evening in September, Cygnus is almost directly overhead for an observer in the United States. The brightest star in this constellation is Deneb.

diagram shows, one counts from the corner of the square to the third star, and moves upward slightly. On a dark night, a faint patch of luminosity can be seen, which is the galaxy of more than 100 billion stars, 2 million light years from us.

In the winter, the most imposing constellation is Orion, which is located somewhat to the south of an observer in the

FIGURE 2.15 The Andromeda Galaxy (M 31), a system of more than 100 billion stars, located about 2 million light years from us, can be found by using the Square of Pegasus as a reference.

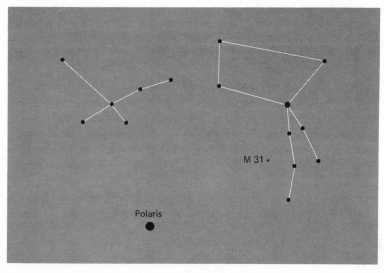

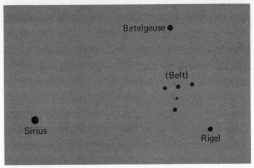

FIGURE 2.16 The Orion region, visible in the evening during the winter months. The three stars forming the belt are
the easiest feature to locate. The middle "star" of the sword
is the Orion nebula.

United States early in the evening in January. The most out-
standing characteristic of this constellation is the three stars in
a line that form the belt of Orion. Hanging down from the belt
are three fainter stars, the sword. The middle "star" of the
sword is not a true star, but is the Great Nebula. Even with the
naked eye it appears diffuse, and using binoculars one can
clearly see it is not a star.

Somewhat above the belt and slightly to the east is a bright
red star, Betelgeuse. This is another red giant, even larger than
Arcturus. Betelgeuse is 700 light years from the sun, and it is
intrinsically 100,000 times as luminous as the sun. Betelgeuse
has a diameter 500 times that of the sun, so that if its center
were placed at the sun, Mars would be inside the star.

Below and to the west of the belt is Rigel, a luminous blue
star. It is a blue giant, larger and more luminous than the sun; it
is also a binary.

A line drawn through the belt stars to the southeast comes to
the brightest star in the heavens: Sirius, in the constellation
Canis Major. It is not intrinsically extremely bright, but it is
relatively near—about 9 light years distant. Sirius is particularly interesting because it has a white dwarf companion. At
one time the companion star must have been the brighter of the
pair, but it has gone through its life cycle and become a very
faint, small, and dense white dwarf, visible now only through a
medium-sized telescope.

The region around the constellation Orion is interesting
because it contains many bright stars and because it includes
easily recognized stars at many stages of development. Within
the Orion nebula are young stars, and others are probably being
formed there now. Rigel is a youthful blue giant, expending
energy at a rate it cannot long maintain. Betelgeuse is an old red

giant. And the companion of Sirius has reached the end; it has
spent its life and is now merely cooling. In Unit 13 we will
discuss in detail the evolution of stars.

QUESTIONS

1 Can an eclipse of the sun and of the moon occur within the
same week? Why or why not?

2 At about what time of day does the full moon rise?

3 Suppose that the axis of the earth were tilted 45° instead of
$23\frac{1}{2}°$. Describe the effects that would be noticeable to us.

4 If an observer is located at 45°N latitude, some stars are always above that person's horizon; these are called the *circumpolar stars* because the observer can see them make
complete circles about the pole. What are the declinations of
the circumpolar stars for this observer?

5 Show that the celestial equator passes through the points on
the horizon directly to the east and to the west.

6 Describe the apparent motion of a star on the celestial equator as seen by a person (*a*) on the earth's equator, (*b*) at a latitude of 45°N, (*c*) at the north pole.

7 What would a person on the surface of the moon, on the side
facing the earth, see during the times that observers on the earth
see (*a*) an eclipse of the moon, and (*b*) a total eclipse of the sun?

3
the solar system

From a consideration of Kepler's laws of planetary motion in Unit 1 one might logically proceed to a discussion of the solar system which is described by those laws or to a further development of the physics which the laws helped inspire. We shall do both; here we consider the solar system, and in the next unit we continue the development of the physics involved in the workings of the solar system.

What we call the solar system, the sun and the objects orbiting around it, includes the six planets known to the ancients (Mercury, Venus, Earth, Mars, Jupiter, Saturn), three planets which they did not know (Uranus, Neptune, Pluto), the satellites of the planets, and a swarm of small planetoids which lie between Mars and Jupiter, as well as comets and a small amount of dust. Since the sun is quite different from the other parts, being in reality a star, we shall defer consideration of it until we are ready to discuss stars in general. We shall give first attention to the planets, those major bodies which orbit the sun.

DISTANCES WITHIN THE SOLAR SYSTEM

It is possible to determine the periods of the planets with high accuracy, and then, from Kepler's third law, to calculate the relative values of the semimajor axes of their orbits. The relative sizes of the orbits can also be determined by direct observation, as shown in Fig. 3.1. Table 1.1 (page 25) gives the semimajor axes (in astronomical units) of all the planets: 0.387 AU for Mercury, 0.723 AU for Venus, 1.52 AU for Mars, etc. For Earth, of course, the value is 1.00 AU by definition. With these values, it is easy to lay out the solar system to scale. The astronomical unit, however, is not a unit used in everyday conversation by most people, and it would be desirable to know its value in miles or kilometers. This brings us for the first time to one of the central problems of astronomy—the measurement of distances.

Triangulation

Many of the measurements in astronomy, including some of the determinations of the length of the astronomical unit, make use of the method of *triangulation*. All one needs to know to

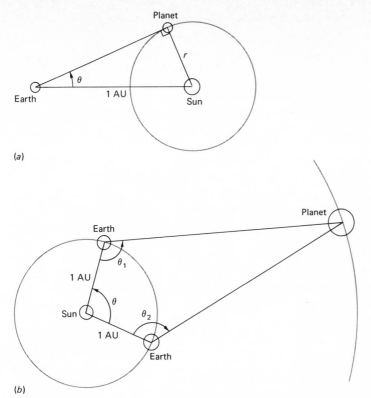

(a)

(b)

FIGURE 3.1 Measurement of sizes of planetary orbits in terms of the astronomical unit. (a) The sizes of the orbits of Mercury and Venus can be computed from observations of the planets at maximum elongation (greatest distance from the sun). The angle θ can be measured, and the distance from the sun to the planet can be computed from the right triangle shown. (b) Computation of the distance to an outer planet can be made by observing the planet at least twice when it is at the same position on its orbit. The earth is not at the same position on its orbit when the two observations are made, and hence angles θ_1 and θ_2 can be measured and θ can be computed from the dates on which the observations are made. Distance from the sun to the planet can then be computed. It was by computations such as this, based on observations made by Brahe, that Kepler determined the orbit of Mars.

understand the principle of triangulation is that a triangle is uniquely determined if two of its sides and the included angle or if two angles and the included side are known. For example, in Fig. 3.2, if one knows the lengths of sides a and b and the value of angle C, one can draw these three parts and the remainder of the triangle is determined. Similarly, if one knows angles A and B and side c, these can be drawn and everything else is fixed. One may find the other sides or angles by methods

FIGURE 3.2 A triangle is uniquely determined if either two sides and the included angle (as a, c, and B) or two angles and the included side (as A, c, and B) are known.

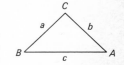

of trigonometry or by making a scale drawing. The important
fact is that the other parts can be found by simple methods.

Astronomical measurements often involve triangles with
two sides very much longer than the third side, with the short
side and some angles known, and in such cases finding the unknowns is relatively simple. Consider the triangles shown in
Fig. 3.3, and suppose that the length of the short side is known
and the length of the long sides is desired. If the long sides are
very much longer than the short side, the two long ones are so
nearly the same length that little error is introduced by treating
them as if they were equal (as if the triangle were isosceles). In
that case the two base angles are equal, and since the sum of the
interior angles must be 180°, knowledge of any one of the
angles yields the other angles immediately. Now compare the
triangle with the segment of a circle, whose radius is the long
side as shown. Again little error is introduced by treating the
base of the triangle (short side) as if it were the arc of the circle.
Now, however, a great simplification has been introduced, for
the arc of the circle is proportional to the angle subtended at the
center of the circle. The proportionality constant depends upon
the unit used to measure angles, but if the unit is the radian, the
constant is 1, so that

$$\text{Arc} = \text{angle in radians} \times \text{radius}$$

If the unit is the degree, the constant is 1/57.3, so that

$$\text{Arc length} = 1/57.3 \times \text{angle in degrees} \times \text{radius}$$

This last equation may be changed to give directly what one
usually wants, the long side of the triangle; thus:

$$\text{Long side} = 57.3 \times \text{base of triangle} \times \frac{1}{\text{angle in degrees}}$$

As an example of the measurement of a distance by
triangulation, consider the problem of determining the distance
to the moon. Suppose that two observers, situated 2,000 mi
apart, simultaneously measure the distance between the vertical and the line of sight to a particular crater on the moon (Fig.
3.4). To make the example simple, we shall assume that they
choose a time when the angles they measure are the same, and
each measures 14.59°. Since the arc of the earth's surface between them has a length 2,000 mi and the radius of the earth is
4,000 mi, the angle at the center of the earth subtended by that
arc is 0.5 × 57.3°, or 28.7°. Since the interior angles of a quadri-

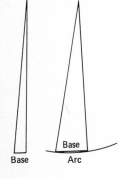

FIGURE 3.3 Thin triangles of the sort used in astronomical triangulation. On the right is shown a comparison of the base of a thin triangle with the arc of the circle whose center is at the apex of the triangle.

lateral must add to 360°, the angle at the apex of the triangle must be 0.48°. If the arc along the surface of the earth is 2,000 mi, the straight line through the earth joining the two observers is 1,990 mi, but the difference is not important since the angular measurements introduce more uncertainty than that difference. If the distance to the moon is now computed from

$$\text{Distance} = 57.3 \times \text{base of triangle} \times \frac{1}{\text{angle in degrees}}$$

it is

$$\text{Distance} = \frac{57.3 \times 1,990}{0.48} = 238,000 \text{ mi}$$

The distance to the moon can be measured with considerable accuracy by triangulation, but distances to the planets are so great that this method does not give good results. Even if advantage is taken of the fact that measurements of direction made at two different times have a base line equal to the distance the earth has traveled in that time, highly precise results by triangulation are impossible. In this sort of calculation, of course, allowance must be made for the movement of the planet during the time interval.

Length of the Astronomical Unit

One part of the solar system is a swarm of objects called *planetoids*, or minor planets. Most of them have orbits between the paths of Mars and Jupiter, but a few have highly eccentric orbits (orbits which depart greatly from circular), and one in particular, Eros, was used in 1932 to make the best measurement that had been made to that time of the astronomical unit. At that time Eros came within about 14 million mi of the earth. Its displacement, as seen from opposite sides of the earth, was observed to be about $\frac{1}{30}°$, or 2 minutes of arc. That means that if an observer had been on Eros at that time, the earth's diameter would have subtended an angle for the observer of 2 minutes of arc. Since that is the apex angle of the triangle having the diameter of the earth as its base, the distance to Eros could be determined in miles. From observations of the motion of Eros over a period of time, however, its position within the solar system relative to the other planets was known, and hence its distance from the earth as measured in astronomical units was known. A comparison of the two types of measurement permitted a determination of the length of the astronomical unit in miles.

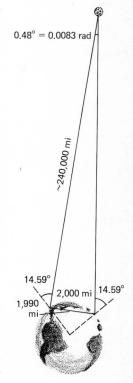

FIGURE 3.4 Measurement of the distance to the moon by triangulation (not to scale).

Another method of finding the value of the astronomical unit so that distances in the solar system can be known in miles or kilometers is to measure the orbital speed of the earth. Observations based on spectrographic analysis (see Unit 5 for a detailed explanation) indicate that the earth moves at an average speed of about $18\frac{1}{2}$ mi/s. The orbit is approximately circular, and the length of the year is known, so that the circumference of the circle and therefore the radius can be calculated directly.

The best measurements of distances within the solar system are made by radar, or in the case of the moon, by visible light. Radar signals have been sent to Venus and the return echo received. Since radar signals travel at the speed of light, 3.00×10^5 km/s, the distance to Venus can be calculated from the time required for the signal to go from the earth to the planet and back with an uncertainty of only a few kilometers.

The first American astronauts to land on the moon left a corner reflector, a device with a number of reflectors which send light back in the same direction from which it arrives. A laser pulse directed toward the moon through a telescope was reflected by the corner reflector and received on the earth again, and measurement of the travel time permitted calculation of the distance from the telescope to the reflector. It is hoped that this sort of measurement will permit determination of the earth-moon distance to within a few centimeters.

From the radar measurements of the distance to Venus, the best value for the astronomical unit is now believed to be 149,587,893 km or about 92,960,000 mi.

Sizes of Planetary Orbits

All the orbits of the planets are approximately circular, a fact which is expressed precisely by saying that the orbits have small eccentricities. Because of this the distances from the sun to the planets are approximately the same as the semimajor axes of their orbits (Table 1.1), and, using the value of the astronomical unit given above, we can compute the distances in kilometers or miles. Knowing that Jupiter is about 483,392,000 mi from the sun will not help people form a very clear picture of the distance, however, unless they have studied the federal budget until such large numbers have some concrete meaning. In order to get across an idea of the scale of the solar system, we shall describe the distances involved in units which may be somewhat more striking.

Suppose that the sun should suddenly cease to shine. How soon would we on the earth know that fact? If you divide 92,960,000 mi by the speed of light, 186,000 mi/s, you will find that the time is about 8 min. Another way of saying that is that

TABLE 3.1

Planet	Mean distance, AU	Mean distance, light time
Mercury	0.387	3 min
Venus	0.723	6 min
Earth	1.000	$8\frac{1}{3}$ min
Mars	1.524	13 min
Jupiter	5.203	40 min
Saturn	9.539	1 h 20 min
Uranus	19.191	2 h 40 min
Neptune	30.071	4 h 10 min
Pluto	39.457	6 h

the earth is 8 light minutes from the sun. The light minute is a unit of distance, as is the mile or the kilometer; it may be difficult to visualize, but at least the number 8 is easier to cope with than 92 million. The distances of the planets from the sun in light travel time are as follows: Mercury, 3 light minutes; Venus, 6 light minutes; Earth, $8\frac{1}{3}$ light minutes; Mars, 13 minutes; Jupiter, 40 minutes; Saturn, 1 hour and 20 minutes; Uranus, 2 hours and 40 minutes; Neptune, 4 hours and 10 minutes; Pluto, the outermost known planet, 6 hours. If the sun suddenly ceased to shine, a resident of Pluto would remain blissfully unaware of the fact for 6 h, whereas the inhabitants of earth would get the word in $8\frac{1}{3}$ min. Table 3.1 gives the mean distance from the sun of all the planets in both astronomical units and light time.

The nearest star is so far away that light requires 4.3 years to reach it. Distances within the solar system, vast by the standards of measurements on earth, are but tiny hops compared to the distances between the stars.

Bode's law Let us rewrite the distances (in astronomical units) of the planets from the sun, rounding off the numbers as follows:

Planet	Distance, AU
Mercury	0.4
Venus	0.7
Earth	1.0
Mars	1.5
Jupiter	5.2
Saturn	9.5
Uranus	19.0
Neptune	30.0
Pluto	39.0

As it turns out, a mathematical progression can be constructed which approximately gives these numbers; this progression is called *Bode's law*. Take the sequence 0, 3, 6, 12, 24, etc., doubling each number after 3 to get the next. Then add 4 to each and divide by 10. Thus $(0 + 4)/10 = 0.4$; $(3 + 4)/10 = 0.7$; $(6 + 4)/10 = 1.0$. This yields the sequence 0.4, 0.7, 1.0, 1.6, 2.8, 5.2, 10, 19.6, 38.3, 77.2. We shall now write these numbers beside the distances of the planets from the sun:

Planet	Distance, AU	Bode
Mercury	0.4	0.4
Venus	0.7	0.7
Earth	1.0	1.0
Mars	1.5	1.6
		2.8
Jupiter	5.2	5.2
Saturn	9.5	10
Uranus	19	19.6
Neptune	30	38.8
Pluto	39	77.2

With the exception of 2.8, to which no planet-sun distance corresponds, the agreement is good up to Uranus; it is fair for Neptune, and is very poor for Pluto.

The name "Bode's law" is inaccurate, for the relation was discovered by Titius of Wittenberg in 1766 and popularized in 1772 by Bode, the director of the Berlin Observatory. To give credit to Titius, the "law" is sometimes called the Titius-Bode law. Of course it is not properly a law in the sense that it accurately describes phenomena; it is approximate, and so far as we know, it may be entirely accidental. When Uranus was discovered in 1781, however, it fit so well into the progression that a search was initiated, for the missing planet at a distance of 2.8 AU. That particular search was unsuccessful, but on January 1, 1801, the Sicilian astronomer Giuseppi Piazzi discovered a small planet between Mars and Jupiter, which he named Ceres, and which was at first assumed to be the missing planet predicted by Bode's law. In March of 1802 Heinrich Olbers discovered another small planet in the same region, and it was named Pallas. During the nineteenth century more than 300 of the minor planets were discovered, and it is now estimated that the number bright enough to leave trails on photographs of the stars is between 50,000 and 100,000. More will be said about them later.

What does Bode's law mean? Why does this progression of

numbers agree so well with the sizes of the orbits of the planets? Is there some physical reason for the agreement or is it fortuitous? No one knows. We know no way to predict where planets should have formed, although it surely was not an entirely random process. At present we can only speculate that the reason Bode's law works to the extent that it does lies hidden in the physics of turbulent motion in the material from which the planets were formed, and that is not understood. Theories of the origin of planetary systems, including our own, will be discussed later, but none of them has been worked out in sufficient detail to know whether Bode's law could be derived from physical principles.

PLANETS AND THE MOON

Now that we have some idea of the size of the solar system, we shall look at some of its components. The study of the planets is an entire field in itself, and many people devote their lives to that study. What will be given here is a superficial description of the gross features of the planets, the sort of information which everyone should know just because these are our neighbors. Anyone wishing more detail about any of the planets can find it in books listed in the bibliography at the end of the book. Many of the numerical facts about the planets are listed in Table 3.2.

The four innermost planets, Mercury, Venus, Earth, and Mars, are often referred to as the *terrestrial* planets. They are comparatively small, they have solid surfaces that we can see or infer, and they are much more like the earth than the next four, which are called the *Jovian* planets.

The Moon

From the standpoint of the universe as a whole, the moon is entirely insignificant (as, of course, is the earth), but residents of the earth cannot ignore it, and we begin with it because it is so near. Exploration of the moon by manned and unmanned vehicles sent from the earth is providing information about both the moon itself and the early stages of formation of the solar system. The investigation of the moon is yielding information about the history of the earth which weathering has long obliterated on our surface.

From the known volume of the moon and its mass, one can compute that its density is 3.34 g/cm^3. (Water has a density of 1.0 g/cm^3.) By comparison, the earth has a density of 5.5 g/cm^3. Since the surface rocks of both bodies are similar in density, one can infer that the moon's core is not as dense as earth's. The

TABLE 3.2 PLANETARY DATA

Planet	Mean dist. from sun, AU	Eccentricity of orbit	Mean diameter, km	Mass (earth = 1)	Mean density (water = 1)
Mercury	0.387	0.2056	4,842	0.05	5.46
Venus	0.723	0.00679	12,822	0.80	4.96
Earth	1.00	0.01676	12,742	1.00	5.52
Mars	1.524	0.09337	6,664	0.11	4.12
Jupiter	5.203	0.04843	139,785	318	1.33
Saturn	9.539	0.05569	115,064	95	0.71
Uranus	19.191	0.04721	47,402	15	1.56
Neptune	30.071	0.008574	43,070	17	2.47
Pluto	39.457	0.24864	5,734	0.10	<5.5

core of the earth is presumed to be iron, but the moon seems to be deficient in iron and all elements more volatile than iron.

The magnetic field of the moon is very small, and one infers that the core of the moon is not molten, although why it has a magnetic field at all is a mystery. The core probably has a temperature in the range 800 to 1000°C.

The moon has no atmosphere. That is to be expected, for it is so small that its gravitational attraction is not sufficient to hold gases at the temperatures reached at its surface. During the lunar day, surface temperatures reach 110°C, slightly above the temperature of boiling water. During the night, temperatures fall to about −170°C.

The most obvious surface features of the moon are the *maria* (so named by Galileo because they look like seas), the mountains, and the craters. The moon has no water and never had enough to form seas. The maria are regions where lava flowed from beneath the surface, probably between 3.1 and 3.4 billion years ago, and formed large, relatively smooth sheets in already existing craters.

The lunar mountains rise above the surrounding surface to heights comparable to those of terrestrial mountains. They are more rugged and sharp than mountains on the earth because of lack of atmosphere-produced weathering. ("Slow weathering" does occur on the moon because of the bombardment of the surface by gas and small dust particles.)

The possibility that some of the moon's craters might be volcanic in origin has often been considered, but it now seems likely that all of them are the result of meteorite impact. Small meteorites still strike the lunar surface, just as they do the earth, but the damage done to the moon is much greater than that done to the earth because the moon has no atmosphere to protect it. When the solar system was much younger, evidently

Surface gravity (earth = 1)	Period of rotation		Period of revolution		Known moons
0.38	59	days	88	days	0
0.87	243	days	225	days	0
1.00	24	h	365	days	1
0.39	24.5	h	687	days	2
2.65	10	h	11.9	yr	13
1.17	10	h	29.5	yr	10
1.05	11	h	84	yr	5
1.23	16	h	165	yr	2
<0.5	6	days	248	yr	0

the moon (and presumably also the earth) was struck by many relatively large objects, creating the huge craters visible on the surface. Bombardment of the surface seems to have been particularly heavy about 4 billion years ago.

Dating of rocks brought back by Apollo astronauts has revealed rocks older than any found on the earth and as old as any meteorites picked up on the earth. Evidently the moon was formed at about the same time as the earth. It is not known whether the moon formed approximately where it is now or whether it formed in another part of the system and was then captured by the earth.

On the moon. Scientist-astronaut Harrison Schmitt is standing beside a huge, split lunar boulder during the Apollo 17 landing at Taurus-Littrow. (NASA.)

Seismographs left on the moon's surface by American astronauts report periodic moonquakes. These occur most often when the moon is nearest the earth and seem to indicate slippage of sections of the crust. It may be that as the moon recedes from the earth, it is becoming less pear-shaped and more nearly spherical. The quakes originate at depths of half the moon's radius, much deeper than quake centers in the earth.

Although there is no volcanic activity on the moon now, there is some evidence that gas is released occasionally from the moon's surface, either vented from below or released by impact of a meteorite.

Most of the moon's surface is covered by a few meters of soil made by the shattering of the original rock by meteorites. The ages of rocks and soil samples returned to the earth, estimated by studying the radioactive elements in the rocks, range from about 3.1 to 4.2 billion years.

Although the United States manned and the Soviet unmanned expeditions to the moon have provided a wealth of new data about that body, some of the most interesting and important questions about the moon are still open for debate.

The most important questions which the expeditions were intended to answer concern the moon's history and its origin. Besides their intrinsic interest, the answers to these questions are important to our understanding of the earth. The geological record of the first billion years of the earth is not available to us, for the first rocks to form either sank back into the hot interior or have been weathered and changed beyond recognition. The moon, with no atmosphere to produce rapid weathering and with less geological activity than earth, still harbors materials from the time when the earth was forming. Rocks have been returned from the moon which solidified as long ago as 4.2 billion years, and it is believed that the moon is not more than one-half billion years older than that.

One of the surprising results of the study of rocks from the highlands, the parts of the moon's crust which solidified first, is the discovery that the highlands rock is basalt, a type which we know results only from lava flows. The conclusion seems inevitable that at one time the surface of the moon was entirely molten. After it cooled huge craters were formed by the impact of large bodies striking the surface, and much later, 3.1 to 3.8 billion years ago, lava flowed again from the interior and filled some of those craters, forming the maria.

Was the moon hot or cold when it first formed? That question is of great interest because the same conditions probably existed on and in the new earth, but this is one of the areas where the data can be interpreted either way, and we are no nearer a solution than we were before men went to the moon.

We believe that the sun, the planets, and the satellites all formed from one large cloud of gas and dust which shrank in size because of gravitational forces and formed a disk about the new sun. Within that disk of gas and dust, chunks began to coalesce as particles collided with one another, and eventually larger and larger pieces came together to form the planets and their satellites (see Unit 15 for a more detailed explanation). One can readily imagine two possibilities. It may be that as a large body began to form, the impact of other pieces falling onto it produced so much heat that the entire body melted and as the planet or satellite reached its final size it was molten throughout. Under such conditions the heavier materials, such as iron, would sink to the center, and a dense core would form. This separation of lighter materials from heavier is termed *differentiation*, and it has occurred in the earth and the moon. The second possibility is that the large bodies in the solar system formed cold, but later became hot enough to melt and cause differentiation. It has been proposed that the moon formed cold, bombardment of the surface by materials of the dust cloud producing enough heat on the surface to melt the surface layers, thus explaining the basalt found in the highlands and mountains. According to this theory, only later did the core become molten so that lava could well up into the maria. In both the moon and the earth, heat released by radioactivity within the interior is assumed to produce the high temperatures required to melt the interior after the body has formed.

Although this controversy has not been definitely resolved, certain indications point to a hot origin for both the earth and the moon.

1 The ages of rocks brought back from the moon suggest that differentiation of the moon occurred within the first billion years of its existence, a time scale consistent with a hot beginning.
2 The oldest rocks found on the earth were formed by heat processes.
3 Some very old meteorites which have struck the earth seem to have been formed by the cooling of molten rock. Since those objects presumably were never part of a body large enough that it would have melted by the release of heat from radioactivity, it is plausible that they formed hot.

Another question to which no definite answer can be given concerns the origin of the moon. Four distinct theories have been proposed and considered.

1 In 1898, the physicist George Darwin, son of the biologist

Charles Darwin, proposed that the moon was torn from the earth because of the earth's rapid spin, and that the Pacific Ocean basin is the hole from which the moon came. Recently this theory has been cited as a possible explanation for the deficiency of iron on the moon. If the separation of the moon from the earth occurred after most of the earth's iron had sunk to the core, the absence of enough iron on the moon to form a heavy core is accounted for. Darwin's theory, however, offers no plausible means by which the tearing apart could have occurred, and it is not widely accepted.

2 As the earth was forming from a cloud of gas and dust, it spun rapidly enough to shed a ring of material which condensed to form the moon. The most obvious problem with that explanation is that such a ring should have been shed at the earth's equator, and consequently the orbit of the moon should lie about the equator; the moon's orbit actually lies more nearly in the plane of the ecliptic than in the plane of the equator. Further, it is doubtful that the earth was spinning rapidly enough to shed such a ring.

3 The earth and the moon formed as a double planet. Our moon has the largest mass of any satellite in the solar system relative to its parent body, so that the earth and the moon are almost like a double planetary system. The difficulty with this hypothesis is that it does not explain the deficiency of the moon in some elements, particularly iron. If both bodies formed at the same time, in the same place, from the same materials, why are their compositions different?

4 The moon formed elsewhere in the solar system as a separate planet and was then captured by the earth. Capture is a rare, unlikely event, but it is possible, and it may be the best of the four explanations.

Mercury

Mercury's maximum angular distance from the sun is 28°, and so it is very difficult to see after the sun sets. Most observations of the planet must be made during the daytime. If observers know where to look, they can see Mercury in the Western sky, if it is above the sun, just after the sun has gone down and before the sky gets dark. By the time the sky is thoroughly dark, the planet is almost certainly below the horizon. Of course it may also be seen in the Eastern sky just before or immediately after sunrise. It is said that Copernicus never saw Mercury; he was not in a good latitude and probably he did not have very good seeing conditions.

Most astronomy books written before 1965 state that Mercury goes around the sun in 88 days and rotates on its axis in the same period so that it always turns its face toward the sun. The belief was that Mercury is tidally locked to the sun with its bulge always facing the sun, just as the moon is tidally locked to the earth. Mercury probably does have a bulge, although it cannot be a large one, but it is possible for a planet with a bulge to be locked in with conditions such that it rotates on its axis in two-thirds the time required to go around the sun, and that apparently is what happens with Mercury. The surprising thing is that if you calculate the way Mercury will look to an observer on the earth, you find that if it has this two-thirds period it always shows us so nearly the same face that the difference is hard to distinguish, given the poor seeing conditions which always plague the observer of Mercury. Because year after year an observer will see almost the same markings, it is not foolish to conclude that the planet is turning in synchronism with its motion around the sun.

In 1965 radar signals were bounced off Mercury, and the shifts in their frequencies indicated that the planet is turning with a period of about 59 days. Since that time it has been found that a two-thirds-period synchronism is possible, and it seems almost certain that Mercury's actual period of rotation on its axis is $\frac{2}{3} \times 88$ days $\approx$ 59 days ($\approx$ means "is approximately equal to").

The pre-1965 books also often stated that Mercury has the hottest and coldest places in the solar system. It was believed, for example, that the side turned toward the sun was hot enough to melt lead, conjuring up interesting pictures of rivers of molten lead. How the lead would return to the "headwaters" is not clear, for surely there would not be a rain of molten lead, but at least the idea makes vivid the notion of a hot region. The other side of Mercury, perpetually turned away from the sun, would be very cold. Measurements of the temperature of Mercury indicate that the earlier ideas were not far from right, for although it does not perpetually turn one face toward the sun, a day on Mercury lasts 88 of our days (176 days from noon to noon), and in that length of time temperatures can rise very high and fall very low. The planet has little if any atmosphere to help transfer heat from the hot side to the cold side. Temperatures on the side facing the sun probably reach 327°C (which is 620°F, the melting point of lead), and on the cold side they may drop as low as −173°C (−280°F).

Since Mercury has no satellite, measurements of its mass had to be made by observations of its effect on other planets until the Venus space probes passed near it. The best estimate now is that its mass is about one-eighteenth that of the earth. Its diameter is about 3,030 mi; it is the smallest and lightest of the planets.

In late March of 1974 Mariner 10 approached Mercury, and for about 10 days it sent back television pictures of the planet—some 2,000 photographs in all. At nearest approach, on March 29, Mariner 10 was only about 420 mi from Mercury.

The Mariner 10 photographs show basins, craters, ridges, and plains strikingly similar to those of the moon; the plains look like lunar maria and have similar cratering. Craters range in size from large basins to those on the edge of detectability.

Because large plains have been flooded by lava which appears to be similar to that which formed the maria on the moon, one can infer that the outer regions of the planet are primarily silicate materials, with a density approximately 3 g/cm³. The average density of the planet is about 5.5 g/cm³, and therefore the planet must be differentiated, with a dense core; probably the core is iron.

Because the cratering suggests that we see some of the results of the final accretion of material as the planet was being formed, we can infer that the differentiation of materials occurred early enough that a solid crust had formed which did not melt before large pieces of material ceased to fall upon the surface. This means that the planet probably formed hot and that the surface was not melted by a rain of large bodies near the end of planetary formation. The existence of such craters on Mercury also indicates that it never had an atmosphere.

Mercury, photographed by Mariner 10 on March 29, 1974, from a distance of 200,000 km. This mosaic of the southern hemisphere is made from 18 pictures. Largest of the craters shown are about 200 km in diameter. (*Jet Propulsion Laboratory.*)

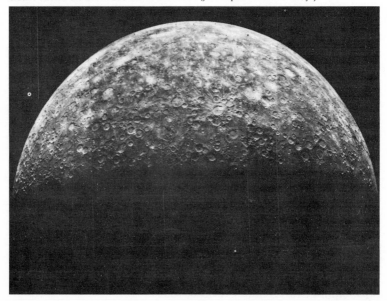

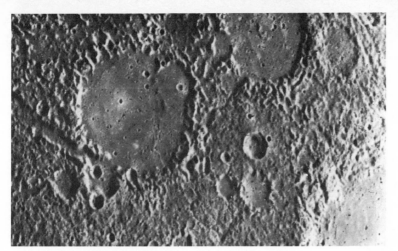

Mercury, photographed by Mariner 10 on March 29, 1974, from a distance of 35,000 km. The picture covers an area 290 by 220 km; the large flat-floored crater near the center is about 80 km in diameter. (*Jet Propulsion Laboratory*.)

It has commonly been assumed that the cratering of the moon and of Mars was severe because of their proximity to the planetoid region, and that cratering on Mercury was much less. That does not seem to be the case; cratering on all three may have been similar. The significance of this observation is not clear, but at the least, we must assume that the early earth suffered bombardment similar to that still recorded on the surfaces of both the moon and Mercury.

Venus

Venus can appear as much as 47° from the sun, and hence it can be bright in the Western sky after the sun has set or in the Eastern sky before sunrise. It is the Evening Star and the Morning Star. Its orbit is the most nearly circular of all the planets, and it is most like the earth in mass and size.

The surface of Venus has never been seen because the planet is always covered by a layer of clouds. The nature of the clouds is uncertain. From both spectroscopic observation and information sent back by the Russian Venera craft which have entered the atmosphere and parachuted to the surface, we now know that the atmosphere is 95 to 96% carbon dioxide, about 2% nitrogen, less than 1% water vapor, and much less than 1% oxygen. Mariner 10 detected hydrogen, helium, neon, and argon in small amounts at the very top of the atmosphere. Whatever the clouds may be, it is clear that they are not water, as was supposed until good data about the atmosphere composition and temperature became available.

Radar signals are regularly bounced off the surface of Venus, and provide enough information for us to make a very rough map of part of the surface. One large mountainous area has been found where peaks reach a height of about 3 km.

From radar observations it is known that Venus rotates retrograde, east to west, in a period of about 243 days. (All the planets except Venus and Uranus rotate west to east.) Most explanations for the origin of the planets can explain why they should all rotate in the same way, but the odd rotations of these two are difficult to explain.

Venus may have begun its life rotating in the same direction as the earth, and perhaps at a similar rate, but because of its nearness to the sun, tidal forces produced by the sun slowed its rotation rate toward the point at which it would have always presented the same face to the sun. It is not tidally locked to the sun, however, and the earth may be responsible for that fact, for Venus rotates at such a rate that it turns the same side toward the earth every time the two planets are closest together. If this explanation is correct, the rotation of Venus is determined by tidal forces produced by both the sun and the earth.

The most detailed information about the temperature and atmospheric pressure of Venus has come from the Russian Venera; the information indicates a surface temperature of about 475°C and a pressure of about 1,300 lb/in². Mariner 10, which passed Venus on February 5, 1974, sent back television pictures of the clouds as well as other data about the planet and its atmosphere, and now data are available that permit an attempt to understand the dynamics of the atmosphere.

Mars

Mars is perhaps the most interesting of the planets; at least it has been the object for the most speculation. We get the best view of it; its surface can be seen without much cloud obscuration; and of all the planets in the system other than earth, it seems the most likely to be hospitable to life.

Speculations that intelligent life exists on Mars go back at least to 1877 when an Italian astronomer named Schiaparelli thought he observed straight lines on the surface, which he called *canali*, or channels. Fine detail can never be seen on the surface of Mars (at its nearest approach it is still 150 times as far away as the moon), but many observers have thought at times that they saw a network of lines, and what seem to be vague lines have appeared on photographs. The notion of artificial canals received support from the fact that when Mars's ice caps melt during the Martian summer, a wave of darkening moves down the planet toward the equator as if water were being directed from the polar regions in a canal system toward the equator, causing foliage to change color. Television pictures sent back from a distance of a few thousand miles show no such lines, however, and it now seems certain that they are optical illusions.

Another fascinating bit of history related to Mars concerns its two moons. After Galileo discovered four of Jupiter's moons, Kepler speculated that Mars should have two satellites, but he had no evidence for it. Then in 1726 Jonathan Swift, in *Gulliver's Travels*, reported that the scientists of Laputa had discovered two satellites around Mars, and he predicted that their periods of rotation about the planet should be 10 and $20\frac{1}{2}$ h. Almost exactly 150 years later, in 1877, two satellites were discovered; their periods are 7 h 39 min and 30 h 18 min. The satellites have been named Phobos and Deimos, meaning *fear* and *panic*. They are so small that their sizes had been estimated only from their brightness until Mariner 9 gave a close-up view of them in December 1971. Both are irregular, shaped somewhat like potatoes. Phobos is 16 km from pole to pole and 23 km wide; Deimos is 9 km through the poles and 11 km across. Both are cratered and pocked.

The Martian day is almost the same as the terrestrial day: 24 h, 37 min, 23 s. This is considered accurate to within a few hundredths of a second. The Martian year is about two of our years. Since the axis of Mars is inclined to the plane of the ecliptic much like that of the earth, it has seasons like the earth does. During the winter in each hemisphere a polar cap forms; during the summer it disappears or almost disappears. Most of the cap is frozen carbon dioxide, but there may be a small amount of water ice which never melts. Temperatures in the equatorial regions tend to run from about 89°F during the day to perhaps −100°F at night. The temperature at the south pole may get as high as 32°F and as low as −150°F, but the temperature at the north probably gets above −100°F only rarely.

In November 1971 the Mariner 9 spacecraft, launched by the United States, went into orbit around Mars. It was the first craft made by man to orbit a planet other than the earth. For $11\frac{1}{2}$ months, until late October 1972, Mariner 9 returned pictures of Mars, as well as the results of observations made with both infrared and ultraviolet spectrometers and an infrared radiometer, and we learned more about the planet during that period than we had in all previous study. The entire surface has been mapped in considerable detail.

Occasionally Mars has severe dust storms, and one of the worst ever recorded was in progress when Mariner 9 reached the planet. That was an initial source of disappointment because the surface of the planet was totally obscured by the

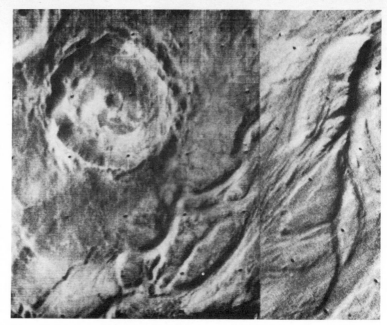

Mars. Photo taken on March 4, 1972, by Mariner 9 of a braided channel sweeping past a crater. Such channels suggest the former presence of fluid erosion of the Martian surface. The crater is 19 km in diameter; both photos in this mosaic were taken from a range of about 1,800 km. (*Jet Propulsion Laboratory*.)

dust, but as the dust settled, viewing conditions became good and the presence of the settling dust permitted observation of wind patterns that otherwise could not have been seen.

Pictures returned from spacecraft which flew past Mars before November 1971 seemed to indicate that the surface was pockmarked, much like the moon, but the detailed pictures of the entire surface returned by Mariner 9 give a quite different impression. It seems certain now that Mars is not a dead and relatively static planet, as had been supposed after examining the earlier pictures, but rather an evolving, geologically active one. It has giant volcanoes and lava flows, faults and rifts suggesting movements of the crust of the planet, and, most surprising, it has evidence of water flow in the not distant past.

Mars is geologically less active than the earth, but it is much more active than the moon, and consequently it has many features which the moon does not have. Both the moon and Mars, as well as Mercury, exhibit the same kind of severe cratering. It has been suggested that approximately 4 billion years ago Jupiter perturbed material in the asteroid belt sufficiently to cause it to spray across the inner parts of the solar system, causing simultaneous bombardment of the surfaces of Mars, Jupiter, and Mercury. Presumably the earth was similarly affected, al-

though we cannot see the effects directly. Craters on Mars appear to be somewhat eroded, but evidently the planet has not
had a heavy atmosphere and large amounts of water for anything like all its life or the evidences of early cratering would have been obliterated. For some unknown reason, the northern hemisphere of Mars is only slightly cratered, but the southern hemisphere is heavily cratered.

The present atmosphere of Mars is thin, and composed almost entirely of carbon dioxide. Very small amounts of other gases, including water and oxygen, have been detected, but the atmospheric pressure and temperature at the surface are such that no liquid water can exist on the surface now. It is possible, however, that large amounts of water are frozen into the polar ice caps. Some of the pictures sent back by Mariner 9 show surface features that can be interpreted most easily as examples of water erosion; they look like deep water channels which were fed by many tributaries. Since meteorite craters are almost entirely absent in the channels, they must have been cut in rather recent times, although they do not all appear to be the same age. No entirely satisfactory explanation for the evidence of water

This mosaic of two photographs of the Tithonius Lacus region on Mars taken by the Mariner 9 spacecraft revealed a canyon nearly 4 times as deep as the Grand Canyon in Arizona when the pictures were compared with pressure measurements taken by the violet spectrometer aboard the spacecraft. The white arrow at left points to the Martian canyon, estimated to be 19,700 ft deep and 75 mi wide. Earth's Grand Canyon is 5,500 ft deep and 13 mi wide. The jagged line at bottom represents the pressure measurements, taken by the ultraviolet instrument, which are translated into distances. (*Jet Propulsion Laboratory*.)

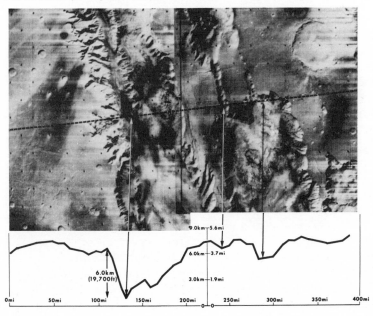

erosion has been offered; one hypothesis is that the Martian climate varies cyclically between a warm and wet phase, in which the atmospheric density is high enough to permit the existence of liquid water, and a cold and dry phase, which we are seeing now. During the cold phase most of the water is locked into the polar caps.

The polar caps appear to be made of two constituents; one melts or disappears during the "summer," and the other always remains in place. The one which appears and disappears is probably carbon dioxide, which changes from vapor to dry ice; and the one which always remains in a solid form is water ice.

Aside from the evidence that water flowed across the surface of Mars at some time in the past, the most striking surface features are a gigantic canyon, which dwarfs the Grand Canyon, and huge volcanoes. The canyon reaches depths of about 20,000 ft and stretches farther than the distance from Los Angeles to New York. One of the volcanoes stands about 26 km (15.5 mi) above the surrounding plain, and the most spectacular, Nix Olympica, is 310 mi across the base and rises almost 14 mi above the plain. All these features are much larger than anything comparable on the earth. Mars seems to have both very old (2 billion years) and relatively young (100 million years) volcanoes.

The questions about the existence of life on Mars are still unanswered, but spacecraft are being planned that will land or

A channel thought to have been formed by running water in Mars's geologic past is seen in this mosaic of three pictures of the planet taken July 1, 1972, by Mariner 9. "Flow" of the channel is northward, from lower left to upper right. This segment of the channel is about 75 km long. (*Jet Propulsion Laboratory.*)

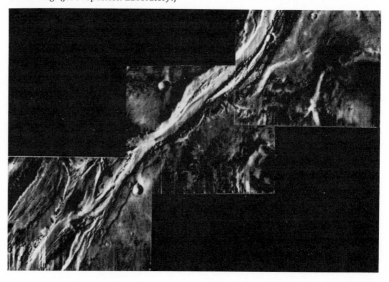

the planet in 1976 in an attempt to find life. The present atmospheric and temperature conditions seem to eliminate the possibility of large animals, but the evidence that liquid water once
existed in large amounts raises hopes that life may have begun
and some forms may have adapted to the present arid conditions.

Jupiter

If we were traveling outward from the sun, after leaving Mars
we would come next to the region where many minor planets,
or planetoids, move. These will be discussed later. Now we
pass on to the next major planet, Jupiter.

Conditions change dramatically by the time we reach
Jupiter. The first four planets are small, the earth, with a diameter of 8,000 mi, being the largest. They have solid surfaces and
relatively thin atmospheres, and the atmospheres are heavy
gases—nitrogen, oxygen, carbon dioxide, etc.

Jupiter is a huge planet, with a diameter of 89,000 mi—11
times that of the earth. Whereas the earth turns on its axis in
24 h, Jupiter turns in 10 h; its rotation rate is so great that it is
noticeably larger through the equatorial plane than through the
poles. It has 13 known satellites. Four of them were seen by
Galileo and hence are called the Galilean satellites; they can be
seen with binoculars, and one is so bright that it would be visible to the naked eye if it were not so close to the planet.

Jupiter's atmosphere is hydrogen, helium, methane, and ammonia. The relative abundances of hydrogen, helium, carbon,
and nitrogen appear to be about the same as in the atmosphere
of the sun, if all the carbon is in the form of methane (CH_4) and
the nitrogen is in ammonia (NH_3). Temperatures measured
from observation of the radiation emitted from the tops of the
visible clouds are approximately $-140°C$ or $-220°F$. Above
the clouds, however, the temperature of the atmosphere appears to rise for a time and then fall again. According to
measurements made with Pioneer 10 as its radio signals
passed through Jupiter's atmosphere, the temperature at one
point reaches a value of about $0°C$, the melting point of ice.

Jupiter is more massive than all the other parts of the solar
system combined except for the sun—318 times the mass of the
earth and about $\frac{1}{1,000}$ the mass of the sun. Its density is only 1.33
times that of water, however, and so we assume that it does not
have much, if any, core made up of stone and metals such as
earth's. (The density of the earth is 5.5 times that of water.)
Probably the interior of Jupiter consists of liquid hydrogen, although at the center hydrogen may be compressed until it is a
metallic solid. The pressure at the center may be 10 million

lb/in², and the temperature is high, perhaps as much as 100,000°F.

Even using a small telescope an observer can see that Jupiter has a banded appearance; the bands apparently are clouds, for they move with differing speeds. The most spectacular feature of the visible surface, which is composed entirely of clouds, is the Great Red Spot. It was seen telescopically at least as early as 1831, and it may have been seen in 1660. It changes in color and position, and somewhat in size, and has been as large as 30,000 mi across. Various explanations have been offered for it—volcanic eruption, floating cloud of particles, etc.—but perhaps the most satisfactory is that it is some kind of vortex in the atmosphere driven and maintained by the atmospheric currents that flow around the planet. Best estimates of the nature of Jupiter suggest that explorers descending through the atmosphere would go through a dense layer of hydrogen, helium, methane, ammonia, and perhaps also nitrogen and the inert gases, until they reached a point where the pressure was so great that the atmosphere had changed to liquid—a liquid sea of hydrogen and methane. Jupiter is almost massive enough to be a small star, and in fact a small amount of heat appears to be flowing up from its interior. It is not becoming hotter, however, and it will remain a cold planet.

It is conceivable that a zone may exist in which temperatures are not extreme and where the atmosphere is similar to that believed to have existed on the earth when life began—an atmosphere with little or no oxygen, but much methane and hydrogen. This possibility has produced speculation that life may exist on Jupiter—some strange form of life not using oxygen and adapted for high pressure. It is an interesting speculation, but we are far from being able to send craft into the atmosphere of Jupiter to try to verify it.

Saturn

The rings of Saturn make it one of the most striking objects in the sky for viewing with even a small telescope. It is the second largest of the planets. One very unusual feature of Saturn is its density—it is less dense than water and so would float if one could put it into a tub of water large enough to hold a planet with a diameter of 72,000 mi. Like Jupiter, Saturn shows bands when observed through a telescope, and since it rotates in approximately 10 h, it is flattened through the poles like Jupiter.

The rings stretch from about 7,000 to 50,000 mi above the surface of the planet. They are made up of many, many particles, most of which seem to be the size of small gravel, and there is evidence that the particles are bits of water ice. Howev-

er, radar reflections indicate that some of the particles are the
size of large boulders. The inclination of the rings to our line of
sight varies, and when we see the rings edge-on, they disappear. The thickness cannot be measured directly, but observations set a limit of 10 to 15 mi, and there are theoretical reasons
for supposing that they may be only feet or inches thick. At
times stars can be seen through the rings.

The temperature of Saturn appears to be about $-145°C$. The
only gases which can be detected in its atmosphere are hydrogen and methane; presumably the ammonia has frozen out. Its
core is presumed to be liquid or solidified hydrogen and
methane.

Saturn is known to have 10 satellites in addition to the rings,
and one of them, Titan, is larger than our moon. Methane has
been observed surrounding Titan, and prior to 1974 it was the
only satellite in the solar system known to possess an atmosphere. In 1974 Pioneer 10 detected an atmosphere around Io, a
satellite of Jupiter.

Uranus

The ancients recognized no planets beyond Saturn, although
they must have seen Uranus. It is visible to the naked eye, looking like a faint star, and between 1690 and 1781 it was plotted
as a star on star charts at least 20 times. It is so faint and its motion is so slow, however, that it was not recognized as a planet
until 1781. Then William Herschel, the incredibly energetic
German-Englishman, who made a living teaching music by day
and observed the stars all night, saw through his telescope that
it looked like a disk rather than a point of light. He thought at
first that it was a comet, but eventually enough observations
permitted a calculation of its orbit, which turned out to be almost circular, lying entirely beyond Saturn.

Uranus is so far from the earth that observations are difficult
and our knowledge of it is not great. It appears to have a diameter of about 30,000 mi and a density about $1\frac{1}{2}$ that of water.
Spectra of the atmosphere give strong indications of methane
and hydrogen, but no indication of ammonia, which probably
is frozen out. The temperatures are thought to be below
$-185°C$.

Uranus has five known satellites, all rather small. In the telescope it appears somewhat greenish, perhaps because of the absorption of light by methane. The rotation period is uncertain
but probably is slightly under 11 h.

The most striking thing about Uranus is that its equatorial
plane makes an angle of 82° with the plane of the ecliptic (the
plane in the sky within which the sun moves). As we have

seen, it is the earth's tilt which produces our seasons; if a planet had no inclination it would have no seasons and the sun would always appear directly above the equator. The tilt of Uranus is so great that the sun at times appears almost directly over the north pole; at other times it appears over the south pole. This must make an interesting mixture of seasons with day and night, although because of Uranus's distance from the sun, seasonal variation in temperature is probably small. Not only does Uranus have a unique orientation with respect to the ecliptic, but, like Venus and the outer four of Jupiter's satellites, it rotates in a retrograde direction.

Neptune

Uranus was discovered by a man with a telescope; Neptune, on the other hand, was discovered by a man with a pencil and paper, or, more accurately, by two such men.

After the discovery of Uranus, its orbit was computed from observations made within the next few years and from older records, for once astronomers knew where to look, they could find it recorded as a star on previous charts of the heavens. By about 1830 it was apparent that the planet was slightly off course —not much, but more than the expected uncertainty in the computations. The calculations of its orbit included allowance for the effect of the sun, holding it in the orbit, and the perturbations produced by the other planets, notably Jupiter and Saturn. One explanation offered for the deviation from the prescribed path was that perhaps the force of gravitation was different at such great distances from the sun. Another explanation was that another planet might lie outside Uranus, and it might attract Uranus forward or retard it, depending upon its position.

The latter suggestion was taken seriously in 1843 by John Adams, an Englishman, shortly after he had completed his work at Cambridge. He began an analysis of the motion of Uranus, trying to explain it in terms of another planet farther out. By 1845 he had concluded that there was indeed another planet, and he computed its position; the results he sent to Sir George Airy, the Astronomer Royal, asking him to look in the part of the sky where his computations indicated that the new planet should be found. Not knowing anything about the young mathematician, Airy sent him a problem designed to test whether he knew enough mathematics that his work should be taken seriously enough to turn the observatory to the search for the predicted planet. Adams ignored the test.

In the meantime, a French mathematician, Leverrier, unaware of the work Adams was doing, tackled the problem and

arrived at approximately the same conclusion. He also sent his
results to Airy, and when Airy asked him to prove his ability by
solving the same test problem, he promptly did so. Airy then
suggested to the director of the Cambridge Observatory that he
search for the predicted planet, but the director, having no up-
to-date star maps of the region, began a slow plotting of all the
stars in the region, hoping that eventually the new planet
would show itself by noticeable movement. (He actually saw
the planet but did not recognize it.)

About a month later, Leverrier suggested to Galle, an astronomer for the Berlin Observatory, that he look for the planet.
Galle had adequate star charts of the region, and on the night of
the day he received Leverrier's letter, he located the new
planet, less than 1° from the position predicted by Leverrier. It
was named Neptune, and Adams and Leverrier share the honor
of discovery.

Neptune requires 165 years to complete one revolution
around the sun, and therefore it will not have returned to the
position in which it was discovered until 2011. It appears to rotate on its axis in approximately 16 h. It has two satellites.

Neptune's mass is about 17 times that of the earth, and its diameter is approximately 28,000 mi. Methane and hydrogen
have been detected in the atmosphere, and helium is very
likely present but unobservable. The temperature must be near
$-210°C$. Neptune probably has a structure very much like that
of Uranus.

Pluto

After the effects of Neptune on Uranus had been computed,
there remained a slight discrepancy between the predicted
position of Uranus and the observed position. It is now known
that the difference was less than the errors of the original observations and so probably was not a real difference, but several
men, among them the Americans W. H. Pickering and Percival
Lowell, began to search for a ninth planet, beyond Neptune,
whose effect would explain the remaining perturbation of the
orbit of Uranus. Lowell computed the position of the new
planet, which he calculated should have a mass about 6.6 times
that of the earth. He searched for it from 1906 until his death in
1916, and he founded the observatory bearing his name in
Arizona for that purpose. In 1930 Clyde Tombaugh, working at
the Lowell Observatory, found the sought-for planet, within 6°
of Lowell's predicted position.

Pluto's orbit has both the highest inclination to the plane of
the ecliptic of any planet and also the largest eccentricity; at its
greatest distance from the sun (aphelion) it is over 4.5 billion

mi away, and at perihelion it is under 2.8 billion mi away.

Pluto probably has a diameter of about 3,700 mi, although that is rather uncertain; it could be somewhat smaller. It has no known satellites, and hence its mass can be determined only by observations of its perturbing effect on Neptune; that uncertain calculation indicates a mass about 0.18 times that of the earth. It varies in brightness with a period of slightly more than 6 days, and that is assumed to be its period of rotation on its axis. Nothing is known of its atmosphere, but at the temperature which it must have, about $-240°C$, only neon, helium, and hydrogen remain gaseous, and helium and hydrogen probably would all escape from such a small body.

When Pluto was discovered and its diameter was estimated, it was calculated that in order to have the mass which Lowell had predicted it must have to produce the observed effects in the motion of Uranus, Pluto must have an unreasonably high density. It is now believed that its density probably is not abnormally high, and that its mass is too small to produce the effects in the motion of Uranus which Lowell used to predict its existence. Almost certainly the discovery of Pluto was accidental. The supposed discrepancies in the orbit of Uranus were produced by observational errors in the past, which made the actual prediction of another planet difficult, if not impossible.

Is Pluto the outermost planet or are there others still unfound? If there are others beyond Pluto, they must be very far out and very small. Tombaugh made a search for other planets after his discovery of Pluto, and he should have found anything as large as Neptune within 270 AU, almost 7 times the distance to Pluto, but nothing was seen.

OTHER MEMBERS OF THE SYSTEM

Minor Planets

In addition to the major planets, there are in the solar system a great number of *minor planets,* or *planetoids.* Originally these were called asteroids because they look like stars in a telescope, but the designation minor planet is coming to be generally accepted. The minor planets lie generally between Mars and Jupiter. The first of them was discovered by Piazzi on January 1, 1801, the first day of the nineteenth century. The second was not discovered until the next year, in March 1802. The standard reference for the minor planets, a Soviet publication called *The Minor Planet Ephemeris,* now lists 1,735 of these bodies. There must actually be thousands more—estimates run as high as perhaps 40,000—that could be easily seen and catalogued, but no one has made the effort. The very small ones are not sufficiently important for anybody to spend the time necessary to calculate orbits for them, or determine enough about them so that they could be definitely identified from one observation to the next.

The largest known is Ceres, 488 mi in diameter. The next in size is called Pallas, with a diameter of 304 mi; Vesta has a diameter of 248 mi, and Juno 118 mi. There probably are fewer than a dozen with diameters as great as 100 mi. There are perhaps a few hundred with diameters more than 25 mi, but most of the planetoids are very small, probably with diameters less than 1 mi.

In some cases, the brightness of these objects varies, and this led observers to conclude that they are not spherical and that they rotate. One in particular, Eros, is shaped like a brick. This was first hypothesized after observing variations in its reflectivity, but on one occasion it came near enough that it could be observed with telescopes and the deduction was confirmed. Eros is 15 or 20 mi long and about 5 mi through the small dimensions, and it tumbles end over end with a period of about 5 h.

Most of the minor planets have orbits which keep them between the orbits of Mars and of Jupiter. There are two interesting characteristics of some of the orbits. The first is that there are gaps in the periods; that is, none of the planetoids have periods of certain specific lengths. These gaps, which are called Kirkwood gaps after the man who first observed and explained them, are simple fractions of the period of Jupiter. Jupiter, which lies just outside the region occupied by the planetoids, is a massive object, and when a minor planet has a period such that each time it comes around, or every second time or every third time it finds itself moving parallel with Jupiter for a while, it experiences a repeated pull which moves it out of that orbit and changes its period. Hence periods which are one-third, two-fifths, three-fifths, or some other simple fraction of Jupiter's period, are not found among the minor planets. (The same effect produces the gaps in Saturn's rings, with the role of Jupiter being played by the satellites of Saturn. In fact, the most recently discovered of Saturn's satellites, Janus, was found in 1966 because it produces an otherwise unexplained gap in the ring system. Since the particles in the ring move in almost circular paths, a gap in periods corresponds to a gap in position.)

The second interesting characteristic of the orbits of some of the minor planets is that they lie at Lagrange points related to Jupiter and the sun. Lagrange, one of the greatest theoreticians in mechanics, discovered that two points on the orbit of a very massive body such as Jupiter, so situated that they, Jupiter, and the sun are at the vertices of equilateral triangles, are *points of equilibrium* (Fig. 3.5). That is, if small bodies move into those positions, they are trapped there; they may oscillate about the positions but they never move far away. There are two such points, one ahead of Jupiter and one behind it, and planetoids are known to be trapped in both positions. These are called the Trojans, and a convention for naming them has been adopted. Those that lie ahead of Jupiter are named for Greek heroes of the Trojan War, and those that lie behind Jupiter are named for Trojan heroes. Unfortunately the convention was not adopted until one planet in each group had been misnamed, so there is one Greek in the Trojan camp and one Trojan in the Greek camp. Each of these groups makes its circuit about the sun in the same period as Jupiter—12 years.

The total mass of the planetoids is not known, but it probably is less than $\frac{1}{1,000}$ that of the earth.

What is thought about the origin of this swarm of small bodies? At one time it was believed that a planet existed in the region and it broke up, but that now seems very unlikely. It probably would not be possible for a planet of any size to form that near to Jupiter because Jupiter would produce such large tidal forces in the embryonic planet that it would be broken up. More likely the influence of Jupiter prevented the formation of a single planet, and perhaps several—maybe a dozen—small planets formed at the time the other planets were forming. Because the attraction of Jupiter would constantly cause the orbits of planets in that region to change, it is possible that a collision occurred at least once between two of the original bodies, and in the collision, they were shattered. If that happened, their fragments probably continued to collide with each other or with the remaining large bodies. The result of such collisions would be to grind the small bodies still smaller, and we assume that such a process is still going on, continuously reducing the size of the smaller objects, but probably not affecting the larger ones much.

Occasionally a long-exposure photograph of the stars, made by causing a telescope to rotate so as to compensate for the rotation of the earth, shows a long streak, made by some object relatively near the earth moving with high speed. Such an object is a minor planet, nearer than usual to the earth. None is known to have come nearer than the moon, but in 1937 Hermes came within 1 million km. Since some of these objects come within

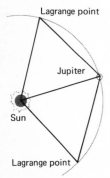

FIGURE 3.5 Lagrange points on the orbit of Jupiter. Each of the two points is located on the orbit such that it, Jupiter, and the sun are at the vertices of an equilateral triangle.

the orbit of the earth, the possibility of a collision cannot be ruled out completely, but it is unlikely. No such collisions are known within historic times nor are any minor planets known to have orbits which would make a collision seem at all probable. One thing is certain, however; if a minor planet of a diameter even as small as 0.5 km collides with the earth, the event will not go unnoticed!

Comets

Since ancient times comets have been considered omens, usually of misfortune to come. A contemporary of Tycho Brahe wrote of comets:

> They are formed by the ascending from the earth of human sins and wickedness, formed into a kind of gas, and ignited by the anger of God. This poisonous stuff falls down again on peoples' heads, and causes all kinds of mischief, such as pestilence, Frenchmen, sudden death, and bad weather.[*]

That particular remark may not have been entirely serious, but the general belief was that comets originated within the atmosphere and were connected with events on earth. Brahe was the first to show that the comet of 1577 exhibited no parallax, and that furthermore it moved so slowly against the background of stars that it must be at least 3 times as far from the earth as the moon.

Undoubtedly the comet best known to the general public is Halley's comet. In 1705, Edmund Halley, friend of Sir Isaac

---

[*]T. L. E. Dreyer, "Tycho Brahe," p. 64, Adams and Charles Black, Edinburgh, 1890.

Edmund Halley. (*Yerkes Observatory photograph.*)

Newton, published a study of the orbits of 24 comets, and he noted that the orbits calculated for the comets of 1531, 1607, and 1682 were almost identical, so that it seemed likely that the 3 comets were really the same body. He predicted that this body would reappear in 1758, and it was discovered on Christmas night of that year. Its period from one appearance to another varies between 74 and 79 years, perturbations by Jupiter causing the variation, but records are known of its observation at each appearance since 240 B.C. It last appeared in 1910 and is expected again in 1986.

About a dozen comets are discovered each year, but few of them are visible to the unaided eye. One can gain immortality of a sort by discovering a comet, for the discoverer's name is attached to it.

When comets are far from the sun, they are small, and consequently cannot be seen. Orbits must be calculated from observations made during the short time the comets are bright, as they pass near the sun, and as a result it is difficult to distinguish between an elongated ellipse, a parabola, and a hyperbola (Fig. 3.6). The ellipse, which has already been discussed, is a closed curve; an object moving on an elliptical path is bound to the sun gravitationally and travels through space with the sun as a permanent part of the solar system. The parabola is the path which an object would follow if it were virtually motionless relative to the sun when it was at a great distance. If, for example, the sun in its movement through space should approach an object moving in the same direction at almost the same speed, the object would be attracted to the sun and would approach it and then leave again on a parabolic path. If an object had some speed different from the sun's, it would approach and then leave on a hyperbolic path. An object which approaches the sun on either a parabolic or a hyperbolic path must come from outside the solar system, and an object which leaves on either of those paths will leave the solar system forever. No comets are known to have approached on parabolic or hyperbolic orbits, but some have been deflected by Jupiter so that they left on such orbits. Some of the orbits of comets within our system may be so long that the period is thousands or millions of years. Some comets have small orbits and low eccentricities, and their orbits can be analyzed well enough so that their motions can be predicted. They have periods ranging from 3.3 years (comet Encke) to 151 years (Rigollet).

Until a comet comes as near the sun as Saturn or Jupiter, it usually is not visible. Probably it is almost entirely solid. A

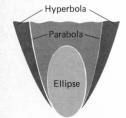

FIGURE 3.6 The three conic sections which may describe the orbits of comets.

Halley's comet, June 6 and 7, 1910. (*Lick Observatory photograph.*)

comet has been described as a chunk of dirty ice, and in the outer reaches of the solar system, the ice is solid. When they can first be observed, in the region of the Jovian planets, comets have spectra like the sun, suggesting that they are solid particles which are merely scattering the sunlight. By the time a comet comes to within about 3 AU of the sun, its spectrum begins to change, indicating that volatile materials are beginning to vaporize, and the spectrum begins to show bands indicating the presence of such gases as C_2, OH, CN, and NH_2. These materials glow by fluorescence; i.e., they absorb the ultraviolet light of the sun and then reemit the energy as visible light. Eventually the comet develops its characteristic *coma*, a small, round, diffuse glow. Sometimes a bright nucleus can be distinguished in the center of the coma, and the nucleus and the coma comprise the head of the comet. Many comets develop tails as they come near the sun. The coma may be of the order of 100,000 mi in diameter, and the tail may stretch for millions of miles. Both are very tenuous gas; stars are dimmed little if any by thousands of miles of coma material.

The tail of a comet generally points away from the sun because of the pressure of sunlight on the material of the tail and because of the pressure of particles which stream out from the sun (solar wind). Because a comet cannot recondense the material which escaped from the nucleus to form the coma or the tail, a periodic comet loses material each time it passes near the sun. As a result the life of such a comet is limited; some make only a few trips around the sun. Eventually a comet loses all its volatile material and can no longer form the coma, and in some cases the head breaks up. When that happens, the solid matter which was in the comet head strings out along the former path of the comet, probably mostly in the form of dust.

Jan Oort, a Dutch astronomer, has suggested that there is a reservoir of comets extending perhaps half way to the nearest star, and that periodically a passing star perturbs some of them enough that they change their orbits and come near the sun, thus becoming visible. In this way he explains the continued presence of comets in spite of the fact that once a comet has become periodic, it cannot live long.

Because comets which have come near to the moons of Jupiter have not disturbed the motion of the moons detectably, one can set a limit on the mass of comets. They cannot be more than one-billionth or perhaps one-trillionth the mass of the earth. Two comets (one of them Halley's in 1910) passed between the earth and the sun, and all attempts to see the solid core silhouetted against the sun were unsuccessful, indicating that there cannot be a solid object as large as 50 mi across in the head. Also, the earth passed through the tail of Halley's comet in 1910, and there were no known effects.

At the time a comet is discovered, predictions of its future performance can be uncertain, a fact well illustrated by the comet Kohoutek, which passed near the sun about Christmas of 1973. In June of 1973, after the orbit of the comet had been computed, predictions were made that it would become as bright as the full moon. Such predictions were based upon the comet's appearance at the time of discovery and upon its computed orbit. It was found far out in the solar system, farther from the sun than comets are usually discovered. Already, however, it was brighter than comets with which it could be compared, comets which later became quite bright. Furthermore, Kohoutek was to pass near the sun, only about 13 million mi from the solar surface, and it was assumed that the effects of solar radiation would produce a spectacularly bright coma and tail.

The predictions were greatly in error, however, for far from becoming as bright as the moon, the comet barely became visible to the naked eye. Observations made from above the atmosphere by American astronauts in Skylab, as well as studies made from the ground, yielded information about Kohoutek which is of great value to students of comets, but Kohoutek apparently lacked the amount of volatile material which many comets have which provide the large coma and tail. It is not expected to become visible again for about 50,000 years.

Meteoroids, Meteors, and Meteorites

The flash of a "shooting star," or *meteor*, is a familiar sight. On a typical dark night several may be seen during any hour. It is estimated that as many as 25 million bright enough to be visible

may strike the atmosphere each day. The flash is caused by a
particle or cluster of particles, probably of mass less than a gram, which burns in the atmosphere, heating the air until the hot, luminous streak of air expands to make a track visible from the earth. Most meteors burn at a height of approximately 60 mi above the earth.

Before the particle enters the atmosphere to produce the luminous flash called a meteor, it is called a *meteoroid*. If it happens to be large enough that it survives the passage through the atmosphere and strikes the ground, it is called a *meteorite*.

When the number of meteors visible becomes much larger than normal, the phenomenon is called a meteor shower. Meteor showers appear to be associated with comets, for they occur when the earth crosses the path of a comet or the path where a comet used to appear. It appears that comets sometimes string particles out along their orbits, and in some cases in which comets have broken up and disappeared as comets, the solid material which was part of them has stretched out over a great distance so that the earth regularly has meteor showers as it passes through the former orbit.

Meteorites are never associated with meteor showers, suggesting that the debris of comets is fine dust. There is reason to believe that meteorites are small planetoids which have drifted far enough from their usual range to be captured by the earth. Meteorites are classed in three general groups: irons or siderites, which are alloys of iron and nickel; stony irons or siderolites, which are mixed iron and silicates; and stones or aerolites, which are primarily silicates and other materials like terrestrial stone. Some meteorites show indication that they were formed under high pressure, and it seems likely that some of them came from deep within the original few bodies in the planetoid region which were later shattered.

In addition to the relatively spectacular materials which have been mentioned, many micrometeorites too small to make a flash strike the earth each day; it is estimated that they add 50 to 100 tons of mass to the earth daily.

Until samples of the moon were returned to earth, meteorites provided our only pieces of extraterrestrial material which could be handled and analyzed. Consequently they have been studied exhaustively. Analysis of their composition reveals that they are composed of the same elements as the earth. The age of meteorites has been computed many times by measuring the degree of decay of radioactive materials in them, and it seems that at least some of the meteorites studied have been in their present form for between 4 and 5 billion years, an age comparable to that of the earth.

It has been known for many years that some meteorites contain carbon and carbon compounds, and in recent years, pieces of what are called *carbonaceous meteorites* have been analyzed with extremely sensitive techniques, and compounds identical with or similar to those produced by terrestrial life have been found. Since much of this work was done on samples of material which had lain in museums for years, however, the possibility could not be ignored that the carbon compounds found were from the earth—that the sample had been contaminated.

On September 28, 1969, a meteorite fell near Murchison, Victoria, in Australia. It was picked up quickly and protected as well as possible from contamination, and then samples were taken from the interior of the chunk and analyzed for carbon compounds. Five of the most common amino acids were found. In this case, however, the nature of the amino acids found indicated that they were not contaminants, for they showed no optical activity. The amino acids of all living things on the earth show optical activity, which means that when a beam of polarized light is passed through them, they rotate the plane of polarization of the light to the right or to the left. The amino acids in the Murchison meteorite do not show optical activity, which indicates that the left- and right-hand forms are present in equal quantities. That situation is common when a material is synthesized without the intervention of life, but it is not normally found on the earth outside the laboratory. It therefore seems highly likely that the carbon compounds found in that meteorite, at least, were in it before it entered the earth's atmosphere.

QUESTIONS

1 Suppose that a planetoid comes so near the earth that if it is observed from opposite sides of the earth, so that one observer sees it on the horizon in the east and the other sees it on the horizon in the west, the angle which the earth's diameter subtends at the planetoid is 0.05°. How far is the planetoid from the earth? *Ans* 9.2 million mi

2 Take the average orbital speed of the earth to be $18\frac{1}{2}$ mi/s, and take the number of seconds in a year to be 3.16×10^7; from these compute the length of the astronomical unit in miles. *Ans* 93 million mi

3 Suppose that a radar pulse is beamed to Venus when the distance of Venus from the earth is 0.276 AU. Take the value of the astronomical unit to be 150×10^6 km and compute the time between transmission and receipt of the reflected signal. *Ans* 276 s

4 A light pulse is sent from the earth to the corner reflector placed on the moon, and the reflected pulse is detected 2.56 s later. How far is the reflector from the transmitting telescope? (Make the computation in metric units.) *Ans* 384,000 km

5 Make a diagram of Mercury and its orbit, with one position on the planet marked by an x, and verify that if it makes one rotation (with respect to the stars) while making two-thirds of a revolution about the sun, the time from noon to noon is 176 of our days.

6 Describe the appearance of day and night and the seasons on Uranus.

7 Make up a table or a graph to permit easy conversion of Celsius and Fahrenheit temperatures over the range used in this unit.

8 It has been suggested that Pluto was once a satellite of Neptune. What facts about its orbit might be better explained by that hypothesis than by the assumption that it formed like the other planets?

9 Make a sketch showing how a comet might be deflected by Jupiter so that its elliptical orbit would be changed to a hyperbola.

10 If a planetoid moved on a circular orbit of radius exactly 2.8 AU, what would be its period?   *Ans*   4.7 years

4
mechanics

Most of the information we receive about the universe outside the earth's atmosphere comes in the form of electromagnetic radiation—light, radio waves, or x-rays. The astronomer looks at the planets or stars and sees angular position—the position projected onto a sphere—but any distance information must be inferred indirectly. If observers are to know anything about the velocities, compositions, temperatures, or other properties of the bodies they study, they must wring that information from the clues brought in the light by which the bodies are seen. We turn now to a consideration of some of the physical principles employed by astronomers in their constant attempts to make better use of those clues and to squeeze the maximum information possible from what little can be seen directly.

BEFORE NEWTON

Kepler's first two laws tell us that the paths of the planets are ellipses, as well as something about how to compute the time when a given planet will reach a particular point. The third law, which involves the periods of the planets and the semimajor axes of their orbits, gives a relationship which holds true for all the planets. But Kepler could not say *why* his laws worked, or give any indication of the range of their validity. They apply to the planets, but do they apply to the satellites of Jupiter? Do they apply to the minor planets, and to the particles in the rings of Saturn? Kepler had discovered experimentally, primarily by working with the planet Mars, that the paths are ellipses, but he had no explanation for nature's choice of that figure. The ancients had said that the paths should be circles for a very good reason: The circle is the perfect figure, and the heavens are perfect, and so obviously the planetary orbits should be circles. But they are not circles, and Kepler did not know why. Furthermore, Kepler could give no reason for the behavior of the planets that would explain either the second or the third law. Early theorists said that something must be *pushing* the planets, perhaps spokes sticking out from the sun, or angels flying around in circles; but these explanations are unsatisfactory because they are *ad hoc*, i.e., invented, without supporting evidence, to explain one problem.

Inertia

Up until the sixteenth century, Aristotelian principles of physics and mechanics offered the best explanation available of motion—terrestrial or celestial. An essential part of Aristotle's concept of motion was the assumption that a body will not continue in motion unless a force acts on it constantly. A dropped stone falls because its natural motion is toward the center of the earth, but a book will not continue sliding across a table unless a force continues to act upon it. According to Aristotle, however, the planets are an exception, their *nature* being to move in circles. Although Aristotle's explanations were known to be unsatisfactory as early as the thirteenth century, no one before Galileo made a significant step beyond them.

Galileo carried out and drew conclusions from two types of experiments which are pertinent to our progression from Kepler's laws to Newton's laws of motion. One of Galileo's activities was to study the motion of falling bodies. He set as his goal simply describing how falling bodies move, without attempting to give an explanation for that motion. The straightforward way to approach that problem is obviously to drop objects various distances and time their falls, then to look for some relation between the times and the distances. Galileo did his work before the invention of the stopwatch (or even of any watch), and so measuring the time required for an object to fall a few feet was impossible. Instead he rolled smooth metal balls down inclined planes, arguing that the only effect of the plane was to "dilute" the motion. Although everything was slowed, the *type* of motion was not changed. In order to time the ball rolling down the plane, he let water run out of a can with a hole in the bottom into a container, then weighed the water after each roll. For short times, the weight of the water was proportional to the time. What Galileo learned from such experiments was that a freely falling body moves with constant acceleration, i.e., that its speed increases by equal increments in equal time intervals. For example, if the ball starts from rest and if at the end of 1s its speed is 2 m/s, then at the end of 2 s its speed will be 4 m/s, after 3 s it will be 6 m/s, etc.

The second set of experiments Galileo carried out involved letting a ball roll down an inclined plane and then up another plane, and he observed that the ball always went almost as high on the second plane as its initial starting point. He reasoned that if the second plane were lowered until it approached more and more nearly the horizontal, the ball would roll farther and farther. In this way Galileo approached the idea of *inertia*—the idea that an object will continue in motion unless a force acts on it to stop it. He still did not have the idea entirely clear, however, for he seems to have thought in terms of motion around

Isaac Newton. (*Yerkes Observatory photograph.*)

the earth, and his concept was that in the absence of a force, a body will continue in motion around the earth at a constant height.

It was Newton who finally stated the principle correctly in the form we now accept: In the absence of an external force, a moving body will continue indefinitely to move in a straight line at constant speed. Of course the Aristotelian idea, that a force is required to maintain motion, is much closer to our experience and intuition. Objects moving across a table or along a track inevitably slow down if no force is pushing them, and it is "faith," not sensory evidence, that leads us to say that if friction were not present the object would not stop. Everyday experience gives little support for the faith. The justification for the confidence physicists since Newton have had in this faith is that it leads to a complete theory of mechanics which can be successfully compared with experience over and over, and Aristotle's presumption did not lead to that result. One of the first and most striking consequences of the acceptance of Newton's concept of inertia was an explanation of Kepler's laws, and that explanation will be demonstrated after some preliminary matters have been discussed.

Velocity and Acceleration

Two terms must be understood accurately if what follows is to make sense: *velocity* and *acceleration*. Everyone knows what is meant by speed: Speed is the rate at which the position of a point or an object is changing. If something is moving at 50 mi/h, it will traverse a distance of 50 mi if the motion continues unchanged for 1 h. And if the speed remains constant, it will travel 25 mi in $\frac{1}{2}$ h or 5 mi in 6 min, etc. Speed is expressed in miles per hour (mi/h), in feet per second (ft/s), meters per second (m/s), etc. All these expressions involve division, for the speed is found by dividing the distance traveled by the time required. If a point moves 30 ft in 5 s, its average speed is 30 ft/5 s = 6 ft/s. If the speed is constant, then the speed is exactly 6 ft/s; if the speed is not constant, then 6 ft/s is the average speed.

Velocity   Velocity differs from speed in that it includes an indication of direction. A car's speed may be 60 mi/h, but its velocity is 60 mi/h in a northward direction. Three people may travel for an hour at a constant speed of 60 mi/h, but the result of an hour of traveling will be quite different if one is going north, one south, and one is traveling on a small circular track. Two automobiles may be in the same lane of the highway, both traveling 60 mi/h. If both have the same velocity, they have no problem, but if one has a velocity westward and the other has a velocity eastward the result may be catastrophic. Clearly speed is not the same as velocity, and the distinction is often of great importance. Velocity is one of a class of quantities which can be represented by an arrow whose length is proportional to the magnitude; with velocity, speed is the magnitude. Such an arrow is called a *vector*, and velocity is called a *vector quantity*. Velocity is thus unlike speed or temperature, both of which are *scalar* quantities, and both of which can be completely described by a number without the arrow. The temperature of the room is 78°F; it is not 78°F eastward or upward!

If the velocity of a point changes, the rate of that change is its *acceleration*. Velocity may change with a change in magnitude (the auto may speed up from 40 to 60 mi/h), or with a change in direction (the auto may go around a corner so that its direction changes from eastward to northward), or both. Regardless of how the change in velocity comes about, a change in velocity divided by the time in which it occurs gives the average acceleration. We shall discuss only two special cases of acceleration.

Linear acceleration   Linear acceleration is the acceleration of a body moving in a straight line. If the direction does not change, the change in velocity must arise from a change in speed. Thus an automobile which accelerates from 40 to 60 mi/h along a straight road in 5 s has an average acceleration of the change in velocity divided by the time, or

$$\frac{60 \text{ mi/h} - 40 \text{ mi/h}}{5 \text{ s}} = \frac{20 \text{ mi/h}}{5 \text{ s}} = 4 \text{ mi/h/s or } 4 \text{ mi/(h)(s)}$$

(See Fig. 4.1.) Since acceleration is a change in velocity divided by a time, and velocity has the units of distance divided by time, acceleration has the units of distance divided by time twice. If a body moves with constant acceleration so that its speed changes from 20 to 80 ft/s in 4 s, the change in speed is

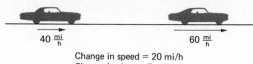

Change in speed = 20 mi/h
Change in time = 5 s
Acceleration = 4 mi/(h)(s)

FIGURE 4.1 Linear acceleration. If the automobile, moving
along a straight line, increases its speed from 40 to 60 mi/h
in 5 s, the change in speed is 20 mi/h, and the acceleration,
directed in the forward direction, is change in velocity
divided by change in time = 20 mi/h/5s = 4 mi/(h)(s).

60 ft/s and the resulting acceleration is 15 ft/s/s, which is
usually written 15 ft/s². That means that the speed changes
15 ft/s every second. If the initial speed was 20 ft/s, at the end of
the first second the speed is 35 ft/s; at the end of the second it is
50 ft/s; at the end of the third it is 65 ft/s; and at the end of 4 s it
is 80 ft/s. A freely falling body, near the surface of the earth,
falls with a constant acceleration of 32 ft/s² or 9.8 m/s².

Centripetal acceleration   The second special case is that of
motion at constant speed around a circle. If a point moves
around a circle at constant speed, its velocity is constantly
changing because the direction is constantly changing; the magnitude of its acceleration is constant. Like velocity, acceleration
is a vector quantity and so it may be represented by an arrow.
For an object moving along a straight line and increasing its
speed, the acceleration arrow points forward; for an object on
the straight line decreasing its speed, the acceleration arrow
points backward; and for an object moving at constant speed on
a circle, the arrow points toward the center of the circle (Fig.
4.2). The magnitude of the acceleration depends upon both the
speed and the radius of the circle; if the speed is represented by
v and the radius of the circle by r, the acceleration is given by

$$a = \frac{v^2}{r}$$

This is called *centripetal acceleration*.

A planet moving in an elliptical orbit has an acceleration at
all times because its direction is constantly changing, but the
acceleration is not a constant because the rate of change of direction, and also the speed of the planet, varies, depending
upon the planet's position in the orbit. In this discussion, we

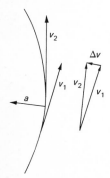

FIGURE 4.2 Centripetal acceleration. A point is moving around a
circular path at constant speed; therefore the velocity vectors are the
same length. But the velocity changes constantly because the direction changes, and the resultant acceleration vector points toward
the center of the circle. Its magnitude is v²/r.

shall approximate the motion of the planets by the motion of a
body at constant speed on a circular path in order to simplify
the mathematics. Since none of the planetary orbits has a large
eccentricity, the approximation is adequate for the planets, although it would not be satisfactory for a body moving in a
highly eccentric orbit, such as a comet.

NEWTON'S LAWS OF MOTION

In the years before 1687, when Isaac Newton published his
masterpiece, the *Principia*, two lines of development had been
proceeding independently. From Copernicus, Brahe, and
Kepler had come some understanding of the motion of the
planets; from Galileo had come a notion of inertia, a statement
of the fact that falling bodies move with constant acceleration,
and an almost accurate statement of the principle that was to
become Newton's first law of motion. Newton combined these
two lines of development in what is termed the *newtonian synthesis* and found one set of physical laws that applies both to
objects on the surface of the earth and to the planets. Almost
until the time of Newton, the idea that the heavens and the terrestrial sphere should be governed by the same physical laws
was unthinkable, and Newton's demonstration that the motion
of falling stones, arrows, projectiles, the moon, and the planets
can all be described or explained by the same principles made
such an impact upon his contemporaries, as well as on the men
of the Enlightenment in the next century, as to elicit comment
such as this from Pope:

> Nature and Nature's Laws lay hid in night
> God said, Let Newton be, and all was light.

The part of Newton's work which is relevant to this discussion can be summarized in four statements: the three laws of
motion and the law of universal gravitation.

First Law of Motion

A body at rest remains at rest and a body in motion continues in
motion in a straight line with constant speed (i.e., it moves with
constant velocity) unless it is acted upon by an external, unbalanced force. A book lying on a table has external forces acting
on it, but they are equal and oppositely directed with the result
that they balance and the book does not move. The earth exerts
a gravitational pull downward on the book, and the table exerts
an upward force (Fig. 4.3). The net, unbalanced force is zero.
An object far out in space, so far from the sun or another star
that its attraction is negligible, will move in a straight line at

FIGURE 4.3 The two external forces acting on the
book are equal in magnitude and are oppositely
directed.

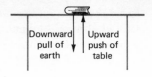

FIGURE 4.4 If the book is moving at constant speed,
the frictional force must be equal to the push supplied
by the hand so that the net force is zero.

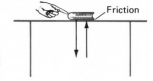

constant speed forever if it begins with that motion. If one
pushes a book so that it slides across a table at constant speed
in a straight line, then the external forces on it must be zero.
The vertical forces are those of gravity and the table's thrust,
and in a horizontal direction, the push of one's hand must be
equal and opposite to the frictional force arising from the sliding motion (Fig. 4.4). The internal forces in a book are the molecular attractions that hold the paper together—the adhesive
force of glue, the tension of thread, etc.—and these, of course,
do not make the book move.

Second Law of Motion

The acceleration of a body is proportional to the unbalanced
external force acting upon it. Since both acceleration and force
are vector quantities, having direction as well as magnitude,

FIGURE 4.5 The only force acting on a
freely falling object is the earth's attraction
for it; hence it falls with constant
acceleration.

FIGURE 4.6 If the applied force (push) is greater than
the frictional force, the book accelerates.

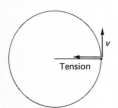

FIGURE 4.7 If a stone tied at the end of a string is being swung in a
circle, tension in the string exerts a centripetal force on the stone.

this statement implies that the acceleration is in the same direction as the unbalanced force and that the magnitude of the acceleration is proportional to the magnitude of the force. A dropped object falls with constant acceleration because the gravitational attraction of the earth for it is constant (Fig. 4.5). The stretched rubber bands of a slingshot apply a force to the stone in the pocket, and it undergoes acceleration. One pushes a book lying on a table with a force greater than the frictional forces and its speed increases (Fig. 4.6). A stone being swung in a circle at the end of a string is pulled toward the center of the circle by the string and hence has a centripetal acceleration, toward the center of the circle (Fig. 4.7).

Third Law of Motion

The traditional statement of the third law is: For every action' there is an equal and opposite reaction. A more easily understood way of stating this is: Forces always come in equal and opposite pairs. The book lying on a table presses down on the table, and the table pushes up against the book with an equal force. The man throwing a shot pushes against the shot, but the shot pushes back against his hand with an equal force. A man leaning against a wall pushes against the wall, but the wall also pushes back against him.

Ramifications of Second Law

Let us now consider the second law of motion in greater detail. Suppose that a 1-kg disk is placed on a frictionless table, and is drawn across the table by a spring. (A frictionless table can be approximated very well by an air table, a table whose top is covered by a grid of small holes through which compressed air is forced. A flat disk placed on such a table floats on a cushion of air and moves with almost no friction.) If the spring is stretched a small amount and the stretch is maintained constant, the disk moves with some constant acceleration, and if the spring is stretched twice as much, the acceleration is twice as great. This is the meaning of the second law: Acceleration is proportional to the applied force, and if the force is doubled, the acceleration is also doubled.

Mass   The magnitude of the acceleration is determined not only by the force, but also by the object being accelerated. Suppose that the spring is stretched enough to produce an acceleration of 3 m/s². If a second, identical, disk is placed on the first, the acceleration produced by that same force is one-half as great—1.5 m/s². And if a third 1-kg disk is added to the stack, the acceleration is cut to one-third—1 m/s².

One's first inclination might be to say that the acceleration depends upon the weight of the object being accelerated, but that is not precisely correct. Weight is the force with which the earth attracts an object, and the weight can be changed merely by changing the latitude or the altitude. In particular, if the object were taken far out in space, away from the earth, its weight would be almost zero. But if a force were applied and its acceleration measured, the same force which produced an acceleration of 3 m/s² on the earth (when the movement was horizontal) would be required to produce that acceleration out in space. The resistance to acceleration depends upon a property of a body which we call its *mass*, and it is entirely different from weight.

The second law, that force is proportional to acceleration, can be stated in symbolic form as

$$F \propto a$$

Any proportionality can be converted to an equation by introducing a proportionality constant, and this proportionality is usually converted to the equation

$$F = ma$$

where m (standing for mass) is the proportionality constant for a given object. The equation implies that there are ways of measuring force, mass, and acceleration, and that when proper numbers are introduced, the equality is true.

If accelerations are measured in meters per second per second, the proper mass unit to use is the kilogram. The kilogram is defined as the mass of a particular piece of platinum alloy kept in Paris, but standards laboratories all over the world have duplicates of the standard kilogram, and all our units of mass are defined in terms of it.

If acceleration is measured in meters per second per second and mass in kilograms, the equation $F = ma$ defines a unit of force which is called the *newton*. One newton (N) will cause a mass of one kilogram to accelerate one meter per second per second.

The relation between mass and force can be illustrated, and a useful relation derived, by answering the question: What is the weight of a 1-kg mass? Weight is a force, and so it is properly measured in newtons. In everyday speech people talk about weight in kilograms, and this is legitimate so long as it is understood that what is meant is the force with which the earth attracts a kilogram mass; if one wishes to be precise, however, kilograms should be used to measure mass and newtons to measure weight.

Suppose a person holds out a 1-kg block, and then drops it. The only force acting on the block is its weight, the attraction of the earth for it. The block accelerates downward at the rate of all freely falling bodies near the surface of the earth, that is, 9.8 m/s². Therefore one can apply the equation relating force and acceleration as

$$F = 1 \text{ kg} \times 9.8 \text{ m/s}^2 = 9.8 \text{ N}$$

Therefore a 1-kg mass weighs 9.8 N.

The acceleration of a freely falling body near the earth is used so often that it is given a special symbol: g.

$$g = 9.8 \text{ m/s}^2 = 980 \text{ cm/s}^2 = 32 \text{ ft/s}^2$$

One may say, then, generally, that a body of mass m has weight mg. With the units we are using m must be expressed in kilograms and weight in newtons. For this use one must take g to be 9.8 m/s²; but other systems of units make use of other equivalent values of g.

Mass and weight Mass and weight have an interesting relationship: Two objects which have the same mass also have the same weight at the same position. That follows, of course, from the fact that all objects, regardless of their size or composition, fall with the same acceleration in the absence of retarding forces such as air friction, and experiments have set a limit of less than 1 part in 10^{11} as the maximum difference between the accelerations of dissimilar materials. That fact is much more important than it perhaps seems. Consider a piece of lead and a piece of wood, and suppose that they have been trimmed until they have the same weight. Of course the piece of wood is much larger, but the two objects weigh the same, which means that the earth attracts them with the same force. There is no a priori reason to suppose that if they are placed on a frictionless track and accelerated so that the masses can be measured by their resistances to acceleration the masses will be the same, but the fact is that they *will* have precisely the same mass. This fact was of key importance in the development of the general theory of relativity, which will be discussed much later; it also provides a means of measuring mass conveniently. Frictionless tables or tracks are not common pieces of equipment, and applying a constant force and measuring accelerations to high precision are all difficult operations. On the other hand, an equal-arm balance, which compares weights, is a convenient device to use and it can be made very precise. Since two objects which have the same mass also have the same weight at the same position on the earth, masses can be compared by comparing weights. Thus one often speaks of "weighing" something when the person really wants to determine its mass.

Centripetal force If any mass moves on a curved path, some force or combination of forces must act on it to hold it to the curve, for its natural tendency (first law of motion) is to move in a straight line. In the particular case of a mass moving at constant speed around a circular path, the acceleration is directed toward the center of the circle, and hence, by the second law of motion, the force must also be directed toward the center of the circle in order to produce that acceleration. Because the force is directed toward the center, it is called a *centripetal force*. Its magnitude can be computed by combining the equation form of the second law of motion with the equation for centripetal acceleration:

$$\text{Centripetal force} = m\frac{v^2}{r}$$

where v represents speed and r is the radius of the circle.

If a stone is tied to a string and swung in a circle, the string exerts a force toward the center of the circle to hold the stone on that path; if the string breaks the stone leaves the circle on a straight-line path. If an automobile goes around a curve which is a segment of a circle, frictional forces between the roadbed and the tires must exert a force toward the center of the circle to cause the auto to change its direction. And if a planet moves about the sun in an approximately circular path, some force must act on that planet, pulling or pushing it toward the center of the circle; if no such centripetal force were acting, the planet would leave its orbit on a straight-line path.

UNIVERSAL GRAVITATION

We turn now to the fourth of Newton's great generalizations, the law of universal gravitation. Newton arrived at the idea by considering the forces ·which must act upon the planets to cause them to move as described by Kepler's laws, but we shall not try to follow the reasoning.

Newton's Statement of the Law

The law of universal gravitation is: Every particle of matter in the universe attracts every other particle with a force proportional to the product of their masses and inversely proportional to the square of the distance between them. In other words, any two particles having masses m_1 and m_2 separated by a distance d attract each other with a force proportional to

$$\frac{m_1 m_2}{d^2}$$

The proportionality can be converted into an equality by the introduction of a constant of proportionality, and the result is

$$F = G\,\frac{m_1 m_2}{d^2}$$

where G is the constant. The statement of the principle of universal gravitation must be regarded as an hypothesis to be tested, and to the extent that experiment agrees with its predictions it must be considered useful. Later we shall discuss some cases in which this principle is inadequate, and when such cases are found the theory must be modified—by the introduction of general relativity. For most purposes, however, the principle as stated is adequate. Calculations based upon Newton's second law of motion and the law of universal gravitation guided the astronauts to the moon and back with amazing accuracy. And when astronauts missed their target area on the moon by a few hundred yards, the fault was not in the laws of physics but in the application of those laws.

If I wish to use the law of gravitation to compute the attraction of the earth for a small object on the surface—my pencil, for example—I seem to have a problem. The statement of the law speaks of "particles," and although I may think of my pencil as a particle, the earth seems rather large to call a particle. What I must do, in order to apply the law, is to imagine the entire earth divided into small particles, and then compute the sum of the attractions of each of those subdivisions for my pencil. The result will be a force which seems to be directed toward the center of the earth. In fact, the force will be precisely the same as if all the mass of the earth were concentrated at its center. In other words, the force of attraction of the earth for my pencil can be computed by using the mass of the pencil for m_1, the mass of the earth for m_2, and the radius of the earth for d. The same argument applies to any body—golf ball, earth, planet, star, etc.—which is spherically symmetric. The gravitational force it exerts on any body outside itself is the same as if all the mass of the spherical body were concentrated at its center.

Mass of the Earth

If we combine the result of the last paragraph with an earlier discussion of freely falling bodies, we can derive an interesting result. The weight of a body of mass m is mg, where g is the acceleration of a freely falling body. That comes from Newton's second law, when we consider the body being accelerated by only its own weight. But according to the law of gravitation, at the surface of the earth the weight W of a body (force of attraction by the earth) is

$$W = G \frac{mM_E}{R_E{}^2}$$

If we combine these two expressions for the weight of the body whose mass is m, we get

$$mg = G\frac{mM_E}{R_E{}^2}$$

Since m occurs on both sides of the equation, it can be divided out. The radius of the earth is known (6.4×10^6 m) and g is 9.8 m/s², and we are left with only the mass of the earth and G unknown. The value of G can be measured in the laboratory, and then the mass of the earth can be computed from the equation.

The value of G was first measured by Cavendish and reported in 1798, more than a century after Newton published his work. Cavendish used a torsion balance, as shown in Fig. 4.8. On each end of the beam was a lead ball, and other balls were brought near in such a way that their attraction for the balls on the beam caused the wire holding the beam to twist. From the amount of twist the force of attraction can be measured, and if the distance between the centers of the spheres is D and the spheres have masses M and m, respectively, then

$$F = 2G\frac{Mm}{D^2}$$

Everything in the equation can be measured except G, which can then be computed. The value accepted now is

$$G = 6.67 \times 10^{-11}\frac{(N)(m^2)}{kg^2}$$

With this value of G, the mass of the earth is computed to be 5.96×10^{24} kg.

Mass of the Sun

If we think of the earth moving in a circular orbit in order to simplify the calculation, we can say that the gravitational attraction between the sun and the earth provide the centripetal force to hold the earth in its path. If the mass of the earth is M_E, its speed in its orbit is v, and the radius of the orbit is a, the centripetal force required to hold the earth on a circular path is $F = M_E v^2/a$. But that same force can be expressed in terms of gravitation, and when the two expressions for the same force

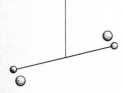

FIGURE 4.8 Cavendish balance. The small spheres on the ends of the beam are attracted by the larger spheres, twisting the suspension fiber. The gravitational attraction between the pairs of spheres can be determined from the angle through which the fiber is twisted.

are equated, the result is

$$\frac{M_E v^2}{a} = G\,\frac{M_E M_S}{a^2}$$

Since the mass of the earth occurs on both sides of the equation, it can be divided out. Using the fact that the speed of the earth is the circumference of the orbit divided by the period, which is the time required to complete one orbit, we have

$$v = \frac{2\pi a}{P}$$

When this is substituted in the equation above, the mass of the sun is

$$M_S = \frac{4\pi^2 a^3}{GP^2}$$

With the value of G given in the preceding section, $a = 1.5 \times 10^8$ m, and $P = 3.16 \times 10^7$ s, we find the mass of the sun to be about 2×10^{30} kg.

Notice that in the expression for the mass of the sun, the a and P can apply to any of the planets; they need not apply to the earth only. What that equation says, then, is that for any planet the ratio a^3/P^2 equals a constant, which involves the mass of the sun. But the fact that the ratio equals a constant is Kepler's third law; thus we have shown that the assumption of universal gravitation leads to Kepler's third law.

Mass of Planets

Let us assume that a planet has a satellite whose mass is much less than the mass of the planet, and it is moving in a circular orbit of radius R with period P. Using the same sort of argument as that in the preceding section—equating centripetal force to gravitational force and replacing speed by $2\pi R/P$—we compute the mass of the planet as

$$M = \frac{4\pi^2 R^3}{GP^2}$$

One can avoid use of constants by dividing the equation for any planet other than earth by the equation written for the earth and its moon and then using ratios of masses, distances, and periods.

Masses of Stars

In discussing the motion of a planet about the sun or a satellite about a planet, we have assumed that the more massive body

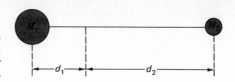

FIGURE 4.9 Center of mass (barycenter) of an asymmetric dumbbell
lies at the point nearer the greater
mass such that $M_1 \times d_1 = M_2 \times d_2$.

does not move. Later we shall wish to consider the motion of double stars, however, and then the two stars usually have masses nearly enough the same that both move. The motion of the earth around the sun is like a child swinging around a post firmly fixed in the ground—the post does not move. But the motion of two stars of approximately equal mass is more like two children with hands clasped and feet close together swinging in a circle. Their feet do not move far, but their bodies both move in circles of about the same size.

If two stars with somewhat different masses revolve about one another, their motion is like that of an asymmetric dumbbell, which rotates when thrown about the center of mass (also called the *barycenter*). The location of the center of mass is shown in Fig. 4.9. If the period of the rotation is P and the total separation $(d_1 + d_2)$ is a, the equations we have been using become

$$P^2 = \frac{4\pi^2 a^3}{G(M_1 + M_2)}$$

Because this equation will be used in the discussion of double stars, we should consider it more carefully. Let us rewrite it so that we can solve for the masses of the stars:

$$M_1 + M_2 = \frac{4\pi^2}{G} \frac{a^3}{P^2}$$

The numerical value of the constant $4\pi^2/G$ depends upon the units used, but it is often convenient to choose units in such a way as to make that constant equal to 1. The choice can be made in this way. Let M_1 equal the mass of the sun and M_2 the mass of the earth. The mass of the earth is negligible compared to that of the sun, and so the equation becomes

$$M_\odot \approx \frac{4\pi^2}{G} \frac{a^3}{P^2}$$

If a is measured in meters and P in seconds, $M_\odot$ comes out to be 2×10^{30} kg, as given previously. But instead of using those units, let us measure a in astronomical units, P in years, and M in solar masses. Then the equation is

$$1 \text{ solar mass} = \frac{4\pi^2}{G} \frac{(1 \text{ AU})^3}{(1 \text{ yr})^2}$$

and since everything in the equation except $4\pi^2/G$ is equal to 1, that constant must be 1 also. Therefore we can write the equation for the sum of the masses of two stars as

$$M_1 + M_2 = \frac{a^3}{P^2}$$

where a is measured in astronomical units, P is measured in years, and M is measured in solar mass units. (A similar simplification of the equation for the mass of a planet given in the previous section can be made by the use of distance in earth-to-moon units, period in sidereal months, and mass in earth masses.)

ENERGY

We turn now from the consideration of forces and accelerations to look at another sort of quantity. Physicists have identified a number of quantities which obey *conservation laws*, which means that in certain kinds of interactions the quantity does not change. One of the most important of these quantities is energy, and it is important because the concept of energy and the principle that it is always conserved is a thread which runs through all of natural science—astronomy, physics, chemistry, biology, geology, etc. Energy changes forms, but it is never created or destroyed. No exceptions to that principle are known in the universe.

No simple definition of energy can be given which will seem clear to a person who is not familiar with the concept, and we shall try to develop some feel for the idea by discussing a number of ways in which energy is observed and the ways in which it changes.

Kinetic Energy

An object in motion has *kinetic energy* by virtue of the fact that it is moving, and the amount of energy is given by the formula $\frac{1}{2}mv^2$ where m is its mass and v is its speed. (Energy is a scalar quantity, not a vector.) Suppose that a book is given a shove and it slides across a table. Energy is expended to make the book move, and that energy is equal to its kinetic energy at the moment the shove ends. As the book slides, however, it slows and eventually stops. Its energy has been converted into heat through friction. Heat, therefore, is another form of energy, and if the heat energy produced by the friction is measured by appropriate methods, it is equal to the kinetic energy lost by the book.

Potential Energy

If a person picks up a book and lifts it, that person is expending
energy in the process. The book gains potential energy because
it has been lifted to a higher position in the gravitational field.
The potential energy it has gained can be calculated from the
formula mgh, where m is its mass, g is the acceleration of a falling body (so that mg is the weight), and h is the height through
which it was raised. If the book is now dropped, it loses potential energy as it falls, and all that potential energy is converted
into kinetic energy. If it falls from a height h, at the time it
strikes the floor the kinetic energy of the falling book is equal to
the potential energy it has lost, so that $\frac{1}{2}mv^2 = mgh$. When it
strikes the floor, all the kinetic energy is abruptly converted
into heat, which is the random motion of the molecules of the
book and the floor.

Formulas for calculating energy and equations expressing
the equalities of those formulas imply that energy can be
measured in definite units, which is true. In the only system of
units we are using, the mks system, the unit of energy is the
joule (J). A mass of one kilogram moving one meter per second
has an energy of

$$\tfrac{1}{2}(1 \text{ kg}) (1 \text{ m/s})^2 = \tfrac{1}{2} \text{ J}$$

A mass of one kilogram lifted one meter gains potential energy
equal to

$$mgh = (1 \text{ kg}) (9.8 \text{ m/s}^2) (1 \text{ m}) = 9.8 \text{ J}$$

Although the term joule may not be familiar, it is closely
related to the watt, a unit everyone uses. The watt is a unit of
power, which is the rate at which energy is used, and one watt
is one joule per second, or one joule/second (J/s). A 100-W light
bulb converts 100 J of energy from an electrical form into heat
and light every second. Something which used power at a rate
of 1 W would use 1 J of energy each second.

By tradition, much of the astrophysical literature uses a unit
of energy called the erg, which is so small that 10^7 ergs $= 1$ J,
but we shall use joules in this book. Another energy unit, used
in discussions of heat, is the calorie.

In discussions of potential energy, the total amount of potential energy is never important, but only changes in that energy.
In fact, the total amount can be made anything one wishes. For
example, if one is discussing lifting a book and thereby giving it
potential energy, the point at which its energy is taken to be
zero can be the table top, the floor, the surface of the earth, the
center of the earth, or any other position. The potential energy
one says the book has depends upon where the zero point is
taken, but the change in that energy if the book is lifted 1 m is

the same in all cases, and only that change is of physical importance.

Imagine two bodies alone far out in space, separated by such a great distance that their mutual attraction is very small. Let them be at rest relative to one another. Since they are at rest, their kinetic energy is zero, and since they attract one another only slightly we can arbitrarily call their mutual potential energy zero. If they attract one another at all, however, as time goes by they begin to move together, and their kinetic energy begins to increase. Since no other bodies are involved, that kinetic energy must come from some store of energy within the system made up of the two objects, and we say that their potential energy is decreasing. Initially the total energy, kinetic plus potential, was zero, and it remains zero for all time. Since the kinetic energy is increasing, the potential energy must be decreasing, but it began at zero and therefore it is becoming negative. (The negative potential energy is a consequence of the initial decision to call the potential energy at a great distance zero; it could have been called 1 million or some other number so that as it decreased the number remained positive, but zero is a more convenient starting point, and the choice is arbitrary and unimportant.)

This kind of analysis can be applied to the planets moving about the sun. Each has kinetic energy because it is moving, and each has potential energy because of the sun's attraction for it. If the potential energy is taken to be zero when the planet is very, very far from the sun, then the potential energy at every point on the actual orbit is negative. Each planet has a definite amount of energy which does not change, and as it approaches the sun so that the potential energy becomes more negative, the speed increases so that the kinetic energy increases. For each planet the following equation is always true:

Kinetic energy + potential energy = constant total energy

For each of the planets, that constant total energy is a negative number.

A body attracted to the sun by a force such as gravitation, which varies inversely as the square of the distance, can have an orbit which is an ellipse, a parabola, or a hyperbola. Which of these orbits the body actually has depends upon its total energy. If the total energy is negative, the orbit is an ellipse. If the total energy is zero, the orbit is a parabola, and if the total energy is positive, the orbit is a hyperbola.

If we think now of two electrically charged particles which attract one another, exactly the same sorts of arguments can be applied. If the potential energy of the charges arising from their attractive force is taken to be zero when they are far apart, then as they approach one another the potential energy becomes negative and the kinetic energy is positive. (Kinetic energy is always positive.) On the other hand, if two electrically charged particles repel one another, their potential energy increases as they are brought close together, for some outside source of energy must push them together.

Radiation

One of the most important methods of transfer of energy from one position to another is radiation. Visible light, infrared, radio waves—all these forms of radiation carry energy. They will be discussed in more detail in the next unit.

Mass-Energy

Prior to the publication of Einstein's special theory of relativity, mass and energy were considered totally separate concepts, and each was believed to be conserved separately. The theory of relativity suggested, and subsequent experimental work has abundantly verified, that mass can be converted into energy and energy can be converted into mass. The formula for computing the amount of energy equivalent to a given mass is the well-known $E = mc^2$ where c is the speed of light. Thus if 1 kg were converted entirely to other forms of energy, it would be equivalent to

$$(1 \text{ kg}) (3 \times 10^8 \text{ m/s})^2 = 9 \times 10^{16} \text{ J}$$

QUESTIONS

1 (a) An automobile is moving 40 ft/s, and in a period of 8 s its speed (along a straight line) increases to 80 ft/s. What is the average acceleration? (b) An automobile moving 40 ft/s maintains a constant speed but goes around a curve which is part of a circle of radius 400 ft. What is the magnitude of the centripetal acceleration? In what direction does the centripetal acceleration vector point?   *Ans* (a) 5 ft/s² (b) 4 ft/s²

2 Distinguish between mass and weight. If a person goes on a diet, does he wish to lose mass or weight?

3 An object of 3-kg mass floating on an air track, where friction is negligible, is pulled with a string so that it has a constant acceleration of 2 m/s². What is the magnitude of the force pulling the object?   *Ans* 6 N

4 A mass of 0.5 kg is lying on an air table where the friction is almost zero. It is swung in a circle and kept in the circular path by a string fastened to a pin in the table. The string is 0.25 m long, and the mass is moving at a speed of 3 m/s. What is the centripetal acceleration of the mass? What force must the string exert on the mass to hold it to the path?   *Ans*   36 m/s²; 18 N

5 An object moving around a circular path is held on the path by a centripetal force. If the speed of the object doubles, how does the required centripetal force change?

6 Compute the mass of the earth from these values: $G = 6.67 \times 10^{-11}$ (N) (m²)/kg², $g = 9.8$ m/s², radius of the earth = 6,400 km.   *Ans* 6.0×10^{24} kg

7 On a planet whose mass and radius are both twice that of the earth, what is g, the acceleration of a freely falling body?

8 The mass of the sun is 2×10^{30} kg; its radius is 6×10^8 m. What is g at its surface?   *Ans*   370 m/s²

9 Suppose that you have a mass of 50 kg (you weigh 110 lb). Compute the gravitational attraction between you and a friend of the same mass when you are 1 m apart. (Neglect the fact that you are not both spheres!)   *Ans* 1.7×10^{-7} N

10 How much mass would have to be converted into energy to keep a 100-W light bulb burning for 10 h?   *Ans* 4×10^{-11} kg

11 Deimos, the outer satellite of Mars, has a period of 1.26 days (0.046 the period of the moon) and a distance from Mars of 23,500 km (0.06 the distance from the earth to the moon). What is the mass of Mars compared to the mass of the earth?

Ans $0.10 \, M_E$

5

optics

Because most of the information we have about astronomical objects comes to us by light (or other electromagnetic waves) it is essential to understand something about the behavior of light and the ways in which it conveys information. We shall consider first the properties of light required to understand the operation of telescopes, and we shall also consider how telescopes are made; then we shall discuss the properties of light which produce such phenomena as diffraction and interference; and finally we shall consider the origins of light and the information it carries about its sources.

TELESCOPES

Telescopes are devices which gather light, radio waves, or x-rays and permit them to be studied. We shall restrict this discussion to optical and radio telescopes.

Optical Telescopes

Visible light can be refracted by such materials as glass formed into lenses, or it can be reflected by mirrors. A telescope whose primary light-gathering element is a lens is called a *refractor*; one whose primary light-gathering element is a mirror is called a *reflector*.

Refractors If a thin pencil of light (usually called a ray) passes from a medium of one density to another of different density—air to glass, air to water, glass to water, etc.—it bends at the interface unless it is perpendicular to the interface. This bending is called refraction, and it is the property of light which makes lenses useful.

The diagram in Fig. 5.1 shows a series of pencils of light

FIGURE 5.1 Rays of light parallel to the axis striking a converging lens come to the focal point.

102

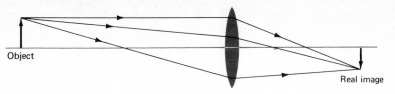

Object

Real image

FIGURE 5.2 Real image formation by a converging lens.

striking a converging lens, that is, a lens which brings parallel rays to a point. The point at which the parallel rays come together is called the *focal point*, and the distance of that point from the lens (assuming that the lens is thin) is the *focal distance*. It does not matter which side the light comes from, for the lens can be turned around and still focus light at the same position.

If a group of rays originates at some point near the axis of the lens but not at a great distance from it, the lens causes those rays which pass through it to come to a point off the axis on the other side, and if many such groups originate from the points of an object, the lens forms an image of the object. If a screen is placed at the position of the image, the image will appear on the screen and can be viewed. Images which can be projected onto a screen are called *real* images.

When a converging lens is used as a magnifier, a *virtual* image is formed. If an observer looks through a lens at an object which is nearer the lens than the focal point, the object will be apparently magnified, but the light reaching the observer's eye will seem to form an image behind the lens. If a screen is placed in that position behind the lens, no image is formed on it; the optical system of the eye must be part of the system in order that the image can be seen.

The object examined with a magnifier need not be a real object, but it can be the real image formed by another lens. If two such lenses are used together, they form a telescope or a microscope, depending upon their focal lengths and the position of the object being examined. Since astronomers are more interested in telescopes than in microscopes, we shall restrict our discussion to telescopes.

FIGURE 5.3 If the object is nearer the lens than the focal point, the eye near the lens sees a virtual image behind the object.

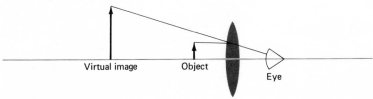

Virtual image   Object

Eye

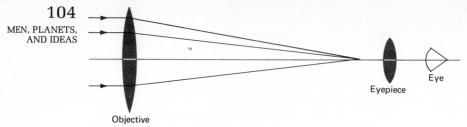

FIGURE 5.4 Schematic drawing of a refracting telescope.

Astronomical bodies are all so far away that the light reaching the earth from them comes in virtually parallel rays, so the operation of a telescope is simple. If light from a star or a planet is focused by a long-focal-length lens (called the *objective* of the telescope), it is formed into a real image of the object. Now if that image is examined by a magnifying glass (called the eyepiece), the combination is a telescope. Since the objective is a lens, and it forms an image by the process of refraction, this kind of telescope is called a refractor.

Many very small telescopes are refractors, but large refractors have so many disadvantages that very large telescopes are never of this type. The largest refractor in the world, in the Yerkes Observatory, has an objective 40 in in diameter.

Reflectors   When the objective of a telescope is a mirror of long focal length instead of a lens of long focal length, the telescope is said to be a reflector.

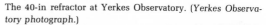

The 40-in refractor at Yerkes Observatory. (*Yerkes Observatory photograph.*)

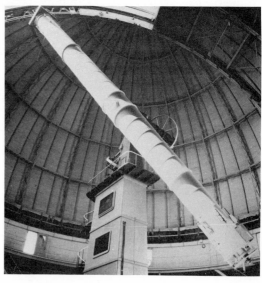

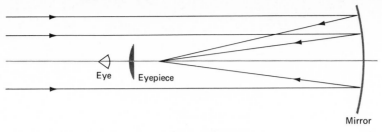

FIGURE 5.5 Schematic drawing of a reflecting telescope.

If a light ray strikes a plane mirror, it is reflected so that the incident and reflected rays make the same angles with the mirror. If a curved surface were covered with small mirrors, parallel rays striking them would be reflected so that they passed through approximately one point. If the curved surface itself is made a mirror, it will act in a similar way on a bundle of parallel rays, bringing them to an approximate focus at the focal point. The diagram in Fig. 5.6 is drawn for a curved surface which is part of a sphere.

Spherical mirrors are unsuitable for use as objectives in a telescope because the light striking regions near the center is brought to focus at a different point from that striking near the edge. This phenomenon is called *spherical aberration*, and it can be eliminated by making the mirror a section of a paraboloid of revolution instead of a sphere. Practically, that means that the mirror is ground to a spherical shape and then

FIGURE 5.6 A concave mirror can bring light rays to a focus.

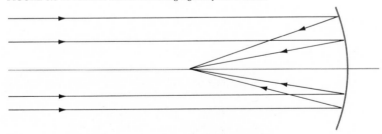

FIGURE 5.7 Spherical aberration. Rays striking the mirror at different distances from the axis are not brought to the same focal point.

FIGURE 5.8 The solid line is a cross section of a spherical
mirror; the dotted line shows a parabolic mirror.

the center is ground slightly more. Figure 5.8 shows a comparison of the cross sections of a sphere and of a paraboloid. All the light coming parallel to the axis and striking a parabolic mirror is brought to a focus at one point, but light coming from objects off the axis is not imaged so well; it forms egg-shaped images when it should form small circular images. That aberration is called *coma* because it makes the images of stars look like comets. Because of the existence of coma, large telescopes with parabolic mirrors are used to look only at small regions of the sky so that they can use light entering parallel to the axis.

Light from a distant object falling onto a parabolic mirror forms a real image at the focal point, just as if it had passed through a lens. Since that image lies in the path of the incoming light, however, some trick must usually be employed to displace it to a more convenient location. If an observer places an eye at the focal point of a 6-in-diameter mirror, all that is likely to be visible is the observer's face, somewhat out of focus. If, on the other hand, the mirror has a diameter of 200 in, such as the Hale telescope at Mt. Palomar, the observer can ride in a

The 200-in Hale telescope, pointing toward the zenith. The mirror
is near the bottom of the picture. (*Hale Observatories.*)

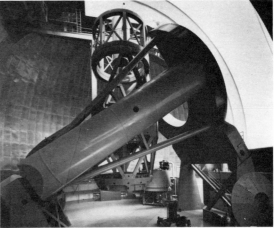

FIGURE 5.9 Newtonian telescope. A diagonal, plane mirror is used to place the real image formed by the mirror outside the tube of the telescope.

small cage at the focal point without blocking a great amount of the light approaching the mirror. That position is called the *prime focus*, and it is usable on some large telescopes.

The most common system for getting the image into a usable location is the Newtonian, shown in Fig. 5.9. A small plane mirror is inserted into the light beam so that it deflects the light before it reaches the focal point and directs it to the side, where it forms a real image which can be observed with an eyepiece. Another system is the Cassegrain, in which a convex mirror sends the light down through a hole in the main mirror to form an image as shown in Fig. 5.10. A third system, the Coudé, directs the light through the support system of the telescope so that the position of the observer does not change as the telescope moves. This is a particularly useful system when heavy pieces of equipment such as spectrographs are being used, for the equipment does not have to be moved, to compensate for the earth's rotation, as the telescope tracks stars. Very large telescopes often can use more than one of these systems.

Comparison of refractors and reflectors There are many reasons for the fact that reflectors are far more widely used than refractors, especially for large telescopes; some are mechanical and some are optical.

A lens must have two surfaces ground; a mirror has only one. Since light passes through a lens, a lens must be ground from a piece of glass with as few flaws as possible. Since the top surface of a mirror is aluminized, light never enters the glass, and bubbles and other defects are of no importance so long as they do not reach the surface. A mirror can have a steel backing to support it, but a lens must be suspended by the edges. Glass is a supercooled fluid which can sag and flow, and this sag sets a limit on the size a lens can be without undergoing so much distortion that its quality deteriorates.

FIGURE 5.10 Cassegrain telescope. A convex mirror is used to send the light through a hole in the center of the primary mirror to form the real image outside the tube of the telescope.

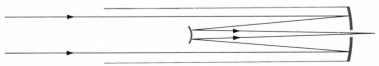

A lens has two optical defects, *spherical* and *chromatic* aberration, which can be eliminated in a mirror. For example, if a mirror is made parabolic, spherical aberration will not occur. No such simple solution exists for a lens, and the problems involved in grinding lenses to shapes other than spherical are so great that as a practical matter lenses large enough to be used in astronomical telescopes are always made with spherical surfaces.

Chromatic aberration arises from the fact that a lens bends light of different colors different amounts, so that the focal point for red light does not coincide with the focal point for blue light. A mirror treats all light the same and hence the problem of chromatic aberration does not arise. Chromatic aberration can be reduced (even eliminated entirely for two or three colors) by the use of multiple lenses; thus a lens consisting of two pieces of glass can be designed to eliminate chromatic aberration at two colors. The difficulties involved in grinding four surfaces rather than two, however, in a lens perhaps 4 ft in diameter are so great that designers of telescopes of that size turn to mirrors.

For some purposes a lens is combined with a mirror. A spherical mirror shows spherical aberration, and a parabolic mirror which is free of spherical aberration has a very small field of view because of coma. A system which combines a wide field of view with low aberrations of all kinds is the Schmidt, which uses a spherical mirror with a thin lens called a corrector plate above it. The best known telescope of this type is the 48-in Schmidt at Palomar, the corrector plate of which is sufficiently thin that the chromatic aberration it introduces is unimportant. This telescope is used only as a camera.

Observers using telescopes to look at objects a block away are most interested in magnification—how much larger things look through the instrument than without it. If one is studying the planets or the moon the magnification of a telescope is also important; but for most astronomical work—looking at stars

FIGURE 5.11 Schmidt camera. The corrector plate is ground to eliminate the spherical aberration which the spherical mirror alone would produce. The film holder is curved.

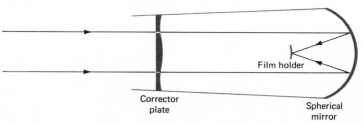

The 48-in Schmidt telescope with a man at the guiding eyepiece. (*Hale Observatories.*)

and other objects in deep space—light-gathering power is more important than magnification. The advantage of a telescope of diameter 200 in over one 100 in is not that the former makes things look larger, but that it gathers 4 times as much light, permitting photographs to be made in one-fourth as much exposure time.

Radio Telescopes

Radio waves are fundamentally the same phenomenon as visible light, differing only in wavelength. Both are electromagnetic radiation, and both obey similar laws of reflection. Like light waves, radio waves can be refracted, but glass has negligible effect on them, and refracting telescopes which use radio waves are not practical.

The fact that radio signals reach the earth from outside the atmosphere was discovered accidentally in 1931 by K. G. Jansky of the Bell Telephone Laboratories. Jansky found that the antennas he was using picked up radio noise which seemed to originate in a particular region of the sky, later identified as the Milky Way. The first antenna made specifically to receive radio waves from outside the atmosphere was built by Grote Reber in 1936, but not until after the Second World War did the study of the radio frequency waves reaching the earth from other objects in the universe become an important branch of astronomy. Such waves have now been received from the sun, some of the planets, some galaxies, certain types of stars, and many other objects.

A radio telescope consists of a metallic reflector which focuses the incident radiation onto an antenna connected to a radio receiver. Since neither the human eye nor photographic film responds to radio waves, one does not "see" with a radio telescope in the same sense as with a visible-light instrument. The output is a tracing on a sheet of paper showing the intensity of the radiation reaching the telescope. If the receiver were connected to a speaker rather than to a recorder, the radio signals being received would sound like static or random noise; even though they are called signals, they do not carry information of the sort on commercial broadcasts.

With a few exceptions, radio telescopes can be divided into three classes: steerable dishes, fixed dishes, and large arrays. *Steerable dishes* are parabolic reflectors which can be pointed and turned to compensate for the earth's rotation so that they always point at the same object. The best known of this type is probably the 250-ft-diameter dish at Jodrell Bank in England. The National Radio Astronomy Observatory in Greenbank, West Virginia, has a 300-ft dish. These reflectors can be made much larger than optical reflectors because the accuracy with which they must be shaped depends on the wavelength of the radiation to be received, and deviations of perhaps an inch can be tolerated in a radio telescope reflector whereas the mirror of an optical telescope must be accurate to within two-millionths of an inch. Since the amount of energy gathered by a mirror is

The 300-ft transit telescope of the National Radio Astronomy Observatory. This is the second largest movable radio telescope in the world; its height is 225 ft, the weight is 500 tons, and the inside area is 1.8 acres. (*National Radio Astronomy Observatory.*)

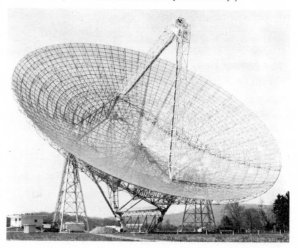

The interferometer of the National Radio Astronomy Observatory consists of three 85-ft telescopes. One is fixed in position, but the other two can be moved along a 5,000-ft base line. (*National Radio Astronomy Observatory.*)

proportional to its area, the advantages of making the reflector large are obvious. On the other hand, the mechanical problems of mounting and moving such a large structure and building it to withstand the wind sets limits on the size. One way of overcoming the engineering problems is to build the reflector in a giant bowl in the ground. It is not movable, but the detector can be moved above the bowl, giving a slight ability to choose the direction in which the instrument will "look." For the most part, however, the telescope looks straight up and depends upon the movement of the earth to swing it across the sky. The best known telescope of this type is at Arecibo, Puerto Rico, and has a diameter of 3,300 ft.

Instead of using a reflector and one detector and directing the reflector toward a source, it is possible to use many detectors and to combine their outputs with appropriate time delays so that the effect is very much the same as if a movable reflector had been used. Arrays of detectors with the "directing" done electronically can be spread out over distances of a mile or more; large arrays like this are in use in many parts of the world.

THE NATURE OF LIGHT

Waves

Visible light, radio waves, x-rays, ultraviolet radiation, infrared radiation, and gamma rays are all part of the electromagnetic spectrum, which means that they are all *electromagnetic*

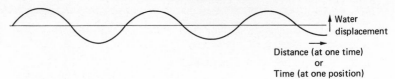

Distance (at one time)
or
Time (at one position)

FIGURE 5.12 Representation of a water wave. This curve may be regarded as a snapshot of the water surface, showing the displacement at one time over some distance, or it may be regarded as a plot against time at one position, showing the variation in height of the water surface as the waves move past.

waves. Let us try to give some feel for that extremely important phenomenon. Imagine an experiment like this: At the edge of a still pool of water a person dips a stick into the water rhythmically—in, out, in, out, etc. Each time the stick touches or leaves the water a disturbance moves out from it, and each disturbance is a wave. All of them together make a wave train, although the distinction between wave and wave train is not always made. Those waves have at least four immediately observable characteristics.

1 They carry energy. They are able to move leaves or corks or the water at some distance from the original disturbance.
2 They have a definite speed of movement through the water. As the first disturbance moves out from the stick it moves with a definite speed, and this is true of the movement of any crest in the train.
3 They have a definite *frequency*, the frequency of the dipping stick. Frequency is the number of waves produced or passing any point each second.
4 They have a definite *wavelength*; i.e., the distance from one crest to the next crest is the same for all waves.

If we designate the frequency of the waves by f, the wavelength by λ, and the speed by v, the three quantities are related by the equality

$$v = f\lambda$$

To see that this is true, imagine the start of the wave train. When the stick dips into the water the first time, a wave moves out, and in 1 s the stick dips f times, producing f waves each separated by a distance λ from the preceding one. At the end of 1 s the first wave has moved out a distance $f\lambda$, and that distance is the wave's speed.

There are other kinds of waves, but the waves on the surface

FIGURE 5.13 Electric field near a positive charge. By convention the field lines are drawn in the direction a positive charge would move if it were placed at the point. Because another positive charge would be repelled by the one shown, the lines point radially outward in all direction (in three dimensions).

of water are as easy as any to see and to visualize. We can represent any such waves by a curve like the one in Fig. 5.12, which
is a plot of the height of the water above its normal position.
The interesting thing about such a plot is that it can show water
height as a function of time or of distance; i.e., it can represent
what happens at one point over a period of time or it can represent what happens at one time and many places.

Electromagnetic Waves

The water wave is a movement of water, and sound in air is a
movement of the particles of the air; but electromagnetic waves
are a movement of nothing. An electromagnetic wave can travel
more easily through a complete vacuum than through matter,
and in this respect it is quite different from water waves. An
electromagnetic wave is made up of fluctuating electric and
magnetic fields, and to have any idea what it is like we must
say something about *fields*.

Everyone knows that electric charges are either positive or
negative, that two positive or two negative charges repel one
another, and that a positive and a negative attract one another.
In order to explain how one charge can affect another some
distance away, we use the idea of field. One visualizes the
space around a charge as being changed in some way, so that
if another charge is brought into the region it interacts with
the changed space and experiences a force. By convention the
field is taken to point in the direction a positive charge would
move if released; the field around a positive charge can be
represented as in Fig. 5.13. If one waves the charge back and
forth along a line, the field which it produces changes constantly, and part of the field is "shaken loose" so that it travels
away from the charge. That traveling field is an electromagnetic wave. Any time an electric field changes, a magnetic field
is produced, so that the two kinds of fields are coupled insep-
arably in the wave. When such a wave passes a wire which
contains charges (electrons) free to move, the charges are
made to oscillate, and the resulting electric current can be
detected and amplified in a radio receiver. The analog of the
diagram showing the height of water in a water wave is a
diagram showing the intensity of the electric (or) magnetic
field, as in Fig. 5.14.

FIGURE 5.14 The electric field is an electromagnetic wave. As in Fig. 5-12, this may
be regarded as a representation of the intensity of the electric field at one time or at one
point.

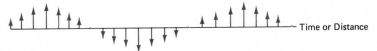

Electromagnetic Spectrum

The electromagnetic spectrum can be discussed in terms of either frequency or wavelength, but we shall use wavelength. One can be easily converted into the other because the speed of propagation of all electromagnetic waves is the same in free space: 3×10^8 m/s. Thus a wave of length one meter has a frequency of $(3 \times 10^8$ m/s)/1 m $= 3 \times 10^8$/s. The symbol "/s" means cycles per second; this unit is now properly called the Hertz (Hz), so that the electric field in a wave of length one meter oscillates 3×10^8 times each second, or it has a frequency of 3×10^8 Hz.

Visible light is made up of waves much smaller than those we see on the surface of water; its wavelengths range from about 4×10^{-7} to 8×10^{-7} m. The short waves are violet and the long ones are red; all the other colors fall between. The yellow light seen when table salt is dropped into a flame has a wavelength of 5.89×10^{-7} m. The standard unit for measurement of wavelengths in the visible range is now the nanometer (10^{-9} m), abbreviated nm; the yellow light of sodium has a wavelength of 589 nm. Before the adoption of the nanometer as the standard unit, wavelengths in the visible portion of the spectrum were most commonly measured in angstroms (Å). One nanometer equals ten angstroms, so that the wavelength of sodium light is 5890 Å.

Radiation whose wavelengths are too short to be seen by the eye, down to perhaps 10^{-9} m, is called *ultraviolet*. No definite cutoff of the ultraviolet range can be given, but somewhere near 10^{-9} m one begins to speak of the radiation as x-rays or gamma rays. As these terms are now used by astronomers, the distinction between x-rays and gamma rays is one of wavelength, with gamma rays having the shorter wavelengths.

From the long end of the visible spectrum out to perhaps one millimeter the radiation is called *infrared*. From one millimeter to several centimeters is the *microwave* region, and beyond that is the *radio* region, extending indefinitely. One can get an idea of the wavelength of the ordinary broadcast band by computing the length of the waves which lie near its center. The AM broadcast band extends 550 to 1,400 kHz (formerly called kilocycles per second, or 1000 cycles per second). Near its center is 1,000 kHz, or 1,000,000 cps, which can be written as 10^6/s. The wavelength can be computed from the relation given earlier which always applies to waves, and we have $\lambda = c/f$, which becomes

$$\lambda = \frac{3 \times 10^8 \text{ m/s}}{10^6/\text{s}} = 3 \times 10^2 \text{ m} = 300 \text{ m}$$

The only essential way in which x-rays, ultraviolet radia-

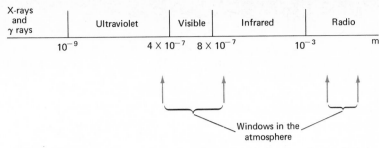

FIGURE 5.15 The electromagnetic spectrum. The windows in the atmosphere indicated are the two regions in which radiation from outside the atmosphere can reach the surface of the earth. The atmosphere is opaque to all other wavelengths.

tion, visible light, infrared radiation, and radio waves differ is in wavelength; all travel at the same speed in empty space, and all are traveling electromagnetic fields. The devices used to detect them differ, including Geiger counters, photographic film, thermocouples, and radio antennas with receivers attached, and the objects and processes which can produce them differ, but all electromagnetic fields can bring information about celestial objects. Not all that information can reach the surface of the earth, however, for the atmosphere is opaque to most of the electromagnetic spectrum. What we can see of the universe outside our atmosphere we must see through two "windows," regions in the spectrum where the atmosphere is transparent. These are from about 290 nm (in the ultraviolet) to the near infrared, and from about 8 mm to 17 m. The visible region, and the ultraviolet and infrared near to the visible, lies in the first window; all radio astronomy must be done through the second window. Since the development of rockets and satellites which carry instruments above the atmosphere, our knowledge of the appearance of the universe as seen by radiation of other wavelengths, notably the shorter radiations, ultraviolet and x-ray, has increased tremendously.

Diffraction and Interference

All waves show two types of behavior which are important in analyzing the data we can get by looking at objects outside the earth: *diffraction* and *interference*. When a wave disturbance of any kind passes through an aperture which limits it, the waves tend to spread out. The wavelength of sound waves from a person's voice is of the order of 3 to 4 ft, and as they pass through a doorway of about that same width, they spread considerably, so that we can hear a person's voice through a doorway even when we cannot see the speaker. Light waves also spread, but we usually are not aware of it because the relative importance of the spreading depends upon the ratio of the wavelength to the size of the aperture. The wavelength of visible light is of the order of 5×10^{-5} cm (0.00005 cm), and its spreading is comparable to that of the sound going through a door only when the light passes through an aperture approximately 0.00005 cm wide. One can demonstrate the spreading of light with slits which are wider than that, but under the best of conditions the aperture must be less than 1 mm for even a very small effect to be seen. The spreading of a wave as it passes through a limiting aperture is one example of diffraction. Other examples occur when an object casts a shadow, blocking out part of the waves. At the edge of the shadow the transition from light to dark is gradual, not instantaneous, because of the bending of some of the waves into the shadow area. Again, this can be seen with visible light only under special, carefully prepared conditions.

If two coherent waves of the same frequency meet, they can interfere with one another to produce what are called interference patterns. The simplest way to produce two coherent waves of the same frequency is often to let the same wave pass through two apertures; if the apertures are sufficiently small, diffraction causes each new wave to spread, and in the overlapping region, the waves interfere with one another. As the diagram in Fig. 5.16 shows, when waves which are approximately circular spread out from two sources which are close together, regions are formed where the combined effect of the two waves is to produce almost no disturbance; here the waves are said to interfere *destructively*. In other regions the crest of one wave coincides with the crest of the other wave, and hence the troughs coincide, with the result that the combination is a more violent wave than either of the original waves alone. This is called *constructive interference*.

The interference of light or radio waves coming from different parts of celestial objects is used to deduce more information about them than could be obtained otherwise; this will be described later.

Resolving power Diffraction sets a limit on the information which can be obtained from a telescope because the edges of the telescope produce diffraction patterns. One result is that stars are never seen as points of light but always as much larger

FIGURE 5.16 Interference pattern produced by two wave sources close together when the two sources are in phase (in step with one another). The lighter lines indicate regions of constructive interference.

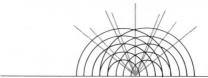

circles than one would expect from geometrical considerations, and the size of the circle of light is determined by the size of the telescope aperture, not by the size of the star. Light passing through a circular aperture produces a series of concentric rings, and the sizes of these depend upon the size of the aperture and the wavelength of the light.

Suppose that one looks with a telescope at two pinpoints of light which are far away. (The telescope is not essential to this description, for the iris of the eye provides an aperture which produces a diffraction pattern on the retina and determines the resolving power of the eye.) Each pinpoint is seen as a small ring or series of concentric rings. Now let the two sources move closer together. As they come sufficiently close their diffraction patterns begin to overlap, and finally, before the sources are actually coincident, the patterns fuse so that an observer is unable to determine whether there is one source or two. The ability of an instrument to distinguish the fact that there are two sources is called its *resolving* power, and the resolving power depends upon the diameter of the limiting aperture and the wavelength of the light being used. Everyone is familiar with the fact that at night the headlights of a distant oncoming automobile look like a single light until the car reaches a certain distance, at which time the fact that there are two lights gradually becomes apparent. The distance at which the two lights become distinguishable depends upon the size of the pupil.

The angle (expressed in radians) subtended at the instrument or eye by two sources which can barely be resolved is approximately 1.22 λ/D, where λ is the wavelength of the radiation being used and D is the diameter of the aperture. We can use this relation to compare optical telescopes with radio telescopes. Let us take as a reasonable estimate of the wavelength of visible light 600 nm, which is yellow light, near the center of the visible part of the spectrum. For the radio telescope, let us use 21 cm, which is an important wavelength for radio signals emitted by cold hydrogen. Now we can ask how large a radio telescope should be to have the same resolving power at 21 cm as a 200-in telescope has at 600 nm (6×10^{-5} cm). Letting the unknown diameter of the radio telescope be D, we can set the two expressions for the minimum angle resolved equal to one another and get

$$1.22 \frac{6 \times 10^{-5} \text{ cm}}{200 \text{ in}} = 1.22 \frac{21 \text{ cm}}{D}$$

Solving for D,

$$D = \frac{21 \text{ cm} \times 200 \text{ in}}{6 \times 10^{-5} \text{ cm}} = 7 \times 10^7 \text{ in} \approx 1,100 \text{ mi}$$

Because there are serious problems involved in constructing a radio telescope 1,100 mi in diameter, radio astronomers with a single telescope must be content to see the universe much as the earth appears to an extremely myopic person without glasses. The radio telescope can see bright spots in the sky, but it cannot tell precisely where they are, or whether they are produced by single or multiple sources.

The example is misleading, for optical telescopes on the earth never come near their theoretical limits of resolution. Turbulence in the atmosphere through which they must peer always distorts the images they produce and reduces their resolution so much that radio-telescope systems using two instruments separated by a moderate distance often have better resolution than the 200-in optical telescope.

The output signals from two or more radio antennae can be combined to yield information unobtainable from a single telescope about positions, size, and structure of radio sources. If two or more antennae are placed along an east-west base line, the distances from the elements to a radio source in the sky overhead changes constantly because of the rotation of the earth. As the relative distances change, the waves being received by the different elements go in and out of phase (as if the waves were in and out of step with each other), and analysis of the resulting combined signal provides information about the source. The signals received by two antennae separated by a great distance can be recorded separately and then combined later; the method is called *very-long-base-line interferometry*.

Two antennae whose separation is variable can be used in a technique called *aperture synthesis*; Sir Martin Ryle of Cambridge University shared the 1974 Nobel Prize in physics for invention of the method.

A large radio telescope may consist of a large reflector which sends the waves striking each part to a common detector, or it may consist of many small detectors spread over a large area, with the outputs of the different detectors combined electrically. Imagine a grid superimposed on either kind of telescope, so that it is divided into squares like a checkerboard. The output of the full telescope is the combination of the contributions from all the squares. Now suppose that the grid is laid out on the earth, and two movable telescopes are placed successively in each pair of squares. In each position their outputs are the same as those of the sections of the larger telescope if it were present, and those outputs can be recorded and combined to produce the effect of the very large telescope. (Actually not all pairs of positions need be used because some sets of pairs yield identical information.) The effect is to "synthesize" the signals that would be produced by a telescope of large aperture by use of two small telescopes.

Photons

Under some circumstances light acts more like a stream of particles than like waves. The "particles" of light are called *photons*.

As the next section will explain in more detail, an atom can acquire energy from its surroundings and then give up that energy in a single pulse of radiation. That flash of light produced by the sudden emission from an atom is called a photon. It can be visualized with some accuracy as a short wave train. We can measure its wavelength, and so it clearly is a wave. On the other hand, if the photon collides with an electron, the collision may take place almost exactly as if the photon were a particle. And if the photon falls on a metal plate, it can give all its energy to one electron, rather than spreading it over a large area as one might expect a wave to do. The energy of a photon depends upon its frequency or wavelength and is given by

$$E = hf = \frac{hc}{\lambda}$$

where h, known as Planck's constant, has the value 6.624×10^{-34} J-s.

To understand diffraction, interference, and polarization we must think of light as waves, but to understand its interaction with matter we must often think of it as particles. One way to visualize something which acts like both waves and particles is to think of the energy being carried by particles, but the particles being directed to their destinations by waves.

The idea of photons is not particularly useful for most applications involving long-wavelength radiation, such as radio waves. In the visible range, one must use both wave and photon concepts, depending upon the phenomenon being studied. And at very short wavelengths, such as those of hard x-rays and gamma rays, the radiation often acts very much like particles, with the wave character being of less importance.

SPECTRA

The light which comes from stars tells not only the direction to its source, but also the composition of the star, its temperature, its velocity toward or away from us, and in some cases such data as the intensity of magnetic fields and the pressure of the gases in the atmosphere. We can often infer something about a star's rotational speed from a study of the light we receive from it. All this information is derived from a study of the spectrum of light emitted by the star, and we turn now to a consideration of spectra and their analyses.

Spectrograph

Light in the visible range, and the near-ultraviolet or near-infrared regions, is analyzed by a *spectrograph*, an example of which is shown in Fig. 5.17. It consists of five basic parts, although certain parts can be combined. Thus the diagram shows two lenses and a prism or a grating; if the grating is ruled on a concave mirror, the two lenses can be eliminated. Light entering the instrument first strikes a narrow slit, which blocks all but a very narrow beam. Beyond the slit is a lens whose focal point lies at the slit so that light coming through the slit and diverging to the lens moves in parallel rays after going through the lens. The third element is either a prism or a diffraction grating; in either case the function of this element is to separate light of different colors, bending light of differing colors different amounts. Next comes another lens, whose function is to focus the light leaving the prism or grating onto the fifth part, which is photographic film or some other detector. For work in the visible or ultraviolet regions, photographic film is almost always used (actually photographic plates, thin glass plates coated with emulsion, are more often used than film), but in the infrared region the detector is likely to be of a different sort, perhaps a thermocouple which is heated by the infrared radiation striking it.

When light strikes an air-glass interface at any angle other than the perpendicular, the light passing through the surface bends, but the amount of bending depends upon the wavelength of the light, i.e., upon its color. Blue light is bent most and red least. If a piece of glass is made into a prism, and if a single beam of white light strikes it as shown in the diagram

FIGURE 5.17 Schematic drawing of a spectrograph, showing, from left to right, the slit, a lens, diffraction grating, lens, and film.

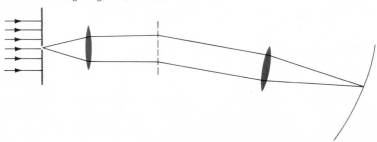

FIGURE 5.18 Dispersion of white light into colors by a glass prism.

in Fig. 5.18, the beam is spread into a fan of colors, with the amount of bending of the emerging light depending upon, but not being proportional to, the wavelength of the light. A diffraction grating produces a bending which is more nearly uniform over the entire visible range than that produced by a prism. The diffraction grating is made by ruling fine lines with a diamond point on a sheet of glass, perhaps as many as 30,000/in. The result is that the glass has alternating regions—those where light is transmitted easily and those where it is not transmitted well—and each of the narrow transmitting regions acts like the slit described in the discussion of diffraction, so that the light which passes through is spread into a diffraction pattern. The diffracted light from many slits interferes to produce constructive interference only in certain directions for light of a particular wavelength. The diagram in Fig. 5.19 shows light of a particular wavelength striking a grating from the left. Only in a direction such that the path difference from one slit to the next is one, two, or some integral number of wavelengths will the emerging light have any appreciable intensity; in all other directions the interference will be destructive. Thus light of one wavelength is concentrated in each direction.

If light of only one wavelength falls onto the slit of the spectrograph, the lenses and the grating produce an image of the slit in that color on the film. If the film is exposed to the light, upon

FIGURE 5.19 The bending of light of one wavelength by a diffraction grating. Each wavelength is turned through a different angle. The angle is determined by the requirement that the path difference from one slit to the next be one wavelength, two wavelengths, or some other integral number of wavelengths. Usually one uses the light for which the path difference is one wavelength.

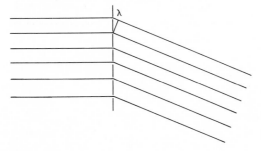

development it will show a single dark line—the image of the slit. If the incident light has several wavelengths which are separated somewhat, an image of the slit is produced by the light of each color, with the result that the developed film shows several "lines." Because of this characteristic appearance of photographed spectra, we speak of *spectral lines*, which represent the discrete wavelengths which make up the spectrum.

Suppose that light is passed through a spectrograph and the spectrum is observed. What does one see? The spectrum will have one of three appearances: It will consist of a series of bright lines; it will be a continuous smear of color from the violet to the red; it will be a continuous smear except that it will be crossed by a series of dark lines. (In some cases it may have either bright lines or dark lines so closely bunched that they appear to be bands of color, but we shall neglect those cases, which arise from energy changes in molecules.) The appearance of the spectrum tells us something about the source of the light, and we shall discuss each of these in detail later.

The continuous spread of color originates from an incandescent solid or liquid or from gases deep enough to be opaque, as for example, those of a star. The hot filament in an incandescent light bulb, a puddle of molten steel, the sun—all these are examples of bodies which give rise to a continuous spectrum. We cannot tell anything about the composition of a source from the appearance of its continuous spectrum because hot tungsten, molten steel, incandescent fire brick, or anything else equally hot give almost precisely the same spectrum. We can, however, tell the temperature of the source, and we shall return to this point later.

Bright-Line Spectra

A bright-line spectrum originates from a gas which is hot or which is excited by some external source of energy, such as an electric current. Common sources of bright-line spectra are "neon" signs (which may or may not be filled with neon), flames, electric arcs, and metallic vapor lamps such as the mercury or sodium lamps often used for street and highway lighting. Any material, when vaporized and heated until it glows, produces a bright-line spectrum, and the particular pattern of lines produced is characteristic of the elements present. Thus hydrogen produces one pattern, helium another, carbon another, and iron something quite different. From a study of the bright-line spectrum, one can determine the elements present and something of their relative abundance. In order to understand how the bright-line spectrum arises, we must discuss certain characteristics of atoms.

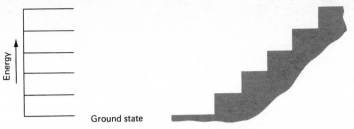

FIGURE 5.20 Energy-level diagram for an object lying on a stairway. The increase in gravitational potential energy from one step to the next is the same, but the object can have only those energies which correspond to positions on steps.

An atom is made up of a nucleus, which consists of protons and neutrons and contains most of the mass of the atom, and of electrons, which are outside the nucleus. For the present we are interested in changes which affect the electrons only, and we will ignore possible changes in the nucleus. Atoms are often pictured as miniature solar systems with the electrons moving in definite orbits about the nucleus, but that picture is too sharp and precise to be accurate. Electrons do not have definite orbits, and we gain little by trying to picture them at all. Much more important is the realization that an atom can absorb and emit energy by rearrangements among its electrons, and that the energy it can absorb or emit is *quantized*, which means that it can take in or give out only certain amounts of energy and never any other amounts. A lifted object has potential energy in the gravitational field, and the amount of energy it can have varies continuously, depending upon the height to which it is lifted. Atoms do not act that way. We can think of their energy in terms of the energy an object can have if it is laid on particular steps on a stairway. If it can be raised one step or two steps or six steps, but not rested between steps, its energy is quantized. The energy such an object might have can be represented on an energy-level diagram as shown in Fig. 5.20, where each horizontal line represents one of the possible energies. The lowest position is called the *ground-state energy*.

The possible energies of atoms are represented similarly by energy-level diagrams. Part of the diagram for hydrogen is shown in Fig. 5.21. An undisturbed hydrogen atom is in the ground state, which means that its energy is as low as possible. If it is struck by a fast electron (perhaps because an electric current is being passed through the hydrogen gas) or otherwise given energy, it may absorb enough energy to be lifted to one of

FIGURE 5.21 The energy-level diagram for hydrogen. The hydrogen atom can have only the energies represented by horizontal lines in this diagram.

the higher levels. Within about 10^{-8} s, however, the atom will give up that energy so that it can fall back to the ground state. In most cases the energy will be given up by radiation; very rarely the atom may collide with another atom and transfer its energy to the other atom instead of radiating it. When the atom radiates away its energy, it may do so in one burst, by jumping from the excited state directly to the ground state, or it may do so in several jumps, coming down by any combination of steps from one level to those below it. In more complex atoms than hydrogen only *some* such sequences of steps are possible, but in hydrogen all combinations may occur. For each jump, whether it covers one step or several, one photon of light is emitted, and it has a wavelength which is inversely proportional to the energy it carries away. (If the energy is known, the wavelength is found from $\lambda = hc/E$.) Thus a small jump, which means small energy, produces long-wavelength light, but a long jump produces short-wavelength light. To dispel any illusion that hydrogen is typically simple, the diagram of sodium in Fig. 5.22 is shown as an example of the energy-level diagrams of more complex atoms.

Because no two elements have the same energy-level diagram, it should be clear why no two elements have the same spectrum. The diagrams are inferred from study of spectra, for the wavelengths of spectral lines tell us the energies which an atom can emit.

Dark-Line Spectra

It is now possible to explain the origin of dark-line spectra, the dark lines crossing a continuous smear of color. This type of spectrum is characteristic of the sun and other stars, but it can be produced in the laboratory, as shown in Fig. 5.23. If the light (continuous spectrum) from an incandescent lamp is passed through a bulb containing cool sodium vapor, and then passed into a spectrograph, the spectrum is continuous except for two dark lines very close together in the yellow. If the light from a sodium vapor lamp is now passed through the same spectrograph, it produces a bright-line spectrum having only two lines—very close together in the yellow. The positions of the bright lines produced by the excited sodium vapor are the same as those of the dark lines produced by the cool sodium vapor. Recalling the energy-level diagrams, we can easily understand what happens. The atom can absorb precisely the same amounts of energy which it can emit; any energy it takes in

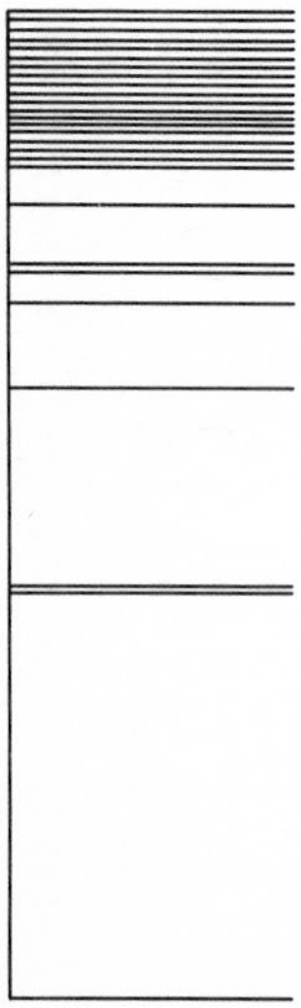

FIGURE 5.22 Partial energy-level diagram for sodium.

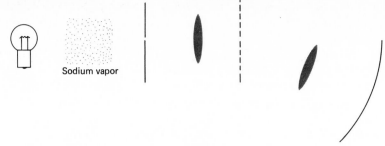

FIGURE 5.23 Laboratory demonstration of the origin of an absorption spectrum. The incandescent lamp produces a continuous spectrum, and after passing through a bulb filled with metallic sodium vapor, the light falls on the slit of a spectrograph. The film records the continuous spectrum except that two lines in the yellow are missing. They are precisely the two lines emitted by a sodium vapor lamp. If the sodium vapor in the bulb were very hot, it would emit yellow light, recorded by the spectrograph as two lines, but when the vapor is cool, it absorbs those same two lines.

must lift it exactly to one of the allowed energy levels. Thus if it is offered energy of any amount, it will take only certain amounts. The white light from the incandescent bulb offered to the cool sodium atoms every possible energy over the entire range represented by the continuous spectrum, but the sodium atoms took only certain amounts of the energy. The white light which continued into the spectrograph still contained all energies, but at those particular wavelengths which the sodium had absorbed, the intensity was reduced so that those regions of the spectrum appeared dark by comparison with the regions nearby. The reader may wonder what the atoms do with the energy they have extracted from the white light. They reradiate it immediately, but since they send it in all directions, the amount which happens to go in the direction of the original white light beam, and thus enters the spectrograph, is insignificant.

One very important difference between the bright-line (*emission*) spectrum of an element and its dark-line (*absorption*) spectrum is that the emission spectrum always has many more lines than the absorption spectrum. An emission line represents the energy released when the atom goes from a high level to a lower one, and an absorption line represents the energy absorbed when the atom goes from a low level to a higher one. If the levels are numbered upward from the bottom, emission lines can be the result of transitions downward that end on, say, level 1, 2, 10, or perhaps level 59. But the transitions upward which produce absorption lines must all begin on level 1, because that is where the atoms are before they absorb. This will be illustrated by a more detailed discussion of the hydrogen spectrum.

FIGURE 5.24 The Balmer spectrum. These are the spectral lines
produced by hydrogen which lie in the visible and near ultraviolet.

If hydrogen light is passed through a spectrograph and the
spectrum is observed with the eye, three or perhaps as many as
five lines are visible: one in the red, one in the blue-green, and
one or more in the violet. These are part of the *Balmer series* of
lines, named for the man who published a formula in 1885 giving the wavelengths of the lines. A photograph of the Balmer
series reveals that there are lines in the ultraviolet which the
eye cannot see. The complete Balmer spectrum looks as shown
in the diagram in Fig. 5.24. If the radiation from hydrogen in
the far ultraviolet is investigated, another series of lines looking
like the Balmer lines shifted to short wavelengths is found; this
series is called the *Lyman* lines. Other groups occur in the infrared, called the Paschen, Brackett, Pfund, etc. A larger part of
the hydrogen spectrum looks something like the diagram in
Fig. 5.25.

Figure 5.26 shows the energy-level diagram of hydrogen, indicating some of the transitions associated with each of the
series. Any transition from a higher level which ends on level 2
produces a line in the Balmer series. The transition from 3 to 2
has the lowest energy of any of those ending on 2, and it
produces the red line in the visible spectrum. The next highest
energy comes from the transition from 4 to 2; it produces the
blue-green. More energetic transitions produce the violet and
ultraviolet, and at the upper levels the spacing becomes so
close that the resulting lines form a continuum in which individual lines cannot be distinguished.

Any transition which terminates on level 1 is part of the
Lyman series, any transition which terminates on level 3 is part
of the Paschen series, and transitions to level 4 produce the
Brackett series.

It is customary to designate the lines of each series by the
first letters of the Greek alphabet, beginning with the longest
line. Thus the red line of the Balmer series is designated Balmer
alpha, or H α; the blue-green line is H β; etc. The longest line

FIGURE 5.25 Part of the hydrogen spectrum, showing four series.

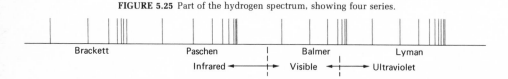

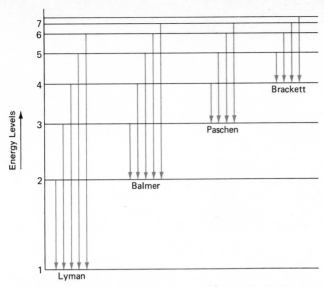

FIGURE 5.26 The energy-level diagram for hydrogen showing the transitions which give rise to four of the series.

of the Lyman series is Lyman alpha. It lies so far in the ultraviolet that the earth's atmosphere is opaque to it, and although the sun and other stars radiate copiously in the Lyman series, none of the radiation can be seen by instruments on the surface of the earth. One of the most important results of astronomical observations made by balloons flying over most of the atmosphere or by rockets and satellites flying above virtually all the atmosphere has been observations made of Lyman radiation.

Suppose that white light is passed through cool hydrogen gas. What absorption lines are likely to be produced? The only lines produced in any strength are the Lyman lines, for they are the only ones which can be absorbed by atoms in the ground state. In order for an atom to absorb light corresponding to a Balmer line, the atom must be in the second energy level; but if an atom finds itself in that condition, it stays only about 10^{-8} s, and the probability that within that time it will absorb light to jump up is small indeed. Only if the hydrogen is so hot that many of the atoms are in the second level much of the time will absorption of the Balmer lines be appreciable.

If a spectrograph is pointed at the sun, the spectrum seen is continuous, with hundreds of dark absorption lines crossing it. Although the sun is entirely gaseous, within a region called the *photosphere* the gas changes from transparent to opaque over a short distance. The gases below that region are largely opaque and hence radiate a continuous spectrum. Before that light can reach the earth, however, it must pass through a deep atmos-

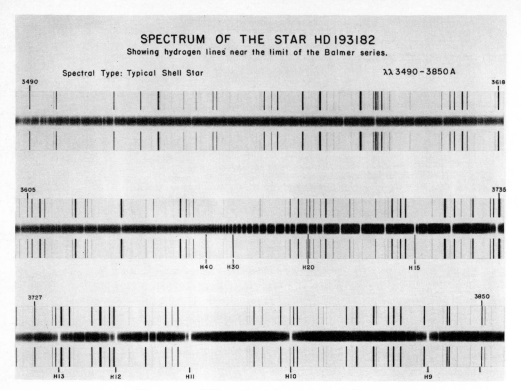

SPECTRUM OF THE STAR HD 193182
Showing hydrogen lines near the limit of the Balmer series.

Spectral Type: Typical Shell Star   λλ 3490 – 3850 A

The spectrum of this star shows the Balmer series of hydrogen from the ninth line (H9) to the fortieth line; beyond that line the Balmer continuum can be seen. The star's spectrum is of the absorption type; an emission spectrum is recorded above and below each segment of the stellar spectrum for comparison purposes. Wavelengths are given in Angstrom units; the corresponding range in nanometers is 349.0 to 385.0 nm. (*Hale Observatories.*)

phere of the sun which is transparent, i.e., a region in which the gases are hot and less dense so that most of the light coming from below passes through. During the passage of light through that region, the atoms of the atmosphere there absorb from the light the energy which under different circumstances they would radiate, and so the absorption lines of the sun reveal that that atmosphere comprises helium, iron, magnesium, sodium, potassium, etc. Thus the absorption lines tell us something about the gases in the atmosphere; the continuous spectrum tells us something about conditions in the deeper layers of the sun.

Continuous Spectra

We know that the continuous spectrum does not reveal composition, for it is almost the same for all kinds of materials. It does reveal temperature.

Let us imagine an experiment in which equipment is set up

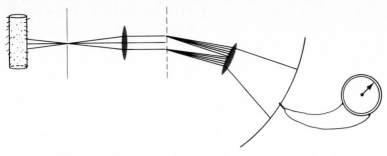

FIGURE 5.27 Measurement of a continuous spectrum. Radiation from the hot object on the left passes through the slit, lenses, and grating and falls on a surface along which a thermocouple is moved. The galvanometer deflection is proportional to the energy striking the thermocouple.

as shown in Fig. 5.27. Light from a hot solid passes through a slit and through a grating, and falls on a curved wall. A thermocouple is moved along the wall, and the thermocouple is blackened so that it absorbs all the energy striking it. (We neglect corrections for the loss of heat which would be made in an actual experiment.) The thermocouple is connected to a meter whose deflection is proportional to the radiant energy striking the thermocouple. Since the thermocouple is two fine wires soldered together, it responds to the energy in a very-narrow-wavelength band, and as it is moved, it indicates the energy in a narrow band at various points in the spectrum. Now if the thermocouple is moved to many different wavelengths and the meter readings are plotted against the wavelength, a curve like Fig. 5.28 is obtained.

The precise shape and height of the curve is affected by the condition of the hot object emitting the light. The object's surface condition—clean, dirty, oxidized, etc.—as well as its composition, can have some effect. In order to eliminate these effects, or to have a standard against which to measure the magnitude of surface effects, physicists discuss continuous-spectrum radiation from an ideal radiator, which is called a *blackbody*.

FIGURE 5.28 Plot of galvanometer reading vs. wavelength from the experiment shown in Fig. 5.27.

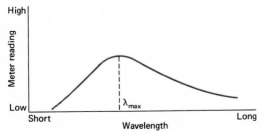

Blackbody The ideal radiator is an object which is totally black, that is, which absorbs all the radiation which strikes it. That fact may not be immediately obvious, but an elegant argument proves that it is true. All the radiation which strikes an opaque object must be either absorbed or reflected. Imagine two objects in a closed box as shown in Fig. 5.29, and suppose that the system has been sitting until all parts are at the same temperature. Since the temperature is not changing, each of the objects must be radiating away energy at the same rate it is absorbing it. The walls of the box radiate onto both objects at the same rate. (If the box and objects are at room temperature, most of the radiation is in the far infrared region of the spectrum.) Let us suppose, however, that one object reflects most of the radiation that strikes it and the other absorbs most of the radiation. Perhaps one is silvery in appearance and the other is almost black. If the temperatures of the two objects are to remain constant, as we know from experience that they do, the one which absorbs more energy must also radiate more, for it must get rid of energy at the same rate it is absorbing it. Therefore, the better the absorber, the better the radiator (and the poorer the reflector). If an object could be found which absorbed all the radiation striking it, it would reflect nothing and would be the best possible radiator. Since such an object would appear black at room temperature, it is called the blackbody.

In practice surfaces which absorb everything which strikes them are not found, although lampblack, platinum black, and similar materials are good approximations. It is possible, however, to obtain the *radiation* from a blackbody without having such a body. Look again at Fig. 5.29 and note that if a perfect blackbody were inside the cavity, the radiation it would emit would be identical with that which already fills the cavity. Therefore the radiation inside a closed cavity is the same as blackbody radiation. If, now, a pinhole is made in the side of the cavity, the radiation which comes out will be blackbody radiation, no matter what the composition or color of the inside walls of the cavity. Experimental work with blackbody radiation is always done actually with *cavity radiation*.

That a small hole punched in a cavity acts as a blackbody

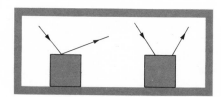

FIGURE 5.29 In a closed box at thermal equilibrium, the object which absorbs more radiation must radiate more; the object which reflects well and absorbs little must radiate little.

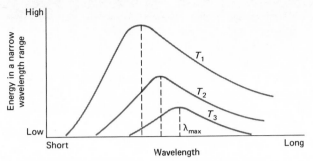

FIGURE 5.30 Radiation curves for a blackbody at three temperatures : $T_1 > T_2 > T_3$. The vertical axis is energy emitted per unit time per area per wavelength region; typical units are watts per square meter per nanometer $[W/(m^2)(nm)]$. Note that as the temperature is raised, the curves become higher, and the position of the peak, λ_{max}, shifts toward shorter wavelengths.

may seem more plausible if the reader will recall the appearance of a small hole, such as might be made by an ice pick, in the side of a pasteboard box which is otherwise closed. Such a hole appears perfectly black because the fraction of light which enters and is reflected out is negligibly small.

An interesting question which might be posed now is: Is the sun a blackbody? Of course the sun does not appear to be black, and the question really is: Does the sun absorb all the radiation which strikes it or does it reflect some? Stated in this way the proper answer, that the sun is *very nearly* a blackbody, seems reasonable.

Now let us repeat the experiment with the spectrograph and thermocouple, using as the source of radiation a very small hole in a cavity in a block of material whose temperature can be varied. For three different temperatures of the cavity, the three curves shown in Fig. 5.30 are obtained.

The difference between the emission of an actual object and the ideal blackbody is described in terms of a number called the *emissivity*. If something radiates 90 percent as much at a particular wavelength as a blackbody at the same temperature, it is said to have an emissivity of 0.90.

Inspection of blackbody emission curves reveals two characteristics immediately: For higher temperatures, the curves are higher and the peaks of the curves are at shorter wavelengths. By suitable calibration of the equipment used to measure the curves, the vertical scale can be made proportional to the energy in a narrow-wavelength band radiated by a unit area of a blackbody. Then the area under the curve is proportional to the total energy radiated, and the first of the observations mentioned above becomes the same as saying that as the temperature of a blackbody is raised, the total energy it radiates increases. That seems obvious, but the exact description of the way its total radiation energy changes is not obvious.

Stefan-Boltzmann law   The exact statement of this relationship is called the *Stefan-Boltzmann law*: The total energy (all wavelengths) radiated per second per unit area by a blackbody is proportional to the fourth power of the Kelvin temperature. In symbolic form,

$$E = \sigma\, T^4$$

where E is energy per second per unit area, σ is called the Stefan-Boltzmann constant, and T is measured in kelvins. (On the Kelvin scale, the freezing point of water is approximately 273 K, the boiling point is 373 K, and room temperature is in the vicinity of 293 K.) The units used for energy and for area determine the units and magnitude of the constant. If energy is measured in ergs and area in square centimeters, then $\sigma = 5.67 \times 10^{-5}$ erg/(cm²)(deg⁴)(s). If energy is measured in joules and area in square meters, $\sigma = 5.67 \times 10^{-8}$ J/(m²)(deg⁴)(s).

Wien displacement law   The law governing the shift of the peak of the radiation curve with change in temperature is called the *Wien displacement law*: The wavelength at which the peak of the curve occurs multiplied by the Kelvin temperature is a constant; that is, $\lambda_{max}\, T = C$, or $T = C/\lambda_{max}$, where $C = 0.2897$ cm-K.

The blackbody emission curves are described by an equation known as *Planck's equation*, which gives the energy emitted per unit area and time as a function of the wavelength and temperature. One can determine the temperature of a blackbody by measuring the rate of energy emission at a particular wavelength; if it is uncertain how nearly the body approximates a blackbody, one can measure at two or three wavelengths and compare with the known blackbody equation.

Using the radiation laws, one can determine the surface temperature of the sun by three methods.

1 The total emission can be computed, from that the emission per square meter can be found, and finally the temperature can be found using the Stefan-Boltzmann equation.
2 The peak of its radiation curve can be located, and the temperature can be determined from the Wien displacement equation.
3 Some pair of wavelengths can be chosen, the relative emission at those positions in the spectrum measured, and the temperature computed from the Planck formula.

All three methods give temperatures which are close to one

another, about 5800 K, confirming the previous statement that the sun is approximately a blackbody.

DOPPLER EFFECT

An optical effect which is of the utmost importance to astronomers is the *doppler effect*: If any kind of wave disturbance is emitted by a source which is moving toward or away from a receiver, the wavelength and frequency of the waves received will differ from the wavelength and frequency of those emitted because of the relative motion of the source and receiver.

Since the magnitude of the frequency shift is dependent upon the ratio of the relative speeds of the source and receiver to the speed of the wave disturbance—i.e., upon the ratio v/c, where v is the relative speed of source and receiver and c is the speed of the waves—the effect is not noticeable with light under normal conditions. Light travels 186,000 mi/s, and that is so much more than the speed of, say, automobiles, that one never notices a change in the color of the headlights or taillights because of the doppler effect. Sound, however, travels approximately 1,100 ft/s, and since an automobile traveling 60 mi/h is moving 88 ft/s, the ratio of the speeds for sound waves is appreciable. As a result, the doppler effect is often observed with sound under common conditions. An observer standing at the side of a highway watching an automobile pass at high speed, blowing its horn as it passes, will hear a drop in pitch of the horn as the auto goes past. The effect is even greater if one is riding in an auto which meets another that is sounding its horn; as the source of sound passes so that it changes from moving toward the listener to moving away from the listener, the pitch drops. The drop in pitch corresponds to a drop in frequency and a consequent increase in wavelength of the waves reaching the ear.

The reason for the doppler effect can be visualized by imagining an experiment such as this: Suppose that one is sitting in a metal rowboat on a lake where waves are moving across the lake at a constant speed toward the boat and slapping the front of the boat. Sitting still and listening, one hears a regular slap, slap, slap. If the listener now begins to move the boat toward the waves, the frequency of the slap increases. Conversely, if one runs the boat with the waves, the frequency of the slapping decreases, since each wave has to overtake a moving boat.

A detailed analysis of the frequency change in the case of water waves or of sound waves is not difficult, and it can be found in any general physics textbook. That analysis is not exactly applicable to the case of light, however, and the differences will be pointed out here and the correct result for the optical case stated. Analyzing the effect for sound, one finds different results depending upon whether the observer is moving or the source is moving. The speed of the sound relevant to the problem is its speed relative to the air, and when an observer moves toward or away from the source of sound, the observer's speed relative to the air changes and hence the speed of that sound relative to the observer changes. The same thing is not true of light, for it always appears to all observers, regardless of their motion, to move at the same speed. To state this matter another way, in the case of sound it makes sense to say that one person is moving relative to the medium carrying the waves (the air) and another person is stationary relative to that medium. In the case of light, no such statement makes sense, for all one can know is that the two observers are moving relative to one another, and there exists no way to determine which one is moving and which is at rest; in fact, such a distinction is utterly meaningless. Consequently the formulas describing the doppler effect for sound are different depending upon whether the observer or the source is moving, but for light the same formula must cover all cases.

If λ_s is the wavelength of light emitted by a source as seen by anyone at rest relative to that source, λ_0 is the wavelength seen by an observer moving relative to the source, v is the relative velocity of the source and observer, assumed to be moving directly toward or away from one another, and c is the velocity of light, then

$$\frac{\lambda_0}{\lambda_s} = \sqrt{\frac{1 + v/c}{1 - v/c}}$$

where v is positive when the source and the observer are receding from one another and negative when they are approaching one another. Thus if an observer looks at a source receding from the observer, its velocity is a positive number and the numerator of the fraction is larger than the denominator, and the ob-

Two spectrograms of Arcturus, taken 6 months apart, showing the doppler effect. The upper spectrogram, taken July 1, 1939, shows a doppler shift corresponding to a radial velocity of $+18$ km/s; the lower, taken January 19, 1940, shows a radial velocity of -32 km/s. The difference of 50 km/s is entirely due to the orbital motion of the earth. (*Hale Observatories.*)

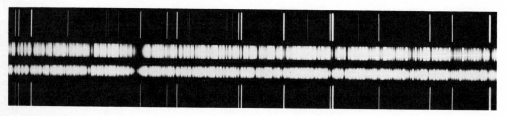

Often v/c is such a small number that an approximation of
this exact expression can be used. If v/c is very small, say less
than 0.1, we can write

$$\lambda_0 \approx \lambda_s \left(1 + \frac{v}{c}\right)$$

where again v is either positive or negative, depending on
whether the source is receding from or approaching the observer.

Often it is useful to speak in terms of the change in wavelength, usually referred to as *wavelength shift*. If a spectral line
has a normal wavelength λ_s, and it is observed to have a wave-
length λ_0, the difference $\lambda_0 - \lambda_s = \Delta\lambda$ is the wavelength shift.
The approximate formula above is equivalent to

$$\frac{\Delta\lambda}{\lambda_s} \approx \frac{v}{c}$$

and this form can be used for low velocities. If an object is moving away from the observer, its spectral lines are shifted toward
the red end of the spectrum; if it is moving toward the observer,
its lines are shifted toward the blue.

Reference has been made repeatedly to a shift of the spectral
lines. The reason is that the shifts observed are usually far too
small to be detected in a continuous spectrum; one does not observe a star which should appear blue to seem red because it is
moving away so rapidly. Instead one looks at spectral lines,
where wavelengths can be measured to a very high degree of
precision, allowing small changes in wavelength to be detected.

QUESTIONS

1 A lens has a focal length of 200 cm for red light, but because
of chromatic aberration its focal length for violet light is only
197 cm. If a color photograph is made of a white object using
this lens, and the film is placed in such a position that violet
light is in sharp focus, how will the photograph appear? (Where
would red light strike the film?)

2 Distinguish between the prime focus, the Newtonian focus,
the Cassegrainian focus, and the Coudé focus of a reflecting
telescope. Radio telescopes never use anything but the prime
focus. Why do they not need to use the others?

3 Sound that corresponds to the note A (above middle C) has a
frequency of 440 Hz. When the speed of sound in air is 330 m/s,
what is the wavelength of that sound?   *Ans* 0.75 m

4 What is the frequency of the light wave which has a wave-length of 600 nm?   *Ans* 5×10^{14} Hz

5 What are the frequency and wavelength of the wave as-sociated with a photon with energy 4×10^{-19} J?

Ans 5×10^{14} Hz; 500 nm

6 When white light passes through a glass prism, red light is deviated less than blue light. But when white light passes through a diffraction grating, red light is deviated more than blue. From the description of the operation of a diffraction grating, can you explain why it bends red light more than blue?

7 Most hydrogen atoms in the atmosphere of the sun are in their lowest energy state. What series of spectral lines do they absorb most strongly? Do we see these absorption lines in the solar spectrum? Why?

8 The continuous spectrum of a star is very nearly that of a blackbody, and the peak of the curve is observed to occur at 700 nm, in the red part of the spectrum. (*a*) Approximately what is the surface temperature of the star? (*b*) How much energy does each square meter of the star's surface radiate each second? (*c*) If another star is twice as hot as the first, where is the peak of its radiation curve? (*d*) How much energy does the second star radiate per square meter per second? [The Wien constant is $C = 2.9 \times 10^6$ nm-K; the Stefan-Boltzmann constant is $\sigma = 5.67 \times 10^{-8}$ J/(m²)(s)(deg⁴).]

Ans   (*a*) 4100 K; (*b*) 17×10^6 J; (*c*) 350 nm; (*d*) 2.7×10^8 J

9 A spectral line with the laboratory wavelength of 500 nm is observed in the spectrum of a star to be at 501 nm. (*a*) Is the star approaching us or receding from us? (*b*) What is the speed of the star relative to the earth?   *Ans*   (*b*) 600 km/s

10 The wavelength of the red line of hydrogen measured in the laboratory is 656.3 nm. At what wavelength is it observed if it is emitted by a star approaching us at 300 km/s?

Ans   655.6 nm

11 The smallest wavelength shift which can usually be mea-sured in the visible region of the spectrum is about 10^{-3} nm. Take 600 nm as a typical value of the wavelengths observed in the visible region and compute the smallest radial velocity which can be measured.   *Ans*   0.5 km/s

6
nuclear energy

The sun and the other stars pour out energy in the form of radiation at a tremendous rate, and that energy must come from some source. One possibility is that the contraction of a star releases gravitational energy, which is the energy being radiated away. When astronomers and astrophysicists first began trying to understand the behavior of stars almost a century ago, the best guess for the source of radiant energy was gravitational energy. Simple computations revealed, however, that the sun could not shine from gravitational energy for more than a few thousand years, and geological evidence has revealed that the earth is far older than that. We now are quite certain that gravitational energy is a negligible factor in the life of a mature star, and that the energy of such stars as the sun comes from changes in the nuclei of atoms and in the elementary particles which make up nuclei.

STRUCTURE OF ATOMS

Atoms are composed of three kinds of particles: protons, neutrons, and electrons. Protons and neutrons make up the nucleus of an atom, and electrons are part of the atom outside the nucleus. Popular pictures show electrons circling the nucleus like planets around the sun, and that picture is near enough to be the best description for our present purposes, which involve distinguishing between the nucleus and the electrons.

A proton has a positive charge, and the number of protons in the nucleus determines the element to which an atom belongs. Hydrogen has one proton; helium has two; lithium has three; carbon has six; etc. The number of protons in the nucleus of each atom is called the *atomic number*; the atomic number of lithium is 3 and of carbon is 6.

Let us consider carbon in more detail. The property that makes an atom a carbon atom is that its nucleus has six protons. If the atom also has six electrons outside its nucleus, it is a neutral carbon atom; if some of those electrons are missing it is still carbon—but an *ion* rather than a neutral atom. In addition to the six protons, a carbon atom has *neutrons* in its nucleus. A neutron has no electric charge, and it has approximately the

137

same mass as a proton; protons and neutrons are about 1,800 times as massive as electrons. The most common carbon atom has 6 neutrons, and so the nucleus consists of 12 particles—6 protons and 6 neutrons. It is hence said to have an atomic mass number of 12. Carbon atoms exist, however, with from 3 to 10 neutrons, and therefore carbon atoms with mass numbers of 9, 10, 11, 12, 13, 14, 15, and 16 exist. Most of those are radioactive, which means they change into something else some time after they are produced.

Isotopes

The forms of an atom which differ only in mass (only in the number of neutrons in the nucleus) are called *isotopes*. Chemists use the symbol C to represent carbon, and a particular isotope of carbon is indicated by a superscript before the C. Often the atomic number appears as a *subscript* before the C, so that some of the isotopes of carbon are designated thus: $^{9}_{6}C$, $^{10}_{6}C$, $^{11}_{6}C$, . . . , $^{16}_{6}C$. The 6 and the C really mean the same thing, for any atom with an atomic number 6 (six protons in the nucleus) must be carbon, and carbon always has the atomic number 6.

PARTICLES NOT FOUND IN ATOMS

The building blocks of the matter of the earth and sun, then, are protons with positive charge, neutrons with no charge, and electrons with negative charge. In 1932 a new particle was found—a particle with the mass of an electron but with a positive charge. It was named the *positron*. Positrons are created by energetic collisions between atomic nuclei, and they may be produced either by cosmic rays from outside the earth's atmosphere striking atmospheric atoms high above the surface of the earth, or by giant particle accelerators. Positrons do not exist long, for within a relatively short time they lose most of their energy and hence most of their speed. After they are slowed almost to a standstill, they combine with an electron and both positron and electron disappear. Their mass disappears and in its place are two very energetic photons called *gamma rays*. We know that when a positron is formed an electron is formed at the same time, and that upon annihilation these particles disappear at the same time. Thus the total charge in the universe remains unchanged.

Antiparticles

The positron is called the *antiparticle* related to the electron. We now know that almost all particles come in pairs, a "normal" particle and its antiparticle. (A few particles are their own antiparticles.) The antiproton is just like a proton, except that it has a negative instead of a positive charge. When an antiproton combines with a proton, the result is a number of other particles which eventually decay and annihilate each other until the final result is a series of gamma rays.

The antineutron poses a problem; it exists, but neither the neutron nor the antineutron has any charge. The difference between the two seems to be the arrangement of electric charge within the particle, for each has positive and negative charge arranged in layers within it. Probably positive and negative layers are interchanged in the two kinds of particles. As with other particles and antiparticles, the neutron and antineutron annihilate one another.

The number of elementary particles has now risen from three to six, and we have only begun. For example, we probably should include as a kind of particle the photon, which is its own antiparticle, having neither charge nor mass.

Neutrinos

In some kinds of nuclear reactions a particle called the *neutrino* is produced. It has no charge, and its mass is very low—probably zero—but it carries away energy. The existence of the neutrino was proposed many years before it was detected experimentally, for its experimental study is very difficult. Neutrinos interact only slightly with matter, and when they pass through the kinds of detectors commonly used by nuclear physicists, they give no indication of their presence. Means do exist to detect neutrinos, however, and it has been found that the neutrino has an antiparticle, the antineutrino. In fact, there are two kinds of neutrinos and two antineutrinos. The symbols for neutrino and antineutrino are ν and $\bar{\nu}$.

Physicists have identified dozens of other particles, bearing such names as pions, muons, lambda particles, etc., but knowledge of these particles is not necessary for our discussion of the energy source of the stars.

The existence of antiparticles suggests an idea that will be discussed in much more detail in a later chapter—the possibility that matter made up entirely of antiparticles, or *antimatter*, might exist. Antiatoms could be made of antiprotons and antineutrons in the nuclei and antielectrons (positrons) surrounding the nuclei. Such atoms would be precisely like the atoms we know, except all positive and negative charges would be interchanged.

NUCLEAR REACTIONS

Atoms, like the "elementary particles," can be created, modified, or destroyed. We shall consider three types of changes which atoms undergo: *decay*, *fission*, and *fusion*.

As an example of decay, we shall consider one radioactive form of carbon, carbon 14. To describe the origin of carbon 14, we must consider another type of nuclear reaction besides decay. Carbon 14 is being constantly produced high in the atmosphere, where cosmic rays, very energetic particles from somewhere in outer space, strike the atmospheric atoms. Some of those collisions release neutrons, and a few neutrons combine with nitrogen to produce carbon 14. The reaction is summarized thus:

$$\,^1_0n + \,^{14}_7N \rightarrow \,^{14}_6C + \,^1_1H$$

The subscripts are given to demonstrate that the sum of subscripts and the sum of superscripts on one side of the equation must be equal to the respective sums on the other side. The subscripts indicate the number of positive charges, and because positive charge is conserved (neither created nor destroyed), the amount that enters the reaction on the left must also be present in the reaction products on the right. The superscripts indicate the number of nuclear particles (protons or neutrons), and that number does not change in this reaction. The particle symbolized by $\,^1_1H$ is a proton, and if it adds an orbiting electron, it can become a neutral hydrogen atom.

From the moment it is formed, carbon 14 begins to disappear. A particular atom may disappear immediately, or it may stay around for 10,000 years, but at some time it will decay into nitrogen 14. The decay process is indicated by

$$\,^{14}_6C \rightarrow \,^0e^- + \,^{14}_7N$$

The symbol $\,^0e^-$ represents an electron, carrying a negative charge and having a mass which is negligible on the scale we are using to measure masses. This electron is sometimes called a *beta particle,* and is symbolized as β^-. It is important to understand what causes the carbon 14 to change into nitrogen 14, for the appearance of an electron is misleading.

Within the nucleus of the carbon 14 atom are six protons and eight neutrons. *There are no electrons within the nucleus.* When the carbon decays, one of the neutrons is transformed into a proton and an electron, and the electron leaves the nucleus. Seven protons and seven neutrons remain in the nucleus, and any atom with seven protons in its nucleus is a nitrogen atom. The fact that the neutron can change into a proton and an electron does not mean that it is simply a combination of those two particles; it is not, and they both come into existence when the neutron disappears.

If the carbon 14 atom is neutral, it has six electrons moving around its nucleus. When it becomes nitrogen, it needs seven electrons to be a neutral atom, and very quickly after the transformation the atom will pick up another electron from those which are commonly wandering around relatively freely.

141

NUCLEAR ENERGY

Carbon-dating

There is no way to predict when one atom of carbon 14 will change into nitrogen, but with a very large number, a statistical prediction can be made. Thus a life insurance actuary cannot predict when a particular man who appears to be in good health will die, but can predict with considerable assurance what fraction of the men at a particular age in the United States will die within a year. If one has a large enough number of atoms of carbon 14, half of them will decay to nitrogen within a period of 5,760 years. It makes no difference how large the sample of carbon 14 is, so long as it is large enough that large-scale statistical analysis is applicable. If one started with 1 g of carbon 14, at the end of 5,760 years 0.5 g would be left; at the end of another 5,760 years 0.25 g would be left; etc. The period 5,760 years is called the *half-life* of carbon 14, that is, the time in which half of any sample will have decayed.

The rate of cosmic-ray bombardment which produces carbon 14 appears to have been reasonably constant for thousands of years, and consequently the rate of production of carbon 14 has been approximately constant. Any living organism is constantly taking in and losing carbon, and part of the carbon it takes into itself is the isotope ^{14}C. So long as the organism is living, a constant exchange of carbon takes place, and the proportion of ^{14}C to ^{12}C remains stable. When the organism dies, the exchange of carbon with the external environment ceases, and the decay of ^{14}C to ^{14}N continues at a constant rate. When an organism has been dead for 5,760 years, the proportion of its carbon which is ^{14}C should be half what it was originally, and half that found in living organisms. By measuring the relative amounts of ^{14}C and ^{12}C in such things as the charred remains of prehistoric campfires or the sarcophagi of Egyptian pharaohs, the age of the wood can be determined.

Carbon 14 has a half-life which is convenient for the archaeologist, and it can be used to date organic materials a few thousand years old. A material with a much longer half-life is necessary if one wishes to determine the age of the earth or meteorites or lunar rocks, however. One such material is uranium 238, which decays to thorium 234 with a half-life of 4.5×10^9 years. Several similarly long-lived isotopes of very heavy metals can be used to estimate the ages of rocks and meteorites.

Fission

A second kind of nuclear process is fission. In decay, a nucleus ejects a particle which may be an electron, a positron, a neutron, or an alpha particle (a helium nucleus—two protons and two neutrons). The ejected particle is always small compared to the nucleus. In fission, however, the nucleus splits into two or three approximately equal pieces. Uranium 235 provides a good example. An atom of uranium 235 can absorb a neutron to become uranium 236. That atom is very unstable, however, and it quickly fissions. The products may be any of a number of nuclei, but most of them have atomic masses in the range 80 to 100. When the uranium nucleus of mass 235 splits into two, three, or more fragments, all approximately equal in mass, a considerable amount of energy is released along with the fragments. That energy is the energy of the atomic bomb and of power reactors which are fueled with uranium 235. Certain other heavy elements, most notable plutonium 239, behave in the same way.

Fusion

During most of the lifetime of a star, a third type of nuclear reaction, fusion, is most important to it. Fusion is the combining of light nuclei to make heavier ones. When very heavy nuclei (mass number above 200) fission, energy is released; when light nuclei (mass number below 56) *combine*, energy is also released. (Atoms with masses between 56 and 200 would release energy if they split, but fission is not observed except in very massive nuclei.) The power of the fusion bomb (the so-called hydrogen bomb) comes from energy released by the combining of light nuclei to form heavier ones, and a tremendous research program is under way in the United States and many other countries to find a way to use the fusion reaction to produce power under controlled, nonexplosive conditions. Thus far, however, only stars have been able to use such reactions continuously.

Mass-energy In all nuclear reactions which release energy, the energy comes from a loss of mass. It is the process discussed in Unit 4, by which mass disappears and energy appears, and 1 kg of mass is equivalent to 9×10^{16} J of energy.

REACTIONS IN STARS

The most important source of the energy of stars is the "burning" of hydrogen to helium. More precisely, the process is one in which four protons are converted into the nucleus of a helium atom (two protons and two neutrons bound together). Two processes are known by which this result may be brought about. We shall consider the simpler first—the *proton-proton* process.

Proton-Proton Cycle

The steps in the proton-proton cycle are shown below in two notations. On the left side p represents proton, d represents deuteron (the nucleus of an isotope of hydrogen containing a neutron in addition to the one proton), e^+ is a positron, and gamma rays and neutrinos are shown as usual. On the right a proton is shown as ^1_1H and a deuteron as ^2_1H.

$$p + p \rightarrow d + e^+ + \nu \qquad\qquad ^1_1H + ^1_1H \rightarrow ^2_1H + e^+ + \nu$$
$$e^+ + e^- \rightarrow 2\gamma \qquad\qquad e^+ + e^- \rightarrow 2\gamma$$
$$d + p \rightarrow {}^3He + \gamma \qquad\qquad ^2_1H + ^1_1H \rightarrow ^3_2He + \gamma$$
$$^3He + {}^3He \rightarrow {}^4He + p + p + \gamma \quad ^3_2He + ^3_2He \rightarrow ^4_2He + ^1_1H$$
$$+ ^1_1H + \gamma$$

To produce a helium 4 nucleus, six protons enter the reactions; two of them are spit out again at the end, still protons. Two positrons are also produced, but they combine with two electrons and their mass is converted into energy. The net effect, therefore, is that four protons are made into a helium nucleus. Now we shall look at the masses of these particles to determine what happens to mass in the process.

The mass of a proton is 1.673×10^{-27} kg; and hence the mass of the four protons used in the reaction is 6.692×10^{-27} kg.

The mass of a helium 4 nucleus is 6.642×10^{-27} kg; this is less than the original mass by 0.050×10^{-27} kg. All that lost mass has been changed into energy. Part of the energy appears in the form of gamma rays, part is carried away by neutrinos, and part becomes the kinetic energy of the particles which are formed.

FIGURE 6.1 A schematic representation of the proton-proton cycle, which produces most of the energy within the sun.

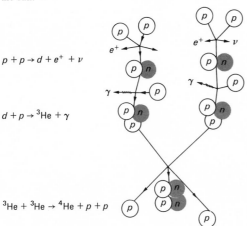

$p + p \rightarrow d + e^+ + \nu$

$d + p \rightarrow {}^3He + \gamma$

$^3He + {}^3He \rightarrow {}^4He + p + p$

This energy loss is real; it is the source of the energy radiated away by the sun and most stars. On the other hand, the fraction of the mass which disappears is small—about 0.72 percent. If 1 kg of hydrogen is converted into helium, the mass of the helium is 993 g; the missing 7 g of mass has been changed into 6.3×10^{14} J of energy, that is, 175 million kWh.

CNO Cycle

The proton-proton process provides almost all the energy for the sun and for stars in its stage of life whose masses are equal to the sun's mass or are less. With more massive stars, however, temperatures are higher, and the carbon-nitrogen-oxygen (CNO) cycle begins to compete with the proton-proton cycle. In very massive stars, the CNO process becomes the more important. The CNO process, proposed by Hans Bethe in 1938, was the first explanation of the source of the energy of the stars made in terms of nuclear transformations. The process begins with a nucleus of carbon 12, which is first changed to nitrogen, then to oxygen, and finally is regenerated as carbon 12. The carbon may be said to function almost as a catalyst in that it is not used up in the process. Obviously this process can operate only in stars which have carbon in their cores; if the stellar interior has only hydrogen and helium, the proton-proton process is the only one which can operate.

The steps of the CNO cycle are:

$$^{12}_{6}C + ^{1}_{1}H \rightarrow ^{13}_{7}N + \gamma$$
$$^{13}_{7}N \rightarrow ^{13}_{6}C + e^{+} + \nu$$
$$^{13}_{6}C + ^{1}_{1}H \rightarrow ^{14}_{7}N + \gamma$$
$$^{14}_{7}N + ^{1}_{1}H \rightarrow ^{15}_{8}O + \gamma$$
$$^{15}_{8}O \rightarrow ^{15}_{7}N + e^{+} + \nu$$
$$^{15}_{7}N + ^{1}_{1}H \rightarrow ^{12}_{6}C + ^{4}_{2}He$$

Because both cycles produce the same net result—the converting of four protons into one helium nucleus—they produce the same amount of energy.

Helium-Carbon Reaction

In the later stages of the lives of most stars, their cores become hot enough to cause other reactions to occur, and a great variety of reactions are probably used to produce energy. The one which becomes important immediately after the burning of hydrogen has converted most of the stellar core to helium is a burning of helium to form carbon, according to the scheme

$$3 \, ^{4}_{2}He \rightarrow ^{12}_{6}C$$

The source of the energy released by this reaction can be seen from this listing of the masses:

Mass of three helium 4 nuclei	19.926×10^{-27} kg
Mass of carbon 12 nucleus	$\underline{19.915 \times 10^{-27}}$ kg
Mass lost in reaction	0.011×10^{-27} kg

That mass which is lost in the reaction is changed into energy and eventually is radiated from the surface of the star.

The structure of atomic nuclei is such that energy is released by fusion until the nucleus produced is iron. To make heavier nuclei than iron, energy must be added rather than extracted, and energy is released when those heavier nuclei are split, or fissioned. The release of energy by fission is probably important in stars only rarely; it may help produce the explosive energy release which tears apart a star to produce a supernova. As we shall see later, however, the current best theory is that the universe began with a composition of almost pure hydrogen plus a small bit of helium. The existence of elements heavier than iron on the earth (e.g., silver, gold, mercury, uranium, etc.) means that reactions capable of building such atoms took place in our vicinity before the solar system formed.

All this may be summarized very briefly: In the interiors of stars light nuclei are combined to form heavier nuclei; in that process part of the mass disappears. The lost mass is changed into energy, and that is why the stars shine!

QUESTIONS

1 Distinguish between the neutron and the neutrino.
2 The sun radiates away energy at the rate of 4×10^{26} W. Compute the amount of mass which must disappear each second to maintain that rate of radiation. Compute how much hydrogen must be converted into helium each second to produce that power. Express the numbers in tons per second. If the sun radiates at about the present rate for 10^{10} years, how much hydrogen will be converted to helium?
    Ans 4×10^9 kg; 6×10^{11} kg, 4.4×10^6 tons/s; 7×10^8 tons/s; 2×10^{29} kg
3 How many hydrogen atoms are converted into helium in the sun each second?   *Ans* 4×10^{38}
4 The mass of our galaxy is estimated to be about 2×10^{41} kg, and its luminosity is estimated to be approximately 1.3×10^{45} J/year. How much mass does it lose each year? How much mass has it lost in 10^{10} years?
                    Ans 1.4×10^{28} kg; 1.4×10^{38} kg

stars
and
galaxies

two

We have not sent men to another star, nor have we sent television cameras near enough to another star to return close-up pictures such as have been received from the nearer planets. What we know about the stars and about the clusters of stars we call galaxies has been deduced from small clues, with the aid of the concepts and the physical methods discussed in the first section.

The ancients made little progress with the study of the stars because they did not have instruments to extend human vision. And even in 1864, after the telescope had been in use for almost two centuries and after the positions of stars and nebulae had been charted with a precision impossible before the time of Galileo, Auguste Comte said, "There are some things of which the human race must forever remain in ignorance, for example, the chemical constitution of the heavenly bodies." But even then the fundamental work on spectral analysis which was to let us determine the chemical compositions and the temperatures of the sun and the stars was 5 years old.

Scientific investigation can aptly be compared to detective work, and the comparison is particularly appropriate in the

2

Spiral reproduced with permission from "The Spiral Way" by Gerald Oster, *Natural History* Magazine, August-September 1974.
1. Solar corona, photographed during total eclipse on August 31, 1932, at Fryeburg, Maine. (*Lick Observatory photograph.*)
2. Solar flare showing magnetic structure, photographed in red light of hydrogen α at Big Bear Solar Observatory on September 17, 1970. (*Hale Observatories.*)

case of astronomy and astrophysics. Few detectives in either fiction or fact work with as few clues as the astronomer, but from their meager information astronomers and astrophysicists determine distances, diameters, masses, compositions, ages, and even genealogies of stars. If they cannot say which is the parent and which the child, they can at least determine to which generation a member of the family belongs.

Like any good mystery story, the story of the discovery of the natures of heavenly objects contains false clues, faulty interpretation of information, and errors which become apparent only slowly. But in one very important respect science is totally unlike fictional detective work and unlike most real detective work. In a mystery story all the answers are revealed at the end; in science one never finds all the answers. As rapidly as one set of answers is found, they suggest more questions, so that the search for clues and the effort to interpret clues is without end. As soon as we determine the compositions of stars, we

3

3. The Lagoon nebula in Sagittarius. This is an emission nebula approximately 2,000 pc from the earth. It is also known as M 8 and as NGC 6523. (*Lick Observatory photograph.*)
4. The Pleiades cluster, showing reflection nebulae around some of the stars. (*Lick Observatory photograph.*)

4

find that the stars are not all the same, and we wonder why. When we observe that some galaxies appear to have spiral arms, we wonder what the arms are. And when we understand the composition of the arms, we are confronted with the problem of understanding how those arms can persist. If science has a last chapter, in which all the threads are untangled and all the questions are answered, we have not yet begun to write even the first paragraph.

In this section we shall consider stars, galaxies, matter between the stars, and many of the other topics which made up the subject matter of astronomy before 1960. Since 1960 other problems, both old and new, have begun to receive a great amount of attention; those will be discussed in the third section. But many problems connected with stars and galaxies remain, and often the only honest way to end discussions of some of their properties is: We do not know.

5. The Horsehead nebula in Orion, photographed in red light with the 200-in telescope. The horsehead is a dark nebula. (*Hale Observatories*.)

6. The Large Magellanic Cloud, an irregular galaxy which is a satellite of our own galaxy. (*Lick Observatory photograph.*)
7. The Whirlpool galaxy in Canes Venatici, also known as M 51 and as NGC 5194-5. (*Lick Observatory photograph.*)

8. A spiral galaxy, NGC 4565, in Coma Berenices, seen
edge-on. This galaxy probably looks very much as our
own Milky Way galaxy would look if it were viewed
edge-on from outside. (*Hale Observatories.*)

9. Spiral galaxy in Ursa Major, known as M 81 or
NGC 3031. This galaxy probably looks very much as
our own Milky Way galaxy would look if viewed from a
great distance at a similar angle of view. (*Lick Observatory photograph.*)

10. Spiral galaxy, NGC 4594, in Virgo. (*Hale Observatories.*)

11. Galaxy, NGC 5128, in Centaurus. This galaxy is a source of radio noise, and in lists of radio sources it is Centaurus A. (*Hale Observatories.*)

12. Cluster of galaxies in Hercules, displaying many different types of galaxies in a single photograph. The photograph was made with the 200-in telescope. (*Hale Observatories.*)

11

12

7
the sun

The sun is of great interest for two reasons. First, with the exception of the energy obtained from nuclear reactions, all the energy available to us comes or came from the sun. When the sun has outbursts on its surface, radio communications may be disrupted on the earth and auroral displays are seen. Small variations in the output of the sun may be the cause of the glaciations which came in the past and may come again.

Second, to the astrophysicist interested in the structure and behavior of stars, the sun is the only star near enough to be observed clearly. All other stars appear as points of light in even the largest telescopes, and are so far away that we cannot observe details about their surfaces. In many respects the sun appears to be an average star, and what we learn of it may be applicable to many of the other stars we cannot study so well.

The determination of the astronomical unit, the distance from earth to sun, has already been discussed. Combining that distance with the period of the earth around the sun provides the sun's mass, 2×10^{30} kg, and combining the distance to the sun with its apparent diameter permits the computation of its actual diameter, 1.4×10^{6} km (approximately 800,000 mi). The sun's diameter is 109 earth diameters. Its volume is therefore approximately 1 million times the volume of the earth. From the volume and the mass one can compute the density, which is 1.4 g/cm³, about one-fourth the density of the earth. From observation of the movement of spots on the surface and also from doppler measurements on the edges of the sun, we know that the sun rotates on its axis in a period of about 25 days, and the equatorial region rotates somewhat faster than the polar regions.

SOLAR ATMOSPHERE AND SURFACE

When one looks at the sun through a smoked glass, or sees the image of the sun projected by a lens, it seems to have a sharp, distinct edge. Thus the statement that the diameter of the sun is about 800,000 mi seems to have some meaning. The sun does not have a solid surface, however, and it does not end abruptly at that appparent surface, and so the meaning of the term "diameter" and the reason for an apparently distinct edge must be considered in more detail.

153

If we could travel down into the sun in a spaceship, well insulated against the heat and intense radiation, we would find no discontinuity. The gases of the sun extend thousands or millions of miles into space around the apparent surface. As the spaceship sank lower and lower, the temperatures outside would change and the density of the gases around would gradually increase, but nowhere would an observer find a sharp change in density. If we continued deep into the interior of the sun, the same observation would hold: density and temperature would increase gradually, but nowhere would we find a surface upon which to stop our descent. The sun is entirely gas all the way to the core. One thing, however, would change rather abruptly. As we began our descent through the outer atmosphere of the sun, we would be able to see the earth and the other planets rather well, and for a great distance that view would be changed little (the planets might become slightly fainter), since the upper atmosphere of the sun is almost transparent to visible light. Rather suddenly, however, we would find our view of the outside space blocked, because we would be in a rather narrow zone (approximately 200 km, or 100 mi) in which the gases of the sun become opaque. That zone, called the *photosphere*, is the "surface" which we see, and the fact that it is thin compared to the diameter of the sun is the reason that the sun seems to have a distinct edge. Light leaving the upper part of the photosphere travels to us with little impediment; light from the bottom of the photosphere is likely to be absorbed and reemitted before it escapes; and light from below the bottom of the photosphere is certain to be stopped before it leaves the sun. The reason for the opacity of the photospheric layer is the formation at that depth of large numbers of negative hydrogen ions—hydrogen atoms that have an additional electron—which absorb all light in or near the visible part of the spectrum.

Even though it is not a surface in the way that the surface of the earth is, the photosphere stops our vision into the sun, and it serves as a natural dividing point for discussions of the sun. What happens in and above the photosphere we can see; what happens below we must infer from study of models of the sun. We shall consider the upper parts of the sun first and then go on to information and guesses about its deeper structures.

Temperature   Since the photosphere appears to us to be opaque, it is the source of the continuous radiation from the sun. If the sun's temperature is determined from the Wien displacement law or by use of the Stefan-Boltzmann law, the result is the temperature of the gases in the photosphere. Such measurements yield values near 5800 K.

Light leaving the bottom of the photosphere is probably almost precisely blackbody in distribution. As it passes through the upper parts of the photosphere and tens of thousands of miles of thinner gases in the outer atmosphere, however, atoms extract from the light certain wavelengths to produce an absorption spectrum, and because of that loss of energy at certain discrete positions, the light which reaches us is not exactly that of a blackbody. The deviation is so slight that for most purposes it can be ignored and the sun can be considered a blackbody radiator. As we have said, determinations of the temperature of the photosphere based upon the Wien law, the Stefan-Boltzmann law, and the Planck law (using more than one wavelength) yield temperatures which differ from one another by less than 300 K. (If the sun were a perfect blackbody, the three methods should give the same results.) At the very top of the photosphere the temperature is only about 4500 K, so the effective temperature, 5800 K, is that of a layer well down in the photosphere.

Granulation Photographs of the photosphere made with good telescopes at times when "seeing" is quite good or made from above most of the earth's atmosphere reveal a granular appearance; the phenomenon is called *granulation*. Regions of perhaps 300 mi in diameter are hotter than the surrounding "walls" and are moving upward at about 1 km/s. Clearly we are seeing the top of convective zones; the brighter regions are places where gas is moving upward, and the dark areas are places where it is sinking back downward. Some idea of the magnitude of this boiling motion can be gained by realizing that if the same sort of motion were to occur on the earth, areas such as Illinois, Wisconsin, Iowa, Nebraska, Kansas, and Missouri would all be rising upward, and material along their borders would be sinking. To make one good granule the New

Solar granulations. This highly magnified section of the sun's surface shows the granular appearance which is the result of convective zones. (*Hale Observatories.*)

England states would have to combine; Texas might make two or three. The pattern of the sun is constantly changing, so that individual granules persist for only about 5 min before they dissolve to be replaced by others.

Chromosphere

The layer of atmosphere directly above the photosphere is the *chromosphere*, so called because during an eclipse of the sun, when the moon blocks out the light from the main disk of the sun, a ring of colored light is seen. The chromosphere extends for about 12,000 km beyond the photosphere. In that distance the density of the gas becomes much less but the temperature rises from 4500 to about 1,000,000 K.

Spectrum When the light from the main disk of the sun is blocked during an eclipse, light from the chromosphere produces a bright-line spectrum. (It is seen for only a very short time and hence is called a flash spectrum.) That light comes from relatively hot gas, glowing because the gas is excited. It is the same gas which produces some of the Fraunhofer absorption spectrum seen in the normal solar spectrum. Under normal conditions, however, the emission of that gas is so much weaker than the continuous emission from the photosphere that the lines are seen as absorption lines. The flash spectrum includes some lines which do not appear in the absorption spectrum; they are produced by ionized atoms in the upper, very hot region of the chromosphere.

Corona

That part of the sun's atmosphere beginning at the top of the chromosphere and extending out as far as it can be identified is called the *corona*. Now the corona can be studied over long periods of time by means of the *coronagraph*, an instrument invented in 1930 by Bernard Lyot; before that invention the corona had been seen only during total eclipses of the sun, which never last more than about 7 min, and are often much shorter. The coronagraph uses a metal disk which blocks the image of the sun and hence serves the same purpose as the moon during eclipse. Because the coronagraph is within the earth's atmosphere, however, it must also use a filter that blocks most of the light scattered by the atmosphere. Although it is a valuable tool for solar astronomers the coronagraph is not a complete substitute for observations made during eclipse.

Temperature The most striking feature of the corona is its temperature—about 1 million K. That temperature is a kinetic temperature, not radiation temperature. This means that it is not determined or demonstrated by the radiation emitted by the

gas, but refers to the average kinetic energy of the gas molecules. If a sample of gas is all at one temperature, the average kinetic energy of the molecules is directly proportional to the
temperature; the relation is

$$\tfrac{1}{2}mv^2 = \tfrac{3}{2}kT$$

where m is the mass of a molecule, v is the average speed, and k
is Boltzmann's constant. If the molecules of a gas have random
motions, their average kinetic energy can be specified by giving
the temperature which corresponds to that energy. (For this use
of temperature to be entirely correct, the number of molecules
having each speed within the possible range should be given by
what is called the *Maxwellian distribution function*.)

What we are saying, therefore, when we say that the temperature of the corona is 1 million K, is that the average kinetic
energy of the atoms of the gas is very high—$\tfrac{3}{2} \times k \times 10^6$. The
value of k is 1.38×10^{-23} J/K, so the average kinetic energy of
each molecule of gas in the corona is of the order of 2×10^{-17} J.
The speed of the typical hydrogen atom nucleus at that energy
is about 1.5×10^5 m/s, but the speed of an electron is more than
40 times higher: 6×10^6 m/s. It is the electrons which give an
indication of this high speed.

At such high energies, light atoms such as hydrogen are completely ionized; they cannot exist for more than the briefest
time as stable atoms. Heavier atoms such as iron or calcium still
have some electrons attached to the nuclei, but they have lost
many electrons and so are highly ionized. The corona comprises a gas made up of protons (hydrogen nuclei), electrons,
the nuclei of such atoms as helium and lithium, and ionized
atoms of heavier elements. Each of the hydrogen atoms present
has contributed one electron (and one proton), but in addition
some heavier atoms have contributed 10 or 12 or even more
electrons; the result is that the coronal gas has more electrons
than any other constituent. Electrons scatter light very efficiently, and they scatter all visible wavelengths equally well.
The corona is too hot for recombination and deexcitation of
atoms to produce very much radiation, and so most of the light
emitted by the corona is light scattered by electrons. Since that
light comes from the photosphere and has passed through the
chromosphere, one might expect that it would be identical to
the light coming directly from the sun—Fraunhofer lines and
all. The fact is, however, that the light is like that of the photosphere but without any absorption lines; it has a continuous
spectrum. During eclipses the corona is often described as
"pearly white." The absence of Fraunhofer lines tells us that
the electrons must be moving so rapidly that as they scatter the
light they give it doppler shifts, and the doppler shifts are
enough to fill in the Fraunhofer lines and obliterate them. For

the electrons to have random speeds that high, they must have energies corresponding to a temperature of about 1 million K.

Spectrum   The corona also shows some emission lines, and because they are not lines often found in laboratory spectra, they were once thought to be produced by an unknown element, perhaps found only in the corona. We now know they are produced by atoms of ordinary elements, ionized to a level not often or easily produced in the laboratory. The most intense of the strange lines is a green line at 530.3 nm produced by iron which has lost 13 of its normal 26 electrons. This is further evidence of high temperature, or, alternatively, of high kinetic energies. Nuclei or electrons must strike the iron atom with enough energy to remove half its electrons, which requires a lot of energy!

Source of energy   The source of the energy is difficult to account for because the corona is a layer of gas at a temperature of 1,000,000 K riding on a layer which has a temperature of 4500 K. Energy certainly does not flow into the sun from outside at a rate which would maintain the corona at such a temperature, nor can conduction from below explain the temperature. (Heat does not flow from a region of 4500 degrees to one of 1,000,000 degrees any more than heat will pass from a cold object to a hot one.) The chromosphere is a region of great turbulence, and jets of hot gases shoot up through it from the photosphere. Some of those jets reach to the top of the chromosphere and rise at speeds of more than 18 mi/s; they are called *spicules*. These gases move at supersonic velocities, producing shock waves (like sonic booms) which rise into the corona. These sonic waves are thought to heat the coronal gases to their astonishing kinetic temperature.

Shape   Photographs of the corona made during total eclipses never show it symmetric about the sun. During times when sunspot activity is at a maximum, the corona is more nearly symmetric than during periods of solar quiescence, but it always extends out farther in some directions than in others.

Sunspots

The sun goes through cycles of activity, and the most apparent indication of its activity is the presence of spots. Very large groups of sunspots can be seen without telescopes. They had been observed before the telescope was invented, but their nature was not realized, and they were usually explained away as optical illusions. When telescopes came into use, observers could follow spots of normal size—not just giant spots—and determine that they really are something associated with the sun. About 1610 the fact that sunspots are "blotches" on the face of the sun was recognized, and their movement across the face of the sun was realized to be an indication that the sun rotates on its axis.

Spots take about $13\frac{1}{2}$ days to cross the visible disk of the sun. In some cases after they have crossed from east to west and disappeared at the edge of the disk, they reappear at the eastern edge again after another $13\frac{1}{2}$ days. It appears at first glance that the sun rotates in a period of 27 days, but when the fact that during that time the earth has been moving also is considered, it is computed that the sun makes one complete revolution in just over 25 days. The time appears to be somewhat different if one is looking at sunspots near the equator or spots at higher latitudes; the equatorial parts of the sun move faster than the parts farther toward the poles.

A sunspot appears dark because it is cooler than the surrounding gas—perhaps by as much as 1500 K in the nucleus of the spot, called the *umbra*. A larger area around the nucleus, called the *penumbra*, has a temperature intermediate between the umbra and that of the surrounding surface of the photosphere. A spot may have a diameter anywhere from a few thousand miles to several tens of thousands of miles; probably the earth could fit into even a small spot.

Magnetic fields In the early twentieth century methods were developed for measuring the strength of magnetic fields on the sun, and it was immediately discovered that sunspots are associated with large magnetic fields. The magnetic field is now believed to be the cause of the spot; the field appears before the visible spot is formed.

Magnetic fields can be measured on the sun (and on some stars) because of an effect that such a field has on light emitted by atoms lying within the field. The effect is called the *Zeeman effect*, named for the physicist who discovered it. When atoms emitting light are in a magnetic field, some of the lines emitted are split into two, three, four, five, or even more components. If the spectrum can be spread out enough, these components may be observed as separate lines; that is, what is one line without a magnetic field may become five closely spaced but separate lines in a magnetic field. If the spectrum cannot be spread enough to separate the components, the effect of the magnetic field is to make the lines appear broader. The magnetic field also causes the components to be polarized in distinctive ways, so that by examining the polarization of a broadened line one can distinguish Zeeman broadening from doppler broadening or broadening produced by such other effects as pressure.

Studies of Zeeman splitting or broadening of lines from the sun have revealed that the sun as a whole has a weak magnetic field, but that in sunspots the field is quite strong. The usual unit for measuring magnetic fields is the gauss (G) and the field in sunspots becomes as strong as 3,000 G. That is as strong as

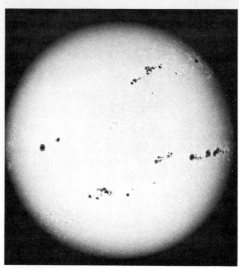

The sun at a time of sunspot maximum on December 21,
1957. (*Hale Observatories.*)

the field of a very strong magnet, and it is about 6,000 times as
strong as the magnetic field of the earth.

Magnetic fields are produced by electric currents, which are
streams of moving electric charges. Probably large magnetic
fields exist deep within the sun and occasionally some of them
rise to the surface to produce sunspots, but the mechanism of
their production and movement is more a matter of speculation
than knowledge. The effect of magnetic fields on the kind of gas
which makes up the photosphere and deeper layers of the sun
is well known, however, and we have a reasonable understanding of the way in which the fields produce a sunspot.

A magnetic field has a direction, and it may be conveniently
represented as a series of parallel lines. Physicists commonly
speak of magnetic fields in terms of "lines," by which they
mean the parallel lines which constitute a picture of the field.
Hot, ionized gas, such as exists in and near the photosphere,
can move with ease along the lines of the magnetic field, but it
can move only with great difficulty across the lines. If
ions—electrons or positive ions—move parallel to the field
lines, the magnetic field has no effect on them and their movement is unhindered. But if the ions try to move across the lines,
a force comes into effect which makes them move in spiral
paths around the magnetic field lines, and their movement is
severely hampered. Thus when a magnetic field breaks through
the photosphere, with its lines sticking up out of the surface, it
inhibits the circulation of the gas within its region, and the
movement of hot gases which normally gives rise to granula-

tion is reduced. When hot gases no longer rise at the usual rate, the gas on the surface begins to cool, and as it cools it becomes a visible sunspot.

Sunspot cycles   The number of sunspots on the surface goes through a cyclic variation; at the same time the position of the spots varies. At times of sunspot *minima*, no spots or perhaps only one or two may be seen at any one time; those spots which are seen may be at latitudes of about 30°. The number of spots increases gradually until at *maxima* dozens of spots, arranged into several large groups, may be seen at once. During the time the number increases, the spots move farther toward the equator, and as maximum is passed and the number begins to diminish, the movement toward the equator continues until finally a few spots are seen at perhaps 8 or 9° latitude. Then the next cycle begins with a few spots appearing at higher latitudes, and the cycle repeats. The entire time from beginning to end of this cycle is about 11 years.

After studying the magnetic field of the spots, astronomers discovered that the cycle just described is only a half cycle. Spots normally occur in pairs, with one spot leading (to the west of) the other spot. One of these spots has the polarity of the north pole of a magnet, and the other has the polarity of the south pole. During one 11-year cycle all the leading spots in the northern hemisphere of the sun have the polarity of a north magnetic pole, and the trailing spots have the opposite polarity. In the southern hemisphere of the sun the polarities are

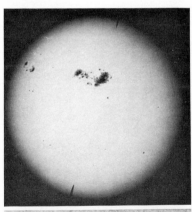

The entire disk of the sun and an enlarged view of an exceptionally large sunspot group. The photograph was taken April, 7, 1947. (*Hale Observatories.*)

reversed, with the leading spots being south poles. In the succeeding 11-year cycle, however, all polarities are reversed, so that where the leading spot had been a north pole, it now is a south pole. The time it takes for everything to return to its original state, then, is 22 years, for in that time the polarities of leading spots have returned to their original values.

Other Surface Phenomena

Although spots are the most obvious indication of the existence of centers of activity on the sun, three other phenomena often appear near sunspot regions during the occurrence of spots or preceding or following them. The three are *faculae, flares,* and *prominences.*

Faculae   Faculae are clouds of gas in the chromosphere which are slightly hotter than the surrounding gas. They are so nearly the temperature of the atmosphere near them that they can be seen in white light only when they are near the edge of the sun's disk. If the sun is observed by the light of a single atom, however, so most of the scattered light from the corona is eliminated, faculae stand out clearly because of their higher temperature. Usually the light of hydrogen or ionized calcium is used. Faculae often appear before sunspots and serve as warnings that spots will soon be visible, and often they persist long after the spots have disappeared.

A large, bright flare on the sun, August 7, 1972. (*Sacramento Peak Observatory, Air Force Cambridge Research Laboratories.*)

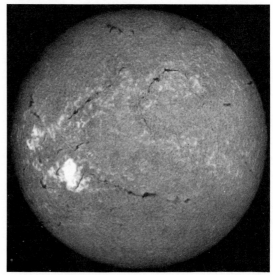

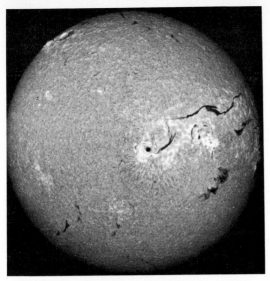

The sun, photographed through a hydrogen α filter. Active regions are shown by bright areas (plages), and one large sunspot near the center of the disk is clearly visible. The dark elongated structures are filaments, which are raised, condensed structures supported by magnetic fields between active regions of opposite polarity. (*Sacramento Peak Observatory, Air Force Cambridge Research Laboratories.*)

Flares   Small regions in the vicinity of sunspots sometimes become suddenly very much brighter than usual; the regions are called flares. During periods of high sunspot activity, as many as 5 to 10 flares may be seen in one day. When a flare is seen at the edge of the disk of the sun it is a jet of gas extending upward several thousand miles, and it may rise hundreds of miles per second. A flare is of short duration—perhaps an hour. During flares x-rays, ultraviolet radiation, and radio waves are emitted in unusual quantities, and particles are ejected into space at high speed. Both the radiation and the particles affect the earth changing radio communication over great distances and disturbing earth's magnetic field; aurorae are visible effects of the particles.

Prominences   In contrast to the flares, which are of short duration, prominences may persist for weeks or months. Photographs taken with only one color of light often show dark lines near sunspots and faculae; the dark lines are called *filaments*. When they are seen at the outer edge of the sun's disk, they often appear as great arches, and then they are called prominences. Prominences can be seen with coronagraphs, and many have been photographed. They characteristically look as if material is falling from the corona onto the disk; apparently

Solar prominence 100,000 mi high,
photographed in red light of hydrogen α
at Big Bear Solar Observatory, June 12,
1972. (*Hale Observatories.*)

material condenses in the corona and flows down along magnetic field lines. The clouds making up a prominence are more dense and cool than other gas in the corona. Occasionally one of the arches will explode upward, expanding in height and length at tremendous speed. A filament will characteristically arch to a height of 20,000 to 60,000 mi, and it will have a length of about 100,000 mi.

All the manifestations of the active sun—spots, faculae, flares, and prominences—probably indicate changes in and

A large eruptive prominence extending out from the disk of the sun. (*Sacramento Peak Observatory, Air Force Cambridge Research Laboratories.*)

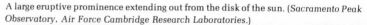

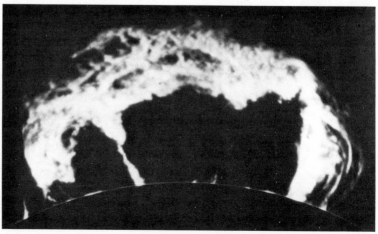

properties of the magnetic field of the sun. Solar astronomers study the face of the sun constantly, and astrophysicists try to understand the origin and convolutions of the magnetic fields within the sun which can produce the effects seen on the surface. Meteorologists are also interested in some aspects of the sun's activity, for the earth's magnetic field, and possibly to some extent the earth's weather, are affected by outbursts on the sun. During the time astronauts have been in space, special watch has been kept for flares, for the x-ray and particle emission from violent flares could be hazardous for men exposed in thin space suits.

SOLAR INTERIOR

We turn next to the interior of the sun, where our information is of a different sort. Since we cannot see below the photosphere, we must take what we know about the sun's gross properties (mass, rotation rate, etc.) plus what we can see at the surface and try to deduce what interior conditions must be. The process is the construction of a model; one imagines how a body of gas like the sun might be constituted and then tries to determine how closely the properties of the model agree with the facts known about the sun.

Models

Scientists constructing models of the sun always begin by neglecting the rotation; it is so small that it must produce only small effects. Models also neglect the magnetic field, less because it is known to be unimportant than because its effects are not known and we do not know how to take them into account. Then the model builder assumes that the sun is in a steady state—that it is not changing in any significant way. Evidence from fossils on the earth indicates that for millions of years the energy output of the sun has been very nearly constant, and one feels confident in assuming that the sun is changing so slowly that the changes can be neglected. We know something of the composition of the outer layers of the sun, and we know the rate at which energy is being brought to the surface and sent into space; with these facts we can begin to make a model.

The steady-state assumption has two implications: The size of the sun is not changing and the rate of flow of energy upward to the surface is constant. Therefore if one thinks of a thin layer of the sun lying somewhere below the photosphere, the pressure at the top of the layer must be great enough to support all the atmosphere above it, and the pressure at the bottom must be enough greater than that to support the additional weight of the layer itself. Furthermore, the energy which enters the layer from the bottom must equal the energy which leaves the top, for the layer must become neither hotter nor cooler. If the sun is thought of as consisting of many thin layers, this sort of reasoning can be applied to all the upper layers. Deep within the sun energy is being released, and the energy which leaves each layer traveling upward must be that which entered it from below plus the energy released within the layer itself. If each layer were totally transparent, energy in the form of radiation would pass through it with no delay, and all parts of the layer could be the same temperature; if the layer were totally opaque, no radiation could pass through and the interior side could be very hot and the outside very cold. The actual material of the sun falls between these extremes, and the opacity determines the rate at which the temperature rises with depth within the sun. The opacity is determined chiefly by the concentrations of elements heavier than hydrogen.

Core Conditions

Depending upon different assumptions about the amounts of heavy elements present in the original material from which the sun formed, and about the amount of mixing occurring in the sun, one can develop different models of the sun. Models which seem to fit observations best, however, assume the following conditions: the temperature in the core of the sun is approximately 15×10^6 K; density at the center is 160 g/cm^3; most of the thermonuclear energy is released within the inner 10 percent of the radius; and convection occurs in a thin zone just below the photosphere, but energy is transported by radiation only (without convection) throughout most of the interior of the sun. The age of the sun probably is near 5 billion years, and it should continue to shine at the present rate for another 5 billion years before significant changes begin to occur as it moves to the red giant stage of its life.

Several models of the sun have been developed, but all give approximately the same picture of the interior. Until recently the only way to test any of them was to examine the accuracy with which they predict the properties of the sun visible from the earth—the temperature at the surface, the composition of the atmosphere, the luminosity, and of course the mass and diameter. But since those observations were the starting points of all the models, they all predicted the observed values very nearly the same. There seemed to be no way to examine the interior of the sun directly in order to further test the models.

Solar Neutrinos

A way does exist, however, to obtain information about conditions at the actual core of the sun. Neutrinos produced by the nuclear reactions there interact so slightly with matter that most of those released escape from the sun immediately. Unlike the light coming from the core of the sun, which takes millions of years to travel to the surface because of many absorptions and reemissions, the neutrinos come as if the outer layers of the sun were not there. Therefore if we can catch and study them, we shall gain information about conditions at the core of the sun now.

Davis's experiment In the mid-1960s Raymond Davis, of Brookhaven National Laboratory, began an experiment a mile underground in the Homestake Gold Mine in South Dakota that was intended to capture neutrinos coming from the sun. The detection equipment is placed far underground to eliminate many of the false counts which might otherwise be produced by cosmic rays. The detector is a tank 20 ft in diameter and 48 ft long filled with 100,000 gallons of perchlorethylene (C_2Cl_4), a common dry-cleaning fluid. The important part of the material is the chlorine, for 25 percent of those chlorine atoms are the isotope ^{37}Cl; the other 75 percent are the isotope ^{35}Cl. The chlorine 37 atom can absorb a neutrino and change into argon 37 and an electron. The reaction is this:

$$\nu + {}^{37}Cl \rightarrow {}^{37}Ar + e^-$$

Because argon is one of the noble gases, it is not bound to the carbons in the molecule, and it drifts away. Periodically Davis and his assistants bubble helium through the tank, and the helium carries out the argon that has been formed. It is frozen out of the helium stream and carried to Brookhaven for analysis.

The amount of argon formed is so small—of the order of atoms per day—that detecting it probably would be impossible except for the fact that it is radioactive. Argon 37 decays to chlorine 37 with a half-life of 35 days; in the decay it captures an electron and emits a gamma ray, which can be detected and counted. It is thus possible to count individual atoms of argon 37, and the experimenters are finding about one argon atom per month.

A neutrino which can produce the change of chlorine to argon must have an energy greater than 0.814 MeV, and most of the neutrinos produced in the sun have less energy than that; consequently only a small number of those produced can possibly be captured. (The MeV, which stands for million electron volts, is a convenient unit for measuring the energy of a single nuclear decay. One MeV is equal to 1.6×10^{-13} J.) In the preceding chapter the proton-proton reactions were listed thus:

$$H + H \rightarrow D + e^+ + \nu$$
$$H + D \rightarrow {}^3He + \gamma$$
$${}^3He + {}^3He \rightarrow {}^4He + H + H + \gamma$$

The neutrino produced in that reaction has an energy lying anywhere below 0.42 MeV, however, and so it cannot be detected by the chlorine 37 reaction. We must consider some reactions which are rare but which produce more energetic neutrinos.

The first step in the series of reactions is the production of a deuterium nucleus from two protons, and 99.74 percent of the deuterium is produced by the reaction just listed: $H + H \rightarrow D + e^+ + \nu$. The other 0.26 percent is produced by the reaction $H + H + e^- \rightarrow D + \nu$. That neutrino has an energy of 1.44 MeV, and can therefore be detected. Another reaction which produces quite energetic neutrinos in significant quantities is the decay of boron 8 to beryllium 8. After ${}^3He$ is produced by the combination of hydrogen and deuterium, most of it is used to form ${}^4He$ by the reaction ${}^3He + {}^3He \rightarrow {}^4He + H + H$. A small amount, however, is used in the reaction ${}^3He + {}^4He \rightarrow {}^7Be + \gamma$. That beryllium 7 combines with another proton to form boron 8 and a neutrino according to the reaction

$${}^7Be + H \rightarrow {}^8B + \nu$$

That neutrino has low energy, but the boron 8 soon decays to beryllium 8 with the release of a very energetic neutrino:

$${}^8B \rightarrow {}^8Be + e^+ + \nu$$

Fewer than 1 percent of the neutrinos produced by the most likely reactions in the sun can be captured by chlorine 37.

When Davis first reported results from his neutrino experiment, he was detecting fewer than half as many neutrinos as expected, but since that time he has improved the experimental methods and apparatus in several ways, and each improvement has lowered the limit on the solar neutrinos being detected. Now he can say with confidence that he is finding fewer than one-sixth the number predicted, and he may be detecting none.

That is a shocking result, and explanations for the discrepancy can be sought in four places. One possibility is poor experimental conditions or experimental design. If the number of neutrinos being detected were too high, one might reasonably suspect that something other than neutrinos were producing events which were mistaken for neutrino captures, but when the number is too low, that sort of explanation is not very convincing. The counters used to detect the decay of argon 37 have been greatly improved since the beginning of the experiment, and in an attempt to reduce the effect of cosmic rays and radioactivity in the walls of the mine, the room containing the tank of cleaning fluid in the mine has been flooded with water. The only effect of all the precautions and improvements has been to lower the possible limit on the number of neutrinos being detected.

A second possible source of error in the experiment concerns the computations of the efficiencies of the various assumed reactions. The ability of chlorine 37 to absorb neutrinos must be computed or inferred from other reactions; it cannot be measured directly. The statement that 0.25 percent of the deuterium is produced by the reaction that released an energetic neutrino is based on a computation of the relative probabilities of the two competing processes; likewise the number of neutrinos expected from the decay of beryllium 8 is based on a computation of the probabilities of all the reactions leading to the formation of beryllium 8. Some of those computations may be wrong. In the several years since Davis first reported neutrino capture rates that are too low, however, the computations have been repeated by various persons, and where possible the measurements on which the computations are based have been repeated in laboratories, with no errors of any significance discovered.

A third possibility is that the nature of the neutrino itself may not be understood. If, for example, the neutrino is unstable so that it decays during the 500 s it requires to come from the sun to the earth, then the particles reaching the earth may be sufficiently different from the originals that they will not combine with chlorine 37. But if the neutrino can change in that way, it must have a mass somewhat different from zero, and attempts have been made to determine its mass with greater precision than had been achieved before. Thus far a mass of zero seems consistent with experiment (although one cannot be sure that it is precisely zero), and no other evidence has been found to indicate that the neutrino is a more complex particle than had been supposed.

The fourth possibility is that the models of the sun are in error. If the core of the sun is considerably cooler than current models predict, then neutrino production will be much lower than that assumed by Davis. But if the core is cooler, one of two things must be true: The models are greatly in error in such matters as their assumptions about the opacity of the sun, or the sun is not in a steady state. The first possibility seems unlikely, and attention is turning to the second possibility. Some arguments offered are these: If periodically the core of the sun mixes with the unreacted materials outside it, the core will be cooled for a time; consequently the neutrino production will decrease. The temperature and luminosity of the outer part of the sun will not change for millions of years, however, because that time is required for the effects of temperature changes at the core to be felt at the photosphere. The neutrino experiment seems to indicate that the core of the sun is relatively cool now, but the outer part of the sun is still at a temperature appropriate for the higher core temperature which existed some millions of years ago. One phenomenon that might be related to a periodic heating and cooling of the sun is the periods of glaciation on the earth, but not enough work has been done on this matter to permit one to say with much assurance where it will lead. One unanswered question is *why* such mixing might occur in the sun.

Until the results of the neutrino experiment conducted by Raymond Davis became well known, most scientists believed that the models of the sun were quite satisfactory and that the sun's future could be predicted with some confidence. We still have no reason to greatly modify our long-run picture of the life expectations of the sun, but confidence in the details of the models has been shaken. Until the source of the discrepancy between the models and the neutrino experiment is discovered, one must put a small question mark beside all conclusions based on models of the sun.

QUESTIONS

1 Describe how the sun would look if the opacity changed gradually and slowly over a distance of one-half the radius of the sun.

2 The rate at which solar energy strikes the earth, called the *solar constant*, is 1.37×10^3 J/(m²)(s). How can that number be used to compute the temperature of the sun's photosphere?

3 Verify the speed given in the text for a proton at a temperature of 10^6 K.

4 Describe the 22-year sunspot cycle.

5 Doppler shifts of the spectral lines in the light coming from the east and west edges of the sun indicate that the radial velocities of the two edges differ by about 4 km/s. From that figure, what is the approximate period of rotation of the sun?

Ans 2.2×10^6 s (= 25 days)

6 Describe Davis's solar neutrino experiment. What hypotheses have been considered to explain his results?

8
the stars— appearances

Before we consider details about the stars, it will be useful to try to gain some perspective on them and on our relation to them. How bright are they? How far away are they? Are they uniformly distributed or are they organized into groups? All these questions will be dealt with in some detail later, but a preview of coming attractions will help the reader avoid becoming lost in details.

SIZES AND DISTANCES

In Unit 3 distances within the solar system were discussed in astronomical units and in terms of light travel time. As we shift our attention to the stars, the distances we must work with become so much greater that only light travel time, or a unit called the *parsec*, are useful. The light year, which is the distance light travels in one year, is a common unit used in discussing distances to the stars. Light travels 3×10^8 m/s or 186,000 mi/s. A year is 3.16×10^7 s, and therefore 1 light year is 9.48×10^{15} m or 5.9×10^{12} mi. Note that a light year is a distance —not a time.

The sizes and distances involved when we begin discussing the stars are so far outside our normal experience that they are difficult to comprehend, but perhaps an analogy will be helpful. Let us suppose that the sun is represented by a basketball with a diameter of almost 10 in. The sun's diameter is almost exactly 100 times larger than the diameter of the earth, and if the sun is represented by a basketball, the earth must be represented by something having a diameter of about 0.1 in. That is approximately the diameter of an apple seed. Now if the sun is a basketball and the earth is an apple seed, how far apart are they? The earth is slightly more than 100 sun diameters from the center of the sun, and so the apple seed should be approximately 100 ft from the basketball. On this scale, Pluto, the outermost planet, is 4,000 ft from the basketball. Now if other stars are represented by spheres of appropriate size, how far 171 away are they? The nearest is almost 5,000 mi away! In fact, the stars in our neighborhood would be represented rather well by balls about 5,000 mi apart on the average. Obviously most of space is empty, and the individual stars are almost alone.

If we wished to represent other stars by balls, would they all be like the basketball which we have used to represent our sun? The answer is "no." Some would be similar, most a little bit smaller, and some slightly larger. Here and there in space would be giant spheres with diameters up to perhaps 700 or 800 ft. About 10 percent of the stars are *white dwarfs*, with diameters only about twice that of the earth, so that they would be represented by very small white peas. In many places we would see two balls close together, revolving about one another. Some pairs almost touch; others are separated by hundreds or thousands of feet. And here and there are more complicated systems—three, four, or more stars forming a small clump, moving about each other. Would planets appear around other stars? We do not know, and representing stars by basketballs and the earth by an apple seed, thousands of miles from any other star, illustrates the difficulties of trying to see other planets.

The sun is one of the approximately 100 billion stars that make up a great system called the Milky Way galaxy, which is approximately 100,000 light years in diameter. We see many other galaxies out in space, and the photographs show two of them (NGC 4565 and NGC 3031) which probably look very much as our system would look if we could see it from outside.

LUMINOSITIES

We must make a distinction between the intrinsic luminosity of stars and their appearance to us. If we observe two identical stars, one near to us and the other far away, the more distant will appear fainter than its twin. A star which appears faint may actually be of low luminosity and near, or it may be very luminous and far away. We cannot tell which is the case until we can determine its distance.

If we take a census of the stars near the sun, we find that most are very faint, so faint that they can be seen only with the use of a telescope. If we assume that the region of space near the sun, which we can catalog rather completely, is representative of all the galaxy, we conclude that faint stars are the most common, although at great distances we cannot see most of these faint stars even with telescopes. The most luminous stars known emit about 160 billion times as much light as the faintest known, and the only stars which can be seen at great distances without telescopes are the rather rare, very luminous members of the stellar population.

The sun is midway between the most luminous and least luminous stars. If one of the most luminous stars were placed where the sun is, the earth would be melted or vaporized. If one of the least luminous were placed in the sun's position, it would give about the same light to the earth as the full moon gives, and the earth's temperature would fall so low that the gases of the earth's atmosphere would freeze into solids.

As Unit 13 will discuss in detail, we believe that all stars follow similar paths of evolution. They are born in a collapse of clouds of gas and dust; they spend much of their lives in the condition the sun is in now, that is, burning their hydrogen; and they die in a variety of ways. The rate at which a star goes through its evolutionary cycle depends upon its mass; massive stars age much more rapidly than light stars. The largest stars that we see are probably stars which were once a size more like that of the sun but which are now in an evolutionary stage in which stars become huge. Some are several hundred times the diameter of the sun, even though their masses are no more than 20 to perhaps 50 times the mass of the sun. At the other extreme are white dwarfs, which are stars which have progressed to one form of stellar death. They have contracted until they have reached such densities that a cubic inch of their material on the earth would weigh tons. A white dwarf the mass of the sun has a diameter comparable to that of the earth. They will be discussed more fully in Unit 13.

MAGNITUDE SCALE

One of the most obvious characteristics of the stars is the fact that they do not all appear equally bright. In the second century B.C. Hipparchus cataloged approximately 1,000 stars and indicated their relative brightnesses on a scale from 1 to 6. The values on that scale are called *magnitudes*, and one speaks of a first-magnitude star, a third-magnitude star, etc. Hipparchus intended the brightest stars he could see to be first magnitude and the faintest he could see to be sixth magnitude. His magnitude system has been modified and made more precise, but it provides the basis for the system we still use.

In the middle of the last century it became possible to make more precise measurements than had been made before of the amount of light reaching a telescope from a star. It was found that the amount of light energy from a first-magnitude star is al-

most precisely 100 times that from a sixth-magnitude star. Telescopes now permit us to see stars fainter than sixth-magnitude
stars, and the magnitude scale can be extended to include
them. The magnitude scale is defined in such a way that a magnitude difference of 5 corresponds to a ratio of light energy of
100 reaching the observer. Thus a star which is $\frac{1}{100}$ as bright as a
sixth-magnitude star is said to have a magnitude of $6 + 5 = 11$,
and a star which is $\frac{1}{100}$ as bright as an eleventh-magnitude star
has a magnitude of 16.

The human eye perceives equal ratios of intensity of light as
equal differences in brightness. (In this respect the eye is like
the ear, which perceives equal ratios of sound intensity as
equal differences of loudness.) Consequently if magnitude
steps are to correspond to what the eye interprets as equal differences in brightness, they must correspond to what really are
equal ratios of energy entering the detector. Hence, the magnitude difference of 5 corresponds to an energy intensity ratio of
100. And a magnitude difference of 1 corresponds to a ratio in
the amount of light received of $\sqrt[5]{100} = 2.512$. The light we receive from a third-magnitude star is 2.5 times that we receive
from a fourth-magnitude star; and the light from a nineteenth-
magnitude star is 2.5 times that from a twentieth-magnitude
star.

The relation between magnitude difference and intensity
ratio is expressed precisely as

$$m_1 - m_2 = 2.5 \log_{10} \frac{I_2}{I_1}$$

where m_1 and m_2 are the magnitudes of stars 1 and 2 whose intensity ratio is I_2/I_1. This equation defines magnitude differences for any intensity ratios whatever, and hence fractional
magnitude differences can be used. The intensity of light
received from a star is now usually measured by photoelectric
measuring devices, and the ratio of light received from two
stars can be measured very accurately; their magnitude difference can then be computed to the same accuracy. For example, let us compute the magnitude difference between two
stars if we receive 24 times as much light from one as from the
other. The intensity ratio $I_2/I_1 = 24$, and the logarithm of 24 is
1.38. The magnitude difference between the two stars is
$2.5 \times 1.38 = 3.45$. The brighter star has the smaller magnitude;
for example, if the brighter star has a magnitude 7.1, the fainter
has a magnitude 10.55.

Although working with arbitrary brightness ratios and magnitude differences usually requires logarithms, we shall select
examples which can be worked by using the table of corre-

Magnitude difference	Intensity ratio
1.0	2.5
1.5	4.0
2.0	6.3
2.5	10.0
3.0	16.0
4.0	40.0
5.0	100.0

The table can be extended by using the given values and the fact that magnitude differences are added but intensity ratios are multiplied. For example, a magnitude difference of 6 (which is 5 + 1) corresponds to an intensity ratio of 250 (which is 100 × 2.5). A magnitude difference of 10 corresponds to an intensity ratio of 100 × 100 = 10,000.

Example 1 If two stars have magnitudes which differ by 8, what is the brightness ratio?

Solution This can be solved in two ways. Without using logarithms, think of 8 as 5 + 3; the ratio corresponding to 5 is 100, and the ratio corresponding to 3 is 16; the ratio corresponding to 5 + 3 is 100 × 16, or 1,600. If we use logarithms, we can write

$$8 = 2.5 \log\frac{I_2}{I_1}$$

$$3.2 = \log\frac{I_2}{I_1}$$

$$\frac{I_2}{I_1} = 1,600$$

The antilog of 3.2 is 1,585, which is 1,600 when it is rounded to two significant figures.

What has been said thus far shows how magnitude differences can be computed, but it does not tell what is used as a starting point. When the magnitude scale was set up for use with modern instruments, Aldebaran and Altair were assigned the magnitude 1.0; since then many stars have been measured carefully and those in a group scattered over the entire sky have been adopted as standards, with their magnitude values adjusted so that they are self-consistent.

Some objects in the sky are brighter than the objects assigned magnitudes of 1, and consequently they have magnitudes which are less than 1. Some magnitudes are negative; for example, the sun has a magnitude of −26.5, the full moon has a magnitude of −12.5, and Sirius has a magnitude of −1.4. The faintest stars which can be seen with the naked eye are about 6.5, and the faintest objects which can be photographed with the 200-in telescope are approximately 23.5.

Until the late 1700s the only way to determine the magnitude of stars was by comparison with others by eye. In the late eighteenth century William Herschel devised a method of making more accurate comparisons by covering the mirror of a telescope by an adjustable diaphragm. If the diaphragm was closed until a star could barely be seen, the light reaching the observer's eye was proportional to the area of the exposed part of the mirror. If two stars were observed in that way, their brightnesses were inversely proportional to the areas of the mirrors when each could barely be seen. Astronomers can also compare brightnesses or intensities by measuring the degree of blackening of a photographic film in the image of a star with an instrument called a *densitometer*. The reading of the densitometer can be calibrated so that magnitude is derived from it. The most accurate measurements now are made with photoelectric instruments in which an electric current is proportional to the intensity of light coming from the star.

COLORS AND TEMPERATURES

Another difference among stars which is visible to casual inspection is color. Some stars appear red, some appear blue, and most appear white. Long-exposure photographs with color film show greater variations in color than the eye recognizes, and the stars range from intense blue to deep red. The only colors between are yellow and white; one does not see green stars. If a star has a temperature such that it radiates strongly in the green part of the spectrum, it also radiates strongly in the red, yellow, and blue, and the eye interprets the total as white or almost white.

Color differences reflect temperature differences, for a star is approximately a blackbody, and the color it shows depends upon the relative amounts of radiation in different parts of the visible spectrum. Figure 8.1 shows blackbody curves for a star which would appear red and for one which would appear blue. The curve for the sun, considered a yellow star, has its peak very near the center of the visible region of the spectrum.

Magnitudes derived by comparing the relative darkening of different star images on photographic plates are not necessarily the same as those derived using the eye. Unless photographic emulsions are specially treated in order to increase sensitivity to long wavelengths, they are most sensitive to the blue end of the spectrum. Thus even if a red star and a blue star appear

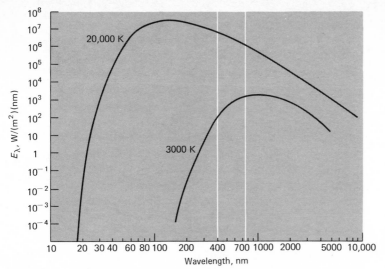

FIGURE 8.1 Blackbody curves for objects at temperatures of 3000 and 20,000 K. The lower temperature corresponds to a cool star, but the actual radiation from a cool red star might be less than the output of an ideal blackbody because of absorption by atoms and molecules in the star's atmosphere. The higher temperature corresponds to a hot star, but not the hottest known. The vertical white lines indicate the limits of the visible spectrum; blue is to the left and red to the right. The units of E_λ are W/(m²)(nm).

equally bright to the eye, the image of the blue star will be darker on a photographic plate, and if the magnitudes are estimated from the photographic image, the blue star will appear the brighter and therefore will have a *smaller* magnitude value. Because the magnitude measured depends upon the method of measurement, two different kinds of magnitude are defined: m_v (v for visual) and m_{pg} (pg for photographic). Because photographic emulsions differ in their response to different colors, and because the photoelectric devices commonly used now for measuring magnitudes have differing spectral responses, it has become standard to use filters which limit the light being received to a rather narrow-wavelength band so that all instruments will yield comparable results. The most common system uses three filters which pass bands of the spectrum centered approximately on 350, 440, and 550 nm; they are designated U, B, and V, for ultraviolet, blue, and visible. The B and V filters give magnitudes which correspond quite closely to the older photographic and visual values. Usually the apparent magnitudes measured using these filters are designated by U, B, or V, and not m_V, m_U, etc.

The colors of stars can be inferred from the intensities of their images on black-and-white film if photographs taken through two different filters can be compared. A convenient way of specifying the color of a star is by giving the difference between any two of the magnitudes measured with U, B, or V filters. The most common difference is one called the *color index:*

$$C = B - V$$

The magnitudes are adjusted so that the color index is zero for a star with a surface temperature of 10,000 K. Such a star, Vega for example, appears white. A hotter star will be bluer, and consequently its blue magnitude will be less than its visual, with the result that its color index is negative. A cooler star has a positive color index. Three examples are these:

Star	Temperature, K	B	V	C = B − V
Spica	23,000	0.7	0.9	−0.2
Sun	5800	−26.1	−26.7	0.6
Antares	3400	2.7	0.9	1.8

The color index is an indication of the temperature of stars which can be measured with high precision, but it is not a linear function of the temperature. If the surface temperatures of stars are plotted along the vertical axis of a graph and the color indices are plotted along the horizontal axis, the corre-

The same star field photographed through blue and yellow filters. The exposures have been adjusted so that most of the stars in the field appear about equally bright in the two photographs, but the brightest star is noticeably different and some faint stars appear on one photograph but not on the other. The difference between two such images can be used to compute the color index of a star. (*Yerkes Observatory photograph.*)

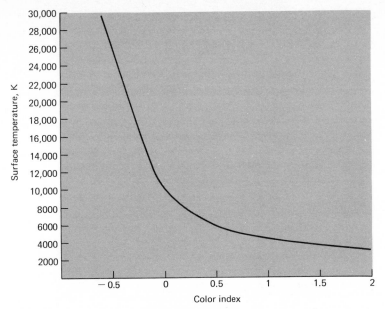

FIGURE 8.2 The color index plotted against the surface temperatures of stars.

sponding points do not fall on a straight line; instead, the graph looks like the one in Fig. 8.2.

Another kind of magnitude is *bolometric magnitude*, or m_{bol}, which is the magnitude a star would have if we could place ourselves above the atmosphere and observe its radiation at all wavelengths. In practice, a correction must be computed that includes the effect of the atmosphere and allows for the fact that a magnitude measured with one of the standard filters is only a sample of the radiation of the star.

STELLAR CLASSES AND TEMPERATURES

Stars may be classified on the basis of their UBV magnitudes, their color indices, or their spectra. Classification based on spectra was proposed in 1863 and reached approximately its present form in the early 1900s. The spectra of all stars show patterns of dark absorption lines, but the patterns vary widely. In some the lines of the Balmer series of hydrogen are strong; in others those lines are very weak. In some the helium lines are strong; in others they are weak. In some the lines of simple molecules can be seen. The spectra can be arranged into classes on the basis of the appearance and relative strengths of lines, and the most common classification scheme has seven main divisions, designated by the letters O, B, A, F, G, K, and M. The proper order of those letters has been learned by generations of

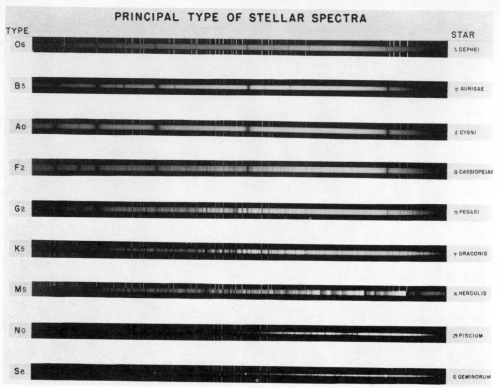

PRINCIPAL TYPE OF STELLAR SPECTRA

TYPE		STAR
O6		λ CEPHEI
B3		η AURIGAE
A0		δ CYGNI
F2		β CASSIOPEIAE
G2		η PEGASI
K5		γ DRACONIS
M5		α HERCULIS
N0		19 PISCIUM
Se		R GEMINORUM

Examples of the principal types of stellar spectra. In each example the spectrum of a star (an absorption spectrum) is recorded between two emission spectra. (*Hale Observatories.*)

students through the mnemonic device "Oh Be A Fine Girl, Kiss Me." (Three other minor classes are designated R, N, and S, and when they are added, the memory device becomes "Oh Be a Fine Girl Kiss Me Right Now, Smack.") Each of the seven main classes is subdivided into 10 smaller divisions indicated by the numbers 0 to 9; thus the sun is a G2 star. (The subdivisions of class O begin with O5.)

Although the fact was not realized when the spectra were classified, the spectral differences used to distinguish classes are temperature indications. Variations in the strength of Balmer absorption lines have little or nothing to do with the quantity of hydrogen in the atmosphere of the star; all stars are mostly hydrogen. The differences arise because in the star with weak Balmer lines the hydrogen is either so cool that most of the hydrogen atoms are in the ground state, from whence they cannot absorb the Balmer lines, or it is so hot that most of the atoms are ionized. All the atoms in cool hydrogen are in the ground state unless they have just absorbed radiation or have been struck by an energetic particle. If the temperature of the hydrogen is raised, however, collisions between atoms gradually become energetic enough to raise them to a higher energy level. Any atom so excited drops back to the ground state within about 10^{-8} s. As the temperature of the gas is increased, the number of atoms in an excited state at any instant increases. At any temperature high enough for excitation to become significant, an equilibrium condition exists in which some fraction of the atoms are at any given moment in one of the excited states. The number in any particular state depends upon the energy difference between the excited state and the ground state, and on the temperature. As the temperature becomes very high, many of the atoms begin to be ionized by collisions, until finally at sufficiently high temperatures most of the atoms may be ionized at any time. The star which shows strong Balmer lines has a temperature which causes large numbers of hydrogen atoms to be found in the second energy level. Thus if a star is too cool or too hot, it shows weak hydrogen lines; the lines are strong only if the temperature is "just right." Similar conditions hold with regard to helium and to the elements that produce other prominent lines. Only cool, red stars can have molecules in great numbers in their atmospheres. As the sequence is given, O to M, it represents temperatures from very high to low, perhaps from 50,000 to

TABLE 8-1 IMPORTANT CHARACTERISTICS OF SPECTRAL CLASSES

Spectral class	Color	Temperature (approximate), K	Salient spectral features	Example
O	Blue	Above 25,000	Strong lines of ionized helium. Lines of doubly ionized oxygen, nitrogen, and carbon.	10 Lacertae
B	Blue	25,000–11,000	Lines of neutral helium are most prominent. Hydrogen lines stronger than in O class.	Rigel, Spica
A	Blue	11,000–7500	Hydrogen lines are most prominent. Singly ionized magnesium, silicon, iron, calcium, and titanium appear.	Sirius, Vega
F	White	7500–6000	Hydrogen lines are weaker than in A stars and neutral metals are stronger.	Procyon
G	Yellow	6000–5000	Lines of ionized calcium are strongest feature. Hydrogen lines are weak. Lines of many neutral and singly ionized metals are present.	Sun, Capella
K	Orange	5000–3500	Neutral metal lines are most prominent. Hydrogen lines are almost nonexistent. Molecular bands are becoming important.	Arcturus
M	Red	Below 3500	Molecular bands are most prominent feature. Titanium oxide bands are very prominent.	Antares, Betelgeuse

3000 K. As in the case of the color index, the relation between spectral class and temperature is not linear. Prominent characteristics of the spectral classes are shown in Table 8.1.

QUESTIONS

1 The visual magnitude of star A is 7; the visual magnitude of B is 4. The light we receive from B is how many times that we receive from A?   *Ans* 16

2 The visual magnitude of Barnard's Star is 9.54; the visual magnitude of Sirius is −1.42. The amount of light we receive from Sirius is how many times that received from Barnard's Star?   *Ans* 2.5×10^4

3 Three stars have the following magnitudes:

	B	V
1	3.1	2.4
2	4.2	4.4
3	7.5	5.9

What is the color index of each? Which is reddest? Bluest? Estimate the surface temperature of each.

4 Balmer lines are prominent in stars with temperatures of about 10,000 K, but weaker in stars both hotter and cooler. Why?

5 Suppose that star A is just visible through a telescope of aperture 4 in and B is just visible when the aperture of the telescope is 20 in. Which star gives us more light? By what factor? What is the magnitude difference between the two? If star A were of visual magnitude 4, what would be the magnitude of B?

Ans A; 25; 3.5; 7.5

6 What would be the magnitude of an object which gave us one-tenth as much light as the full moon?   *Ans* −10

7 What is the magnitude of an object which gives us one-hundredth as much light as a star at the limit of naked-eye visibility?   *Ans* 11

8 If you studied the spectrum of what appeared to be a single star and found lines of both ionized helium and molecular bands, what conclusion could you draw?

9 From a star of magnitude 1.0, we receive energy at approximately the rate 1×10^{-8} J/(m²)(s). At what rate do we receive energy from a star of magnitude 6.0?   *Ans* 1×10^{-10} J/(m²)(s)

9
the stars–distances and intrinsic properties

In Unit 8 we discussed the colors and apparent magnitudes of the stars. Before we go much further in a consideration of their properties we must determine how far away they are, and that brings us again to the constant problem of the astronomer—the determination of distances.

ANNUAL PARALLAX

A necessary consequence of the facts that the earth moves around the sun and that not all stars are the same distance from the earth is the possibility of seeing *annual parallax*, an apparent motion of nearer stars against the background of more distant stars because of the earth's motion. Suppose that all the stars are motionless. Then as the earth moves around the sun, a nearby star which lies in the plane of the earth's motion should move back and forth along a line; a star which lies in a direction perpendicular to the plane of the orbit should appear to move in an elliptical path which is a duplicate of the path of the earth; and a star lying somewhere between the perpendicular and the plane of the orbit should appear to move in a path which is a flattened ellipse (Fig. 9.1). If the stars are motionless, the parallax of a nearby star should be observable in one year; if the stars are in motion, more than one year's observations will be required to separate parallactic motion from true motion. The latter is the actual situation.

Annual parallax was not actually observed until 1838, although William Herschel had made a careful attempt to detect it somewhat earlier. In that year F. Bessel detected the parallax of the star 61 Cygni, and F. Struve detected the motion of Vega. Both men used specially equipped telescopes to measure the

FIGURE 9.1 Annual parallax. Because of the earth's annual motion around the sun, a nearby star above the earth's orbit appears to trace a small ellipse against the background of more distant stars. A star in the plane of the earth's orbit appears to move back and forth along a line projected against the distant stars.

separation between the star being observed and some background object; the work was tedious and difficult. Now photographs are taken of the same area of the sky throughout the year, and from measurements of the movements of the images of nearby stars their parallaxes can be determined.

If the parallax of a star can be measured, the distance to the star can be computed by the method of annual parallax, or *trigonometric parallax*. The trigonometric parallax of a star is defined as the angle subtended at the star by one astronomical unit. That is, we use the displacement of the star during half a year. Figure 9.2 shows the triangle in which the base is the astronomical unit and the angle at the star is the trigonometric parallax; if that angle is known, the distance to the star can be computed easily.

The Parsec

Instead of computing stellar distances in terms of kilometers, or some similar distance unit, astronomers have defined a new unit—the *parsec*, a combination of "parallax" and "second." One parsec (pc) is the distance to an object whose trigonometric parallax is one second of arc. That is, if the angle p in Fig. 9.2 is 1 second of arc, then the distance is 1 pc. Since the angle varies inversely with the distance, the general expression for the distance to an object whose parallax is known is

$$D = \frac{1}{p}$$

where p is measured in seconds of arc and D is parsecs. Thus a star whose parallactic angle is 0.25 second of arc is 4 pc distant. One parsec is equal to 3.26 light years.

Limits of the Method

No star is as near as 1 pc, so all parallactic angles measured are less than 1 second of arc. The nearest stars are a triple system, Alpha Centauri, and the main star there has a parallax of 0.76 second of arc, making it 1.3 pc away. That angle of about 0.75 second of arc is the angle subtended by a penny a little more than 3 mi away! Trigonometric parallaxes can be measured for stars to about 100 pc, but at that distance the accuracy is not high. Parallaxes for almost 7,000 stars have been measured; for only the 700 within 20 pc are the distances known to within a 10 percent accuracy. Knowledge of the distances to those 700 stars provides the foundation upon which all measurements of distances to stars and galaxies are based. They are used to calibrate other, indirect, methods which are used for objects much

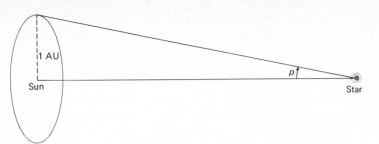

FIGURE 9.2 The angle of parallax p is the angle subtended at the star by a distance of 1 AU. The distance from the sun to the star in parsecs is $1/p$ when p is measured in seconds of arc.

too distant for trigonometric parallax to be observed. Several of the indirect methods will be described later.

INTRINSIC LUMINOSITY

The magnitudes described in Unit 8 are *apparent* magnitudes—measures of how bright stars appear to be, or how much light from the stars reaches the earth. Differences in apparent magnitude can arise because of differences in the *intrinsic luminosity* of the stars and differences in their distances from us, and because of the presence of dust in space which dims the light of distant stars. The last effect will be ignored until a later chapter. If two stars are identical in intrinsic luminosity, but one is farther away than the other, the more distant will appear the fainter. If the distance to a star is known, the intrinsic luminosity of that star can be determined.

Because all stars are so distant from us, no matter how large a star may be it appears as a point source of light to us, and the light from a point source decreases with distance in a simple way. All the light which leaves a star passes through any sphere drawn around the star, but because the area of a sphere increases proportionally to the square of its radius, the light per unit area on a section of any of the spheres decreases as the square of the distance from the source, as shown in Fig. 9.3. If the intensity of illumination is measured in terms of the energy striking a given area, that intensity varies with distance from

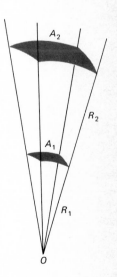

FIGURE 9.3 Inverse square variation of light intensity. Because the areas A_1 and A_2 are cut from spheres of radii R_1 and R_2, the areas are proportional to the squares of the radii. If a light source is at O, the same light passes through A_1 and through A_2, and consequently the intensity (energy per unit area per unit time) is less at A_2 than at A_1. The intensity is inversely proportional to the area through which the light must pass and therefore is inversely proportional to the square of the distance from the source.

$$I \propto \frac{1}{R^2}$$

where R is the distance from the source. Thus when two stars which are intrinsically identical are compared, if one is twice as far from the earth as the other, the light we receive from the more distant star is one-fourth what we receive from the nearer. (The more distant star will therefore appear 1.5 magnitudes fainter.) If a star could be moved from its present position outward, until the distance had doubled, its apparent brightness would be one-fourth its original value, and when the distance had tripled, the brightness would be one-ninth the original value. If the star were moved 10 times as far away as it was originally, it would appear $\frac{1}{100}$ as bright. Conversely, if the distance were reduced by one-half, the brightness would be 4 times its original value.

Absolute Magnitude

A convenient way to compare the intrinsic luminosities of the stars is to imagine them all placed at some standard distance from the earth. Thus we refer to the *absolute magnitude* of a star as the *apparent magnitude it would have if it were at* 10 pc. For example, if a star is seventh magnitude, at a distance of 100 pc from the earth, one can reason as follows: If the star were moved from 100 to 10 pc, its distance would decrease by a factor of 10, and consequently it would appear brighter by a factor of $10^2 = 100$; but if it were brighter by a factor of 100, its magnitude would be less by 5. Therefore it has an absolute magnitude of 2. If the sun were moved out 10 pc, it would appear as a fifth-magnitude star; its absolute magnitude is therefore 5.

Absolute magnitude is often indicated by a capital M, and if the magnitude is measured on the UBV system, it is given as M_U, M_B, or M_V. The absolute visual magnitudes of stars range approximately from $M_v = -9$ to $M_v = 19$. With an absolute magnitude $M_v = 5$, the sun is almost precisely in the middle of the range. Stars exist which are 14 magnitudes brighter and 14 magnitudes fainter; 14 magnitudes means a ratio of luminosity of 400,000.

If the sun were replaced by one of the faintest stars known, the apparent magnitude of -26.5 would be reduced to -12.5, which is just the apparent magnitude of the full moon, so we can imagine what such a star would look like. If there were no other source of heat, temperature on the earth's surface would drop to a few kelvins. Imagining one of the brightest stars in the position of the sun is more difficult, for we have no experience with such a tremendous outpouring of light. We can estimate its effect in this way, however: If the radiation striking the earth increased by a factor of 400,000, the radiation the earth emits would have to increase by the same factor, for it would eventually reach an equilibrium condition in which as much radiation was emitted as received. The radiation emitted by a blackbody varies as the fourth power of its temperature, and if the earth were a blackbody, the increase in radiation would cause its temperature to increase by a factor $\sqrt[4]{400,000} = 25$. The average temperature of the earth is somewhat less than 300 K, and thus the temperature of our hypothetical blackbody should approach 7500 K. The example is far-fetched, both because the earth is not a blackbody and because before its temperature reached anything like 7500 K it would vaporize completely and begin to disperse into space. It does seem safe to say that if the sun were replaced by a star 400,000 times as luminous, life on earth would become impossible very quickly.

Magnitude and Distance

Let us define L as the luminosity of a star, measured in joules per second of energy emitted in a particular wavelength region, and I as the intensity of radiation reaching the earth in the same wavelength band, measured in joules per second per square meter. Then if the distance to the star is R meters,

$$I = \frac{L}{4\pi R^2}$$

If two stars are compared,

$$\frac{I_1}{I_2} = \frac{L_1}{L_2}\left(\frac{R_2}{R_1}\right)^2 \tag{1}$$

The same relation is true no matter what units of measure are used, and we turn now to measurements in parsecs. To reduce the chance of confusion, we will use d to indicate distance measured in parsecs. If a star giving us radiation of intensity I is at distance d, and an identical star is at 10 pc, at which distance the intensity of its radiation is I_s (s for standard), the equation becomes

$$\frac{I}{I_s} = \left(\frac{10}{d}\right)^2 \tag{2}$$

If we imagine a star producing intensity I at distance d moved to a distance 10 pc, at which position its intensity is I_s, then the ratio I/I_s corresponds to the difference between apparent and absolute magnitude. If the apparent magnitude of the star is m and its absolute magnitude is M, we have, from the definition given in Unit 8 of magnitude difference,

$$M - m = 2.5 \log \frac{I}{I_s}$$

If we substitute from Eq. (2), this becomes

$$M - m = 5 \log \frac{10}{d} = 5 - 5 \log d \qquad (3)$$

or

$$m - M = 5 \log d - 5 \qquad (3a)$$

The uses of the three numbered equations will be illustrated by two examples.

Example 1 A star of the sixth magnitude is located 40 pc from the sun. What is its absolute magnitude?

Solution This problem can be solved by substituting numbers into Eq. (2) or (3), and both these methods will be demonstrated, but first we will solve it by "talking" it through. The question can be restated as: How bright would the star appear if it were moved in from 40 to 10 pc? If it were moved from 40 to 10 pc, its distance would have decreased by a factor of 4, and its apparent brightness would have increased by a factor of 16 ($4^2 = 16$). But if the brightness has increased by a factor of 16, its apparent magnitude has decreased by 3, and hence its absolute magnitude is $6 - 3 = 3$.

Substituting into Eq. (2) gives

$$\frac{I}{I_s} = \left(\frac{10}{40}\right)^2 = \frac{1}{16}$$
$$I_s = 16I$$

But if the brightness at standard distance is 16 times the original brightness, then the magnitude at standard distance is 3 less than the original magnitude, or the absolute magnitude is $6 - 3 = 3$.

Substituting into Eq. (3) rewritten in the form

$$M = m + 5 - 5 \log d$$

yields

$$M = 6 + 5 - 5 \log 40$$

The $\log_{10} 40$ is 1.6, and therefore this equation is

$$M = 6 + 5 - 5 \times 1.6 = 3$$

Example 2 A star of second magnitude is at 5 pc. What is its absolute magnitude?

Solution If the star were moved from 5 to 10 pc, its distance would be doubled and its apparent brightness would be decreased to one-fourth its original value. A change of a factor of 4 in apparent brightness corresponds to a difference in magnitude of 1.5, and since the star would decrease in brightness, its magnitude would increase, from 2 to 3.5.

Substituting into Eq. (2),

$$\frac{I}{I_s} = \left(\frac{10}{5}\right)^2 = 4$$

$$I_s = \tfrac{1}{4}I$$

But if the brightness at standard distance is one-fourth times the brightness at the original position, the magnitude at standard distance is 1.5 more than the original magnitude. Therefore the absolute magnitude is $2 + 1.5 = 3.5$.

Substituting into Eq. (3),

$$\begin{aligned}
M &= m + 5 - 5 \log d \\
&= 2 + 5 - 5 \log 5 \\
&= 2 + 5 - 5 \times 0.7 = 3.5
\end{aligned}$$

The absolute magnitude of a star can be computed from these equations only if the distance is known. It appears, however, that if the absolute magnitude could be found by some indirect means, the equations could be used to compute the distance to the star. Astronomers have found many methods of estimating the intrinsic luminosity of objects—stars, clusters, and galaxies—and distances for most objects beyond the 700 stars for which trigonometric parallax can be used are computed by comparing the apparent brightness with the intrinsic luminosity estimated by one method or another.

Without considering how intrinsic luminosity (or absolute magnitude) can be estimated, we shall work an example to show how the distance can be computed if both apparent and absolute magnitudes are known.

Example 3 A star has an apparent magnitude of 8 and an absolute magnitude of 0. How far away is it?

Solution This problem is closely related to the problems in Examples 1 and 2, and we shall first follow our "verbal" method. The star in its present location is eighth magnitude, but if it were at 10 pc it would be 0 magnitude. Therefore if it were brought in from its actual position to 10 pc it would appear brighter by 8 magnitudes; but 8 magnitudes corresponds to a brightness ratio of 1,600 ($8 = 3 + 5$, which corresponds to 16×100). If the brightness is to increase by a factor of 1,600, the distance must be decreased by a factor of $\sqrt[2]{1,600} = 40$. If the distance decreased by a factor of 40 when it became 10 pc, the distance now must be 400 pc.

Using Eq. (2), and the fact that a magnitude difference of 8 corresponds to a brightness ratio of 1,600, we have

$$\frac{I}{I_s} = \frac{1}{1,600} = \left(\frac{10}{d}\right)^2$$

$$\frac{d}{10} = 40$$

$$d = 400 \text{ pc}$$

Using Eq. (3),

$$\begin{aligned}
M - m &= 5 - 5 \log d \\
0 - 8 &= 5 - 5 \log d \\
5 \log d &= 13 \\
\log d &= 2.6 \\
d &= 400 \text{ pc}
\end{aligned}$$

THE HERTZSPRUNG-RUSSELL DIAGRAM

In the early years of the twentieth century the Danish astronomer Ejnar Hertzsprung and an American, Henry Norris Russell, independently found a way to correlate two kinds of information about stars. The result, known as the *Hertzsprung-Russell diagram* (more commonly, the H-R diagram), is one of the most useful devices yet developed for organizing information about the stars.

Construction

In the H-R diagram, each star is represented by a point. The horizontal axis of the diagram is labeled with the spectral classes (O, B, A, F, G, K, M), and the vertical axis is labeled with absolute magnitudes. Thus a star which is spectral class O and is very luminous is represented by a point at the upper left corner of the diagram. The sun, which is spectral class G and of moderate luminosity, is represented by a point near the center of the diagram. Color index or any other function of surface temperature can be used on the horizontal scale instead of

FIGURE 9.4 The H-R diagram.

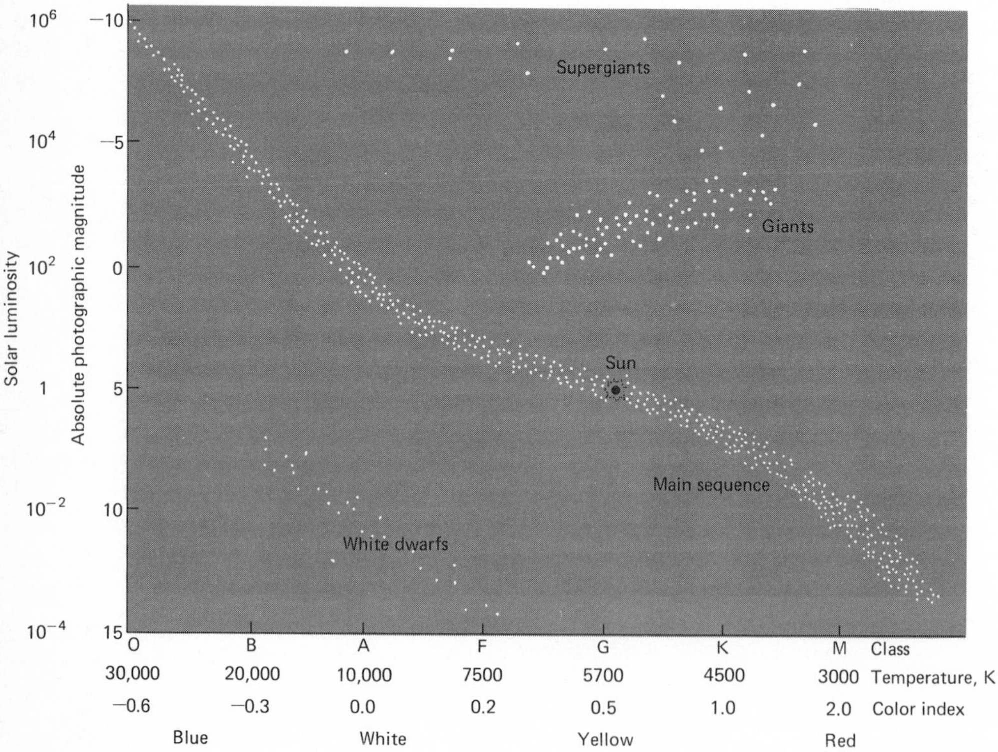

stellar class to generate essentially the same graph. In general terms, then, the H-R diagram is a plot of intrinsic luminosity (or absolute magnitude) against surface temperature.

Figure 9.4 shows such a diagram. Most of the stars fall on a broad line running roughly from upper left to lower right; that band is called the *main sequence*. In the lower left are white dwarfs, and to the right of the main sequence and above it are giants and supergiants. Stars at the left side of the diagram are hot and blue or white; those in the center are yellow, and those at the right are red. Stars near the bottom of the diagram are intrinsically faint, and those near the top are very luminous.

A star cannot be plotted on the H-R diagram unless its absolute magnitude is known, which means that its distance from the sun must be known. The basic material for construction of the diagram, therefore, is the data on the 700 stars for which accurate distances have been determined, and Russell originally worked with stars near the sun for which such distance measurements had been made. Hertzsprung, on the other hand, worked with stars in clusters. By looking at clusters of stars in which all the stars are at almost the same distance from the earth, he was able to determine relative luminosities of the stars. Even though he did not know their distance and therefore could not determine absolute magnitude, the fact that they were all at about the same distance implied that the stars which appeared brightest were actually the most luminous. Now data from observations of both clusters and near neighbors are combined in the construction of H-R diagrams.

Spectroscopic Parallax

The relations between temperature and intrinsic luminosity shown by the H-R diagram seem to apply to all stars, and the diagram is used in computing distances to stars too far away for annual parallax to be applied. Distances obtained by the use of the H-R diagram are called *spectroscopic parallaxes*.

The method can be illustrated by an example. The star Spica is a BO star with a visual magnitude of 0.9. From the H-R diagram we estimate that it has an absolute visual magnitude of about $M_v = -2$. Now we can substitute the apparent and absolute magnitudes into Eq. (3) and compute the distance. We can also reason as follows: If the star were at 10 pc, it would appear to be of -2 magnitude. It actually is approximately 3 magnitudes fainter, which means that it is fainter by a factor of 16. If it were moved from 10 pc out until the brightness decreased by a factor of 16, it would be 4 times its original distance, or 40 pc out. Therefore the star must be approximately 40 pc from the sun.

The method of spectroscopic parallax is used to determine the distance to stars too far away for annual parallax to be observed, and it is used to help calibrate other methods which measure even greater distances. It is most reliable when applied to a cluster of stars so that many points can be fitted to the main sequence line and the best common distance for all the stars determined.

One difficulty involved in using the H-R diagram is deciding whether a star of the G or K class is on the main sequence, is a giant, or is a supergiant. Fortunately, the distinction can be made with reasonable certainty because main-sequence, giant, and supergiant stars have recognizable differences in spectra. The densities of the gases in their atmospheres where absorption takes place to produce the dark lines are vastly different for the different species of stars. The masses of the stars vary over a rather small range, but their radii and hence volumes vary tremendously; the result is that the densities of their atmospheres vary greatly. Spectral lines are never infinitely narrow, and their width is determined by several factors, one of which is the pressure of the gas producing the line. The lines from a supergiant, whose low-pressure atmosphere is very thin, are therefore sharper than the lines from a normal main-sequence star. Other minor effects also can be used to distinguish between the various types of stars. It is believed that in our region of space about 90 percent of the stars are main sequence; 10 percent are white dwarfs, and fewer than 1 percent are giants and supergiants.

DIAMETERS OF STARS

Because of their great distances from us, all stars appear as points of light. The size of a star's image seen in a telescope or produced in a photographic emulsion is determined by the diffraction of light in the optical system or by scattering of light in the emulsion. In no case does the size of the image on a picture indicate anything about the diameter of a star, and what we know about diameters must be determined by indirect methods.

Diameters Deduced

Some interesting conclusions about the diameters of stars can be drawn from their positions on the H-R diagram. Consider, for example, a white dwarf with a surface temperature of 12,000 K and an absolute magnitude of 10. The star is $\frac{1}{100}$ as luminous as the sun, but its temperature is twice that of the sun. From the Stefan-Boltzmann law we may infer that each square meter of the star's surface radiates 16 times as much energy as a square meter of the sun's surface (radiation per unit area is proportional to T^4). If it is to be only $\frac{1}{100}$ as luminous as the sun, it must have only $\frac{1}{1,600}$ the sun's surface area. That implies that it has $\frac{1}{40}$ the sun's diameter. Indeed, white dwarfs are that size or less.

At the other extreme, we may consider a supergiant which has a surface temperature of 3000 K and an absolute magnitude of −5. It is 10,000 times as luminous as the sun, but each square meter of its surface radiates only $\frac{1}{16}$ as much as a square meter of the sun. It must, therefore, have an area 160,000 times the area of the sun, and its diameter must be 400 times the diameter of the sun. In fact, Mira has a diameter 420 times that of the sun (if the sun were placed at its center, the earth would be inside the star). It is clear, then, that white dwarfs, because they are white and hence hot but are of low luminosity, must have small radii. On the other hand, red giants, because they are very luminous even though they are red and hence cool, must have large radii.

The physics which has been used in the preceding two examples can be expressed in an equation. If the luminosity of a star, expressed in watts (joules per second) being emitted, is L, the radius of the star is R, the surface temperature is T, and the Stefan-Boltzmann constant is σ, these quantities are related as

$$L = 4\pi R^2 \sigma T^4$$

If the luminosity of a star is compared to the luminosity of the sun, the constants can be dropped from the ratios. The use of the Stefan-Boltzmann law in the preceding examples means that all the radiation from the star must be considered when the magnitude is specified, and consequently bolometric magnitude must be used.

For most stars, the method described above is the only way to estimate diameter or radius. Two other methods exist, however, for estimating the diameters of stars, and one will be examined now; the other applies only to binary stars and will be discussed in a later section.

Diameters Measured

In the 1920s Albert Michelson constructed what is known as a *stellar interferometer* on the 100-in telescope on Mount Wilson, and with it he was able to measure the diameters of seven stars. The interferometer is made up of a beam across the front of the telescope carrying mirrors whose distance from the axis can be adjusted. Light from the star striking the mirrors is redirected by other mirrors into the telescope itself, as shown

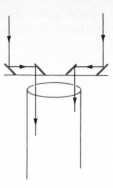

FIGURE 9.5 The arrangement of four mirrors above a telescope
to form the Michelson stellar interferometer.

in Fig. 9.5. When a star which was either very large or very near
or both was examined, the difference in path length for light
from the two edges of the star produced interference fringes in
the eyepiece of the telescope, and as the adjustable mirrors
were moved, the fringes shifted. An analysis of the fringes
yielded information about the angular diameter of the star as
seen from the telescope. That angular diameter was combined
with the known (or estimated) distance to the star to give an actual diameter. The stars examined by Michelson ranged in diameter from 23 solar diameters (Arcturus) to 750 solar diameters (Betelgeuse at its greatest size—it is variable).

Barnard's Star in Ophiuchus, the star with the greatest known
proper motion. The two photographs show the change in position
in 22 years. (*Yerkes Observatory photograph.*)

A different type of instrument, the *stellar intensity inter-ferometer*, has recently been put into operation in New South Wales. This instrument uses two telescopes, and combines the output of the photoelectric sensing units to obtain information comparable to the fringe patterns of the Michelson instrument; it has given information about the diameters of a few more stars. Both types of instruments are limited to very near or very large stars.

MOTION OF STARS

Although the stars appear not to change position, and during the lifetime of a human their movements are undetectable to the naked eye, they are in fact moving. That fact was first discovered by Edmund Halley in 1718, when he found that some of the prominent stars were not at the positions they had been when they were charted by the ancient Greeks. The usual way of detecting and measuring the movement of stars now is to compare photographs taken years or decades apart. If the photographs are taken on the same day of the year, the earth is at the same position on its orbit both times and thus parallactic effects are eliminated.

Proper Motion

If a star appears to be moving across our line of sight, its motion can be described in terms of the rate at which its angular position is changing. A star's *angular speed*, designated μ, is called its *proper* motion and is measured in seconds of arc per year. The star with the largest known proper motion is Barnard's Star, which is moving across the celestial sphere at the rate of 10.3 seconds of arc per year. At that magnificent speed it will change its position by an amount equal to the diameter of the moon in 175 years.

Tangential Velocity

If we know the angle through which a star appears to move in 1 year, and the distance to the star, we can find the actual distance perpendicular to our line of sight that the star moves in 1 year. Usually that is changed to a velocity *across* the line of sight, expressed in kilometers per second, and it is called the *transverse velocity* or the *tangential velocity*, designated v_T. The most common transverse velocities are 20 to 25 km/s, but a few stars have much higher velocities. The greatest is that of Kapteyn's star, which is moving relative to us across our line of sight at 163 km/s.

Transverse velocities can be measured only for nearby stars, since very distant stars show no apparent motion, no matter how fast they may be moving, even over a period of centuries. *Radial velocity*, however, that is, motion *along* the line of sight, can be measured for any star for which a good spectrum can be obtained, for it is determined by the doppler effect. About as many stars are observed to be approaching us as receding from us, and the most common speed is close to 15 km/s. Radial velocity is designated v_R.

Stars move not just across or along our line of sight, and transverse and radial velocities are only the components of a star's true velocity measured in those directions. The true velocity, called the *space velocity V*, can be computed by applying the pythagorean theorem to the right triangle in which the legs are v_T and v_R and the hypotenuse is V:

$$V = \sqrt{v_T{}^2 + v_R{}^2}$$

MOTION OF THE SUN

The motions discussed so far have been motions relative to the sun, but the question inevitably arises: Is the sun itself moving, and are some of the motions we see relative to it influenced by its motion? The answer is that the sun is moving, and, relative to the stars *near* it, it is moving about 20 km/s toward Vega and away from Orion. (We shall later consider the fact that the sun's actual motion relative to other galaxies at great distances is somewhat more than 20 km/s.)

We can compute our speed relative to the nearby stars by comparing the apparent motions of the stars in different directions. All the stars have random motions, and if the sun were not moving and the motions were truly random, we could expect the average of either transverse or radial velocities in any direction to be approximately zero. What we find is that if we look at the stars in the direction of Vega, the average of the transverse velocities is zero, but the radial velocities have an average suggesting that we are moving toward them at about 20 km/s. We also find that if we look in the opposite direction from Vega, toward Orion, we appear to be receding from those stars at about 20 km/s. In a direction 90° to the line joining Vega and Orion, the average of radial velocities is approximately zero, but the average of transverse velocities is about 20 km/s toward Orion, suggesting that we are moving in the opposite direction.

QUESTIONS

1 Verify that 1 parsec is 3.26 light years.

2 The star 40 Eridani has a visual magnitude 4.5; it is 5.00 pc from the sun. What is its absolute visual magnitude?

Ans 6.0

3 Achernar has a visual magnitude of 0.51 and an absolute visual magnitude of −1.0. What is its distance from the sun?

Ans 20 pc

4 The absolute visual magnitude of Vega is 0.5; the absolute visual magnitude of Betelgeuse is −5.5. The difference in the actual light emission is what factor? Which is the more luminous?

Ans 250

5 The absolute visual magnitude of Antares is −4.5. Which is the more luminous, Antares or Betelgeuse? By what factor?

Ans 2.5

6 Star A, which is known to be 30 light years from the sun, sends us twice as much energy as star B. The two stars are known to have about the same intrinsic luminosity. How far away is star B?

Ans 42 light years

7 Two stars have the same visual magnitude. One is a faint star, located 20 light years away; the other is 10,000 times as luminous. How far away is it?

Ans 2,000 light years

8 A main-sequence star of type A has an apparent magnitude 10. Use the H-R diagram to estimate the distance to the star.

9 The sun has an absolute magnitude 5 and a surface temperature of approximately 6000 K. If another star has the same surface temperature but an absolute magnitude of 0.0, how does its diameter compare with the diameter of the sun?

Ans 10 times as great ($D_S = 10\ D_\odot$)

10 If a star has a temperature half that of the sun but is 10,000 times as luminous, how does its radius compare with that of the sun?

Ans 400 times as great ($R_S = 400\ R_\odot$)

11 If a star has a temperature twice that of the sun but is $\frac{1}{10,000}$ times as luminous, how does its radius compare with that of the sun?

Ans $R_S = \frac{1}{400}\ R_\odot$

12 Use the stars listed in Appendix 4 and Appendix 5 to construct an H-R diagram. Let the horizontal axis run from 0 to M 9 and the vertical axis from +15 to −10. Use a dot for each star in Appendix 4 and an x for each star in Appendix 5. Compare your plots with the H-R diagram given in the chapter. Why are they somewhat different?

10
double stars

At least two reasons can be cited for studying double stars. First, they provide information about masses, and sometimes radii, that we can get in no other way. Second, they are common; more than 50 percent of the stars in the neighborhood of the sun are members of multiple systems. Of the 32 stars within 4 pc of the sun, 11 are alone, 4 have unseen companions that may be stars or planets, 14 are in pairs, and 3 are in a triplet. It is possible that the 11 stars listed as being alone actually have companions which we have not been able to detect, for solitary stars seem to be the exception rather than the rule.

One question which arises when one sees two stars seemingly very close together is whether they are actually very far apart, but seem close because they appear in the same line of sight. When a pair of stars has been observed sufficiently that the motion of one about the other can be *seen*, the question can be answered unequivocally: They are gravitationally bound into a system. Without that kind of observational information, one must fall back on probabilities. The probability that two stars will accidentally appear to be as close together as binaries is very, very small, and astronomers have concluded that most of the 40,000 pairs of stars which *appear* to be binaries are physically bound systems.

In 1650, when the telescope was still a rather new instrument, the Italian astronomer J. B. Riccioli noticed that Mizar, in the handle of the Big Dipper, appeared to be two stars. He had discovered the first double star. In the eighteenth century William Herschel noticed that one of the stars of a pair was often brighter than the other. On the assumption that the fainter star was farther away, he began a program of observing double stars in an attempt to detect annual parallax, which he assumed would be easy to observe as the stars moved nearer together and then farther apart. Between 1782 and 1821 he published three catalogs of double stars, listing 800 pairs, but he was not able to detect annual parallax by observing any of them. Herschel did realize that not all of his pairs just seemed close together by accident, for in 1804 he observed that one of the stars of the pair that is Castor was moving around the other. John Herschel, his son, continued cataloging double stars, and he listed more than

10,000 systems that are two, three, or more stars. Doubles are
the most common, but larger systems are not rare. Castor, for in-
stance, is actually six stars, and Mizar is four, each of the visible components being a double that telescopes do not resolve.

Two stars which appear close together but actually are not
are called an *optical binary*; their true nature can sometimes be
determined from measurements of their velocities. Alcor and
Mizar, in the handle of the Big Dipper, are such a pair.

True binary stars are classified on the basis of the observations which reveal their binary nature; the three most important types are:

Visual binaries: Pairs in which both stars can be seen in the
field view of a telescope.

Spectroscopic binaries: Pairs in which the doppler shifts of
spectral lines reveal the motion of one star about another.

Eclipsing binaries: Pairs in which the stars alternately
eclipse one another because the plane of their motion lies almost in our line of sight.

These types are not mutually exclusive, and the same system
could meet all three criteria so that it would be a visual binary,
a spectroscopic binary, and an eclipsing binary.

VISUAL BINARIES

Of the 40,000 visual binaries known, about 300 have been studied in sufficient detail and over a sufficiently long time that the
characteristics of their orbits are known. In true binary systems, both stars move about the center of mass, but it is more
convenient to treat them as if one of the stars (by convention,
the brighter) is fixed and the other is moving about it. Over a
period of time the fainter star appears to move about the
brighter, and if its distance from the brighter and its direction
are measured, its position can be plotted. If the system is
watched for a complete period, or a substantial part of a period,
the length of the period can be determined. And if the distance
to the pair and the angular separation of the two stars are
known, their actual separation can be computed. One can then
plot the apparent orbit of the fainter about the brighter in terms
of actual distance, perhaps in astronomical units.

Observations of visual binaries involve different calculations, however, depending on whether the orbit is being seen
from directly above or obliquely. If the orbit is observed from
directly above, and one star is taken to be fixed and the other is
regarded as if it is a planet moving about the fixed one, Kepler's
first two laws both apply. The orbit of the moving star is an

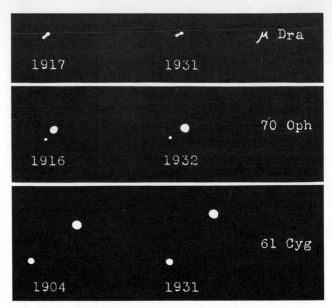

Orbital motion of μ Draconis, 70 Ophiuchi, and 61 Cygni during 14, 16, and 27 years, respectively. (*Yerkes Observatory photograph.*)

ellipse, with the fixed star at one focus, and the law of constant areal velocity is obeyed. If the orbit is seen obliquely, however, it still appears to be an ellipse, for an ellipse viewed from any angle is still an ellipse, and the law of constant areal velocity is still satisfied; the oblique viewpoint changes all areas in the same way. But in this case, the fixed star is no longer at a focus, and that fact can be used to determine the angle at which the orbit is being viewed.

Suppose that an orbit is actually a circle, and the fixed star is at its center. If that orbit is viewed from an angle perhaps 30° above its plane, it will appear to be a rather flat ellipse. But the brighter star will still be at its center, far from the foci, as shown in Fig. 10.1.

By a mathematical analysis too complicated to describe here, it is possible to determine from the apparent orbit the actual eccentricity of the orbit, the angular size of the semimajor axis, and the inclination to our line of sight. If the distance to the system is known, the angular size of the semimajor axis can be changed into a linear distance, and since the period is known,

FIGURE 10.1 A circular orbit viewed obliquely. The brighter star appears in the center instead of at one of the foci.

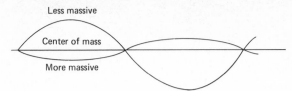

FIGURE 10.2 The center of mass of a pair of double stars moves on a straight line. At any time the distances of the two stars from that line are inversely proportional to their masses.

the sum of the masses of the two stars can be found from

$$M_1 + M_2 = \frac{a^3}{P^2}$$

where a is the semimajor axis, measured in astronomical units, P is measured in years, and M_1 and M_2 are measured in solar masses.

Example 1 Suppose that the period of a double star system is 40 years and the semimajor axis of the orbit of the fainter star about the brighter is 22 AU. Then the sum of the masses of the two stars is

$$M_1 + M_2 = \frac{(22)^3}{(40)^2} = \frac{10,648}{1,600} = 6.65 \text{ solar masses}$$

The division of the sum of the masses between two stars can be found only if the center of mass of the system can be located, and that can be done only if the motion of the system against the background stars can be observed. The center of mass moves on a straight line, and both members of the pair move on wavy lines which cross the path of the center of mass (see Fig. 10.2). If such wavy motion can be observed, the masses of the individual stars can be computed. A few binaries are known because the only visible star of the system moves across the sky in a wavy line.

Periods

The median period for visual binaries is about 70 years, and it is unlikely that any system with a period less than a year could be seen as a visual binary. The reason is quite simple: In order for two stars to be resolved by a telescope so that they can be seen as separate sources of light, they must have an angular separation of more than about 0.5 second of arc. Such an angular separation implies that they are either very near the earth or quite far apart. If they are far apart, their period is long, and periods probably range up to hundreds or thousands of years. However, observations have not been made for a sufficient

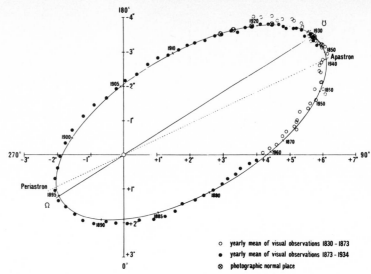

Diagram of orbit of 70 Ophiuchi, whose period is 88 years. (*Yerkes Observatory photograph.*)

○ yearly mean of visual observations 1830 - 1873
● yearly mean of visual observations 1873 - 1934
⊗ photographic normal place

period of time for periods of several hundred years or more to be determined.

The members of visual systems which have been studied well tend to be either of class B or A. Class B and A stars are both massive and luminous, and it may be that they appear common only because they are easily seen.

SPECTROSCOPIC BINARIES

Spectroscopic binaries are pairs that reveal their double nature by the doppler effect. Sometimes one star is so much brighter than the other that its light covers the light of the fainter and only one set of lines is seen, but the lines are observed to shift

An example of a double-lined spectroscopic binary. (*Hale Observatories.*)

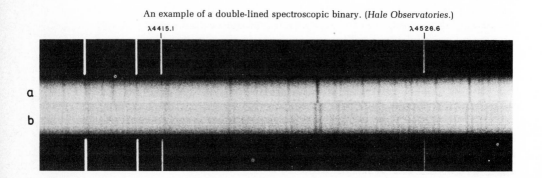

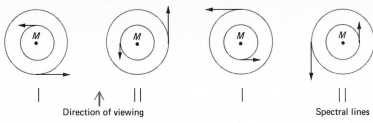

Direction of viewing   Spectral lines

FIGURE 10.3 Spectroscopic binary viewed from the direction indicated. Each star moves about the center of mass, indicated by M. A spectral line from each star is shown; they coalesce when the stars are moving across the line of sight so that the doppler shifts are zero, but they separate when the stars are moving with components of velocity toward and away from us.

back and forth, revealing a periodic motion toward and away from us. Such a system is called a *single-lined spectroscopic binary*. If both stars are approximately the same luminosity, the spectral lines of both are seen, and the system is a *double-lined spectroscopic binary*. In the spectrum the lines coalesce into one set, then gradually separate into a double set, then move back together into singles, etc. Figure 10.3 shows the relative positions of the two stars compared to the spectrum. About 1,000 spectroscopic binaries are known, and orbits have been determined for about 400. In addition, about 20 percent of the stars for which radial velocities have been determined repeatedly exhibit variations in their velocities, suggesting that they are single-lined binaries.

Because the doppler shifts cannot be observed unless the stars have velocities along the line of sight greater than about 1 km/s, the periods of spectroscopic binaries are always short. Periods range from hours to weeks, with the median period being about 9 days.

Velocity Curve

From the doppler shifts one can calculate the component of the velocities of the stars along our line of sight and from those velocities plot a velocity curve. In general the entire system may have a velocity toward or away from us, and so the veloci-

An example of a single-lined spectroscopic binary, α^1 Geminorum. (*Lick Observatory photograph.*)

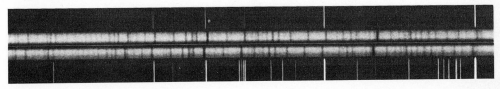

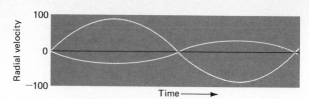

FIGURE 10.4 The radial velocities of a spectroscopic binary pair relative to their center of mass. (The radial velocity of the center of mass has been subtracted from all other velocities.) The spectroscopic lines reach maximum separation when the radial velocities are maximum.

ty of the center of mass must be computed and the velocities of the stars relative to it plotted. Figure 10.4 shows the velocity curve for the simplest possible situation—two stars in circular orbits about the center of mass, with the plane of the orbits passing through us.

Example 2 We wish to find the individual masses of the stars in Fig. 10.4, and their distances from the center of mass.

Solution Inspection of the curves gives the period of rotation about the center of mass. Each star moves in a circle about the center of mass. Let us call the mass of one star M, its distance from the center of mass R, and its orbital velocity V. Let the other star's mass be m, its distance from the center of mass be r, and its velocity by v. The velocities V and v are the maximum observed velocities relative to the center of mass, that is, the actual speeds about the orbits, which are the largest velocities we ever observe. If star M travels at speed V for a time P, it travels a distance VP; that is the circumference of its orbit. But if the radius of that orbit is R, then the circumference is $2\pi R$, and the radius of the orbit is $VP/2\pi$. Similarly the radius of the orbit of the other star is $vP/2\pi$. Because R and r are measured from the center of mass, $MR = mr$, or $M/m = r/R$. Therefore the ratio of the two masses are known. But their sum can be known also, for the distance between the star $a = R + r$, and then

$$M + m = \frac{a^3}{P^2}$$

And if both the sum and the ratio are known, the individual masses can be easily determined.

In this example we assumed that our line of sight lay in the orbital plane so that the maximum velocities we saw were the actual maximum velocities of the stars. Usually we have no way of knowing that we are seeing the orbit edge-on, and we

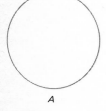

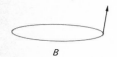

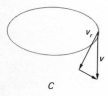

FIGURE 10.5 Maximum radial velocities seen for three orientations of the orbit of a member of a spectroscopic binary pair. If the orbit is seen from directly above as in A, no radial velocity is seen. If the orbit is seen from the edge as in B, the greatest radial velocity seen is the true velocity of the star. If the orbit is seen obliquely as in C, the true velocity V is never directed toward us, and we see only the component V_r which is in our line of sight.

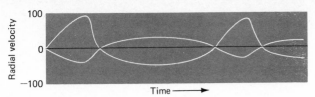

FIGURE 10.6 Distorted radial velocity curves produced by stars moving about their center of mass in elliptical orbits. Compare with Fig. 10.4, which shows the radial velocity curves for circular orbits.

must assume that we know only the components of the orbital speeds which are in our line of sight. For example, to take the extreme case, if we see the orbit from directly above, we will see no doppler shift, for the stars never approach us or move away from us. (In a case like that we may still be able to recognize that we are seeing a binary if the two stars are of quite different stellar classes. For example, if one star is type A and the other is type F, the spectrum we see will be a mixture of the two types, and a spectroscopist may recognize that two spectra have been combined. A double recognized in that way is called a *spectrum double*.) If we see the orbit from directly above, we see no orbital velocity; if we see it edge-on we see the true orbital velocity; at all other orientations we see something less than the true velocity. When we measure velocities by the doppler shift, then, we know that the actual velocities may be equal to or greater than the ones we see. If the velocities are greater than our measurements indicate, the masses of the stars are greater than our computed values; consequently the masses we compute represent a lower limit of the actual masses.

If the orbits of the two stars are not circles, the velocity curve is distorted from the simple shape shown in Fig. 10.4, and it looks more like that in Fig. 10.6. Analysis of this curve yields the same kind of information about the orbits as that discussed in Example 2. If an observer knows that a particular orbit is being viewed edge-on, the sum of masses can be computed. If that information is not available, the situation is the same as for the circular case: one finds a lower limit on the sum of the masses.

ECLIPSING BINARIES

If we see a binary system almost edge-on, one star partially eclipses the other on each revolution, and it is called an *eclipsing binary*. In that case we do have information about the inclination of the plane of the orbit to our line of sight. Thousands of eclipsing binaries are known. Any double pair is an eclipsing

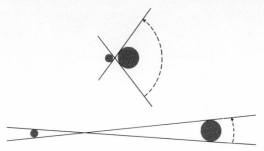

FIGURE 10.7 Comparison of the regions within which an
eclipse can be seen for two stars very close together and the
same two stars with greater separation. The region within
which the eclipse can be seen is a band on the celestial
sphere of the angular width indicated by the arrows.

binary if it is seen from the proper direction, but unless the
stars are very close together the region in which the eclipses
can be seen is so small that the probability of our finding ourselves in that region is low. Figure 10.7 shows the geometry for
observing eclipse for stars that are close together and stars that
are farther apart. In visual binaries, the stars are quite far apart,
and the probability that we will be situated to see the eclipse is
very small. Few if any visual binaries are also observed as
eclipsing binaries. On the other hand, if the stars are close
enough together so that we are likely to see their eclipse, they
also move with high orbital velocities and are hence spectroscopic binaries. Most eclipsing binaries are also observed as
spectroscopic binaries.

In actual practice a number of factors may complicate the
analysis of eclipsing binaries: The eclipse may be only partial;
the orbits may be elliptical; one star may cause a hot spot on the
other directly opposite it; gas may stream from one to the other;
etc. We will ignore such complicating matters and consider a
few simple cases to illustrate what information can be determined and how it is obtained.

Light Curve

Let us suppose that a small, hot star is moving around a large,
cooler star, and that we observe the system so that their equa-
tors coincide (we see a central eclipse). When we can see both
stars, the luminosity we observe is the sum of their effects, but
when one passes behind the other, the light drops. The light
curve compared to the orbital motion of the stars is shown in
Fig. 10.8. We are assuming circular orbits, and consequently
the times between minima are the same. The deeper minimum
is called the *primary eclipse,* and the shallower one is the *sec-*

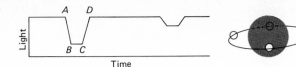

FIGURE 10.8 Light curve for eclipsing binary when the eclipse is central. The points labeled *A*, *B*, *C*, and *D* indicate first, second, third, and fourth contacts.

ondary eclipse. If the smaller star is the hotter, the primary eclipse occurs when it passes behind the cooler star; if the smaller star is cooler, primary eclipse will occur when it passes in front. *First contact* is the condition when the eclipse begins (point *A* in the diagram), second contact is the condition when the eclipse reaches minimum (point *B*), and third and fourth contacts are the corresponding points (*C* and *D*) when the eclipse is ending. From an inspection of the light curve one can draw some conclusions about the stars and their orbit. The ratio of the time between first and second contacts to the time between first and third contacts gives the ratio of the diameters of the two stars. The ratio of the times between contacts to the time for the entire orbital cycle gives the ratio of the diameters of the stars to the circumference of the orbit. From the ratio of the depths of the minima we can determine the ratio of the temperatures of the two stars; in each case the same area is blocked from view, and the decrease in radiation from that area is dependent upon the Stefan-Boltzmann law.

Combination of Types

If our only information is the light curve, then we cannot go beyond ratios of diameters for the stars and the orbit. If the system is also a spectroscopic binary and absolute velocities and the time of revolution are known (from the light curve), we can find the circumference and the diameter of the orbit. From that we can compute the diameters of the stars in meters. From the individual velocities and the separation of the stars we can find the mass of each star. Complete analyses of this sort have been carried out on a few dozen stars, giving us information about their diameters and masses which could have been obtained in no other way.

Algol

An eclipsing binary of particular interest is Algol, the Demon Star (also known as Beta Persei). In 1669 G. Montonari observed that Algol varies in apparent brightness with a period of

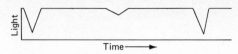

FIGURE 10.9 The light curve of Algol. The period is 2.87 days and the maximum change in brightness is slightlymore than one magnitude.

about 3 days—actually 2 days, 20 h, 49 min—and in 1783 John Goodricke suggested that the behavior could be explained if the star is an eclipsing binary. In 1890 H. Vogel classified Algol as a single-lined spectroscopic binary. The light curve of Algol is shown in Fig. 10.9. The fact that the minima do not have flat bottoms indicates that the eclipses are partial; at the lowest part of the primary minimum the larger star is blocking the light from about 70 percent of the brighter and slightly smaller companion. The brighter of the two is of class B8; it is 5 times as massive as the sun and has a radius 3 times that of the sun. The fainter star is of class K0; its mass is equal to the sun's, and its radius is 3.2 times greater. The two stars are about 11 million km, or $\frac{1}{14}$ AU apart. Because the stars are so large and so close together, the view of one from the surface of the other would be spectacular. The brighter is about 25 times as luminous as the fainter, and reflection causes the fainter star to appear much brighter when we see it illuminated by the brighter star than when it is between us and the brighter star.

About a century ago it was observed that the period of Algol varies slightly. It has since been determined that the radial velocity of the center of mass of the pair varies with a period of 1.87 years. Recently the spectrum of a F2-type star has been found associated with Algol, and it seems certain that a third star is part of the system. It is estimated to have a mass about 1.3 times the sun's and to lie about 3 AU from the other stars. As it and the eclipsing stars move about one another, the eclipsing pair oscillates toward and away from us, and hence the spectra show a variable doppler shift. Also, when they are away from us the distance their light must travel to reach us is slightly greater than when they are near, and this difference in light travel time produces a variation in observed period of the eclipses.

Data going back about two centuries indicate that the period has another variation, with a periodicity of almost 200 years. That variation is believed to be caused by a fourth member of the system, which probably lies 60 or 70 AU from the others and has a mass several times that of the sun. The fourth member of the system has not been observed with a telescope, although from what is known of it, we speculate that it is visible.

Sirius

Sirius is the brightest of the stars, and it is the nearest (8 light years) star visible to the naked eye in most of the United States. (The closest star to the sun, Proxima Centauri, is not visible this far north.) In 1844 Bessel discovered that Sirius moves across the sky with a sinuous motion, having a period of 50 years, and he attributed the wavy motion to the presence of an unseen companion. In 1862 Alvan Clark, an American lens and telescope maker, observed the companion, a white dwarf now known as Sirius B.

MASS-LUMINOSITY RELATION

The only star for which we can make a simple, straightforward mass calculation is the sun, but good data are available for computing the masses of about 100 other stars, which are members of binary systems. For stars on the main sequence (not white dwarfs or red giants) mass seems to be related to absolute luminosity, so that when the stars are plotted, as in Fig. 10.10, the points representing the stars lie approximately on a line. This plot is called the *mass-luminosity relation*. One should expect that the theories of stellar structure would explain why stars with greater masses should have greater luminosities, and indeed the models of stars do give some indication why this is true. Because this relation is a result of the operation of physical laws within stars, it is presumed to

FIGURE 10.10 The mass-luminosity diagram for stars on the main sequence.

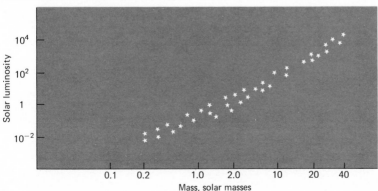

apply to all main-sequence stars, and hence one can estimate the masses of stars from observations of their luminosities.

QUESTIONS

1 Find the combined mass of the two stars in a binary system whose period of mutual revolution is 2 years and for which the semimajor axis of the relative orbit is 2 AU.   *Ans* $2\,M_\odot$

2 A few stars are both visual binaries and spectroscopic binaries. Why are such systems rare?

3 From the definition of the parsec, show that the semimajor axis of the orbit of a visual binary system, measured in astronomical units, is equal to its angular size, measured in seconds of arc, times the distance to the system in parsecs.

4 If a binary pair has a parallax of 0.125 second of arc, a semimajor axis of 2 seconds of arc, and a period of 50 years, what is the sum of the masses of the stars?   *Ans* $1.6\,M_\odot$

5 An eclipsing binary system consists of a small star which moves about a larger star, producing central eclipses. The orbital speed of the small star with respect to the larger is 6×10^4 mi/h. The time from first contact to second contact is 1 h; from second to third is 3 h; from third to fourth is 1 h. Sketch the light curve. What are the diameters of the two stars?   *Ans* 6×10^4 mi; 24×10^4 mi

6 From the mass-luminosity diagram estimate the mass of a star whose luminosity is 100 times that of the sun.

7 Why does the primary minimum in the light curve of an eclipsing binary occur when the star of greater surface brightness is eclipsed?

8 Explain why the analysis of spectroscopic binaries often gives only a lower limit on the sum of the masses.

variable stars

Eclipsing binaries often appear to be a single star that varies in light output. There also exist stars which *actually* fluctuate even though they are single; these are called *intrinsic variables*. Usually the adjective *intrinsic* is omitted, and stars whose light output varies in time are simply called variables. Variables are divided into two main categories—the pulsating variables and the eruptive or explosive stars.

PULSATING VARIABLES

Pulsating stars have variations in light which resemble in some ways the variations of eclipsing binaries, but they have certain features which clearly differentiate them from binaries. One difference is that since the light output of a pulsating star is constantly changing, its light curve has no flat spots like the light curve from a binary. The pulsating star changes temperature and consequently stellar class. Spectral lines from the surface of pulsating stars show periodic doppler shifts, indicating that the star is alternately swelling and shrinking.

The first pulsating variable found was observed in 1596 by David Fabricus, who noticed a star which was not on existing star charts. Its proper designation is Omicron Ceti, but it has long been known as Mira the Wonderful. It is a long-period (11 months) variable of a type that will be discussed later. The second variable discovered was Delta Cephei, found in 1784, which has given its name to the class of stars that resemble it. Although Cepheids are not the most common type of variable, they are one of the most important, and we consider them first.

Cepheids

Cepheids typically vary in brightness in a very regular way. If their magnitude is measured repeatedly, the variation with time can be plotted to produce a light curve such as that in Fig. 11.1. The period is the time from maximum back to maximum; periods of Cepheids vary from less than 1 to about 50 days, but the most common period is about 1 week. More than 600

211

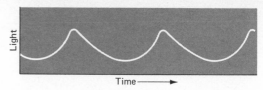

FIGURE 11.1 The light output of a typical Cepheid variable plotted against time. The curve is asymmetric; the brightening phase lasts for about 30 percent of the period.

Cepheids have been found in our galaxy, and they are concentrated along the disk of the galaxy. As the Cepheid varies in light output by approximately 1 magnitude, its temperature varies by something like 1000 K, and because of that its class changes somewhat. The typical Cepheid is an F- or G-type star, yellow in color like the sun but with an absolute magnitude somewhere between −3 and −5. The absolute magnitude of the sun is 5, and so Cepheids may be 10,000 times as luminous as the sun. As a consequence, they can be seen for great distances. Cepheids do lie above the main sequence on the H-R diagram and generally to the left of the giants. They are now thought to represent a particular phase—perhaps a short one—in the life of a star.

Period-magnitude relation   One reason that the Cepheids are important is that they can be used as distance indicators. In the early 1900s Ms. Henrietta Leavitt, working at the Harvard College Observatory, discovered numerous Cepheid-like stars in photographs of the Small Magellanic cloud. The Clouds of Magellan (Large and Small) are two small galaxies, probably satellites of our galaxy, which are so far south that they cannot be seen from the United States. (Their names derive from the fact that they were described by the survivors of Magellan's trip around the earth.) Ms. Leavitt noticed an interesting thing about the variable stars that looked like the Cepheids in our own galaxy: the brighter one of the stars is, the longer is its period. In fact, if the period is plotted against the apparent magnitude, most of the stars fall along a narrow band (Fig. 11.2).

At the time of Ms. Leavitt's discovery, no one knew the distance to either of the Magellanic clouds, but it was clear that the distance to the Clouds is much greater than the distance across one of them. Therefore all the Cepheids seen in the Smaller Magellanic cloud are approximately the same distance from the earth, for even the distance between two stars at different ends of the Cloud is insignificant compared to the distance each is from the earth. If all the stars are approximately the same distance from us, the ones that appear brighter actually are brighter, and a plot of apparent magnitude against period is the same as the plot of absolute magnitude against period, except that the numbers on the magnitude scale are different. If all the stars are at the same distance, their absolute magnitudes can be found by subtracting some constant number from their apparent magnitudes, but the number depends upon the distance, and the distance was not known when Ms. Leavitt's work was first reported. Since the periods of Cepheids are related simply to their absolute magnitudes, however, it was known that if the absolute magnitude of a few Cepheids could be determined, then the magnitudes of all the others could be read from the graph.

If we knew what numbers to place on the magnitude scale shown in Fig. 11.2 that would make it an absolute magnitude scale, then we could look at a Cepheid, determine its period, and read off from the diagram its absolute magnitude. And if the absolute magnitude and apparent magnitude of any object are known, its distance can be computed. Thus a determination of the period of a Cepheid plus the observation of its apparent magnitude could rather directly give its distance. Unfortunately, no Cepheid lies near enough for its distance to be measured by trigonometric parallax, and indirect methods must be used to calibrate the magnitude-period plot. The most accurate calibration now is carried out by the observation of Cepheids in clusters of stars where many of the stars are normal, main-sequence stars, whose absolute magnitudes can be estimated from the H-R diagram. If the clusters are relatively small, the distance to the Cepheids is the same as that to other stars in the cluster, and so the absolute magnitudes of the Cepheids can be determined.

Harlow Shapley was one of the first astronomers to realize the possibilities of using Cepheids as distance indicators, and he began the task of trying to determine the absolute magni-

FIGURE 11.2 When the apparent magnitudes of a group of Cepheids, all of which are at about the same distance from the earth, are plotted against their periods (actually against the logarithms of their periods), the points fall within a narrow strip.

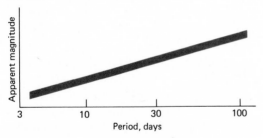

tudes of the Cepheids in our own galaxy. One of the most important early uses of distances derived from observations of Cepheids was Hubble's estimation of the distance to the Andromeda galaxy in the 1920s. At that time it was not known whether the Andromeda galaxy (and other objects which we now know are also galaxies) was a part of the Milky Way or whether it was something outside. Hubble observed a number of Cepheids within the Andromeda galaxy and from them deduced that it is about 1 million light years away, clearly outside the Milky Way. (That distance is now believed to be 2.5 million light years.)

Until the early 1950s the calibration of the period-luminosity relation for Cepheids was rather crude, and it was not recognized that two different kinds of Cepheids exist. After astronomers found that Cepheids come in two models, estimates of distances to other galaxies were drastically changed. When it turned out, for example, that the Cepheids being seen in galaxies such as M 31 (the Andromeda galaxy) were actually 4 times brighter than had been thought, estimates of the distances to those galaxies had to be doubled to explain why the stars looked as faint as they did.

The two types of Cepheids are *Type I* and *Type II*; these are respectively Population I and Population II stars, distinctions which go much further than just Cepheids. Population I stars are found near the plane of the galaxy; most of the objects in our neighborhood are Population I objects. Population II objects are found in all parts of the galaxy, but they are the dominant objects in the globular clusters and near the nucleus of the galaxy. (We use the term "objects" instead of "stars" because such things as nebulae, and some clusters, may be included in the categories.) Figure 11.3 shows the magnitude-period relation for Population I and Population II Cepheids, and for RR Lyrae stars, which will be discussed later.

FIGURE 11.3 The period-luminosity diagram for Cepheids and RR Lyrae stars.

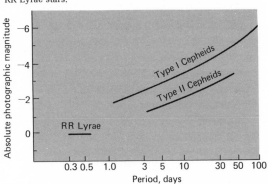

Everyone knows one Cepheid, although few realize that it is variable. It is Polaris, the North Star, which varies by only about 0.1 magnitude, not enough to be noticeable to the eye. Its period is approximately 4 days.

Pulsation mechanism As the luminosity of a Cepheid varies, the temperature of the surface changes; doppler shifts reveal that the radius of the star also changes. The radial velocity of the surface can be measured and plotted as a function of time, and one can compute the distance the surface rises and falls during a cycle. The change in radius is of the order of 10 percent; Delta Cephei, for example, changes its radius by almost 1 million mi, and the radius at the midpoint of the motion is approximately 12 million mi. The radius does not change enough to cause much change in the luminosity, and so the luminosity change is produced primarily by the temperature change. Although one might reasonably suppose that when a star is at minimum size its gases will be compressed to a maximum and will therefore be hottest, and hence the luminosity will be at a maximum, this is not the case. The greatest luminosity occurs when the velocity of expansion is greatest, after the star has passed minimum size and begun expanding outward. It is believed that the compression of the star during contraction does indeed cause heating, but the light produced by that extra heating takes some time to reach the surface and escape. When it does become visible, the temperature of the surface of the star may increase by about 1000 K, but by that time the expansion has begun and in fact has reached greatest velocity.

The Cepheid variable is believed to be a relatively brief phase in the life of many stars. The nature of the evolutionary processes in stars in the variable stage is not understood well enough for astronomers to compute the time they will remain in that phase. Perhaps 1 million years is a reasonable guess, since if the period were much greater, we should see more Cepheids than we actually observe.

What causes a star to begin oscillating is not known with any certainty, but we do have a theoretical model that explains something of what maintains the motion. As the star contracts, the temperatures of layers not far below the photosphere rise; the layers are relatively opaque, so that they trap the radiation their greater temperature has produced. The trapped radiation exerts a pressure against layers farther out, and finally begins to force those layers up. As the star begins to expand, its opacity decreases and radiation begins to stream out. The pressure below decreases, but the layers moving outward have enough momentum to continue moving, and they overshoot their equilibrium position. As a result of those layers overshooting the place where the pressure would just hold up the upper layers,

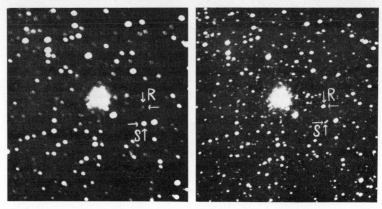

The variable stars R and S Scorpii. (*Yerkes Observatory photograph.*)

the star expands enough so that when it falls back it repeats the cycle.

RR Lyrae Stars

Like Cepheids, RR Lyrae stars are useful in determination of distances. They are more common than Cepheids, and about 3,000 of them have been identified in our own and nearby galaxies. They are Population II objects, with periods of less than 1 day. As inspection of Fig. 11.3 will show, they all have approximately the same luminosity: about 0.5 magnitude. RR Lyrae stars are found so often in globular clusters, those spherical clusters containing as many as 1 million stars each which lie outside the main disk of our galaxy, that they are often called *cluster variables*.

Suppose that in a globular cluster several RR Lyrae stars are observed to have an apparent magnitude 15.5. We can assume that the stars have absolute magnitudes of about 0.5, and we can hence reason that the distance is such that they appear 15 magnitudes fainter than if they were at 10 pc, or that they are fainter by a factor of 1 million. If a star were moved from 10 pc outward until its brightness decreased by a factor of 10^6, its distance would have increased by 10^3; it would have been moved to 10,000 pc. That is the distance to the globular cluster.

Long-period Variables

The most common of the variables are the *long-period* or *Mira-type* variables, several thousands of which are known. They are red giants or red supergiants of spectral classes M, R, N, or S, and their periods of variation range from about 70 to 700 days.

Mira is of apparent magnitude about 8 most of the time, but every 11 months it brightens to magnitude 2.5 or more; once it reached 1.2. The variations of the long-period variables are not as regular as those of the Cepheids, and consequently they are not useful as distance-measuring aids. Mira, for example, has an average period of 331 days, but it varies as much as 25 days from the longest to the shortest. (Mira, incidentally, is one of the stars whose diameter has been measured by the stellar interferometer. Its diameter is 420 times that of the sun, and it varies by about 20 percent during Mira's oscillations.)

The amplitude-of-magnitude variation of the long-period variables can be quite high. Some vary by as much as 7 magnitudes, which means that their light output in the visible region changes by a factor of about 600. Their temperatures do not fluctuate greatly, however, and an explanation of the large magnitude change must take into account the shape of blackbody curves. Long-period stars have surface temperatures in the vicinity of 2000 K, and most of their radiation falls in the infrared part of the spectrum. If the temperature of the surface increases by a few hundred degrees, the increase in total radiation is not great; in other words, the *bolometric magnitude* does not change greatly. However, when the temperature rises even a small amount, the peak of the blackbody curve shifts toward the visible enough to cause a tremendous increase in the visible radiation emitted. And since we usually observe only the visible, the star appears to brighten far more than we might expect from a naive consideration of the increase in radiation predicted by the Stefan-Boltzmann law.

ERUPTIVE VARIABLES

Explosive stars, such as *novae* and *supernovae*, suddenly increase their brightnesses by tremendous factors and then fade over long periods of time. Such an increase may be a once-in-a-lifetime event for the star, or it may occur over and over.

Novae

The name *nova* means new, but a nova is not a new star; on the contrary, it is probably very old. The name was given before astronomers realized that a star which suddenly becomes prominent, blazing brightly where it has not been noticed before, has really been a faint star in that same position all the time. Before outbursts, and after they have returned to normal, years or decades later, novae are small, hot stars. Eight stars are known to have flared as novae more than once; they are called *recurrent novae*. The average time between eruptions is about 30 years.

When a nova begins to brighten, its luminosity may increase by a factor between 10^4 and 10^5 within a day or two. It then begins to decrease, and may be down by a factor of 50 within a week. The decrease from then on is gradual, lasting years. Spectra of novae indicate that a shell of matter, at the most a few percent of the mass of the star, is ejected from the star, traveling outward at a speed of about 1,000 km/s. We know that a shell of matter is ejected because during the outburst spectral lines are shifted toward the blue, indicating that the surface of the star is approaching us. After some time the shell becomes thin enough so that it is no longer opaque, and then we begin to see emission lines in the spectrum, lines emitted by the tenuous gas of the shell as it releases energy it has absorbed from the ultraviolet radiation of the star. In at least one case the expanding shell became visible in the telescope some months later.

The causes of novae explosions are not known with certainty, but current theory suggests that the causes are external to the star. In at least some observed cases, novae are members of very close binary pairs, and it is suspected that *all* novae are in such pairs. Very close binaries sometimes exchange mass, with matter flowing from one to the other as their radii change. Imagine two stars quite close together, one in the white dwarf

Direct photographs of Nova Herculis 1934, showing the large change in brightness between March 10 and May 6, 1935. (*Lick Observatory photograph.*)

stage and the other beginning to swell to become a red giant. If

the ballooning star expands enough, some of its gases will cross
the line at which its attraction becomes less than the attraction
of its white dwarf companion, and the gases will begin to fall
into the white dwarf. A white dwarf is very dense, however,
and its surface is somewhat unyielding, and as gas from the
large star rains down upon that surface, the temperature of the
gas shoots up abruptly as its kinetic energy from the fall is
changed into thermal energy. As the temperature rises rapidly
to an extremely high point, the gases falling onto the dwarf expand so rapidly that part of them are blown away from the system, forming the expanding shell which we see.

About 200 novae have been discovered in our galaxy, and we
may assume that many more are hidden from our view by the
dust which pervades our region of space. Observations of the
Andromeda galaxy suggest that if our own galaxy is as much
like Andromeda as we suppose, approximately 25 novae occur
in our galaxy each year.

If the double-star explanation of novae is correct, it is clear
that we can learn little about the structure and evolution of
stars in general from observation of novae. Supernovae, however, are believed to mark a possible end of the evolution of some
kinds of stars, and if we can understand supernovae, we will
know more about the internal changes of many stars.

Supernovae

A nova rarely increases in brightness by more than a factor of
10^5, but a supernova increases by 10^8 or more. At peak
luminosity, most supernovae are more than 100 million times
as luminous as the sun. Such a star may become as bright as the
entire galaxy it is in, and the galaxy may have something of the
order of 100 billion stars. Supernovae are rare events, however,
and Nature has not been kind to astronomers on earth in
providing examples to be studied. During the last millenium
three supernovae are known to have occurred within our galaxy
and within a distance permitting easy study, but all three occurred before the invention of the telescope or spectrograph.
Supernovae in other galaxies have been observed, and what we
know of them comes from observations made across intergalactic distances.

On July 4, 1054, a supernova was recorded by the Chinese,
who called it a Guest Star. The suggestion has been made that
certain pictures on the walls of caves made by Indians in the
American southwest represent that event, but we cannot be
sure that interpretation of the pictures is correct. No record of

notice of the star has been found in Europe, although at its peak it was brighter than Venus. In 1572 Tycho Brahe observed a supernova, now known as Tycho's nova, which at one point became bright enough to be seen in the daytime. In 1604, Kepler's nova, so named because Kepler studied it, became bright enough to be observed; like the others, it faded within a few months. Remnants of other supernovae have been found in our galaxy, and it is estimated from the frequency with which supernovae are observed in other galaxies that an average of one per century should occur in ours.

Like a nova, a supernova blows mass away into space in the form of an expanding shell, but the supernova differs from a nova in at least two important respects. First, the mass is ejected at much higher velocities—up to 4,000 km/s. Second, a substantial part of the star is shot out into space; a supernova must be drastically and permanently changed by the explosion. The causes of supernovae are discussed in Unit 12.

Crab nebula   The Chinese Guest Star of 1054 appeared in the region of sky we identify as the constellation Taurus. In Taurus now, in the same location, as nearly as one can estimate from descriptions in the old records, we see a structure called the Crab nebula. The Crab nebula is one of the most intensely studied objects in the sky; it has been studied particularly carefully since the discovery a few years ago that one of the stars embedded in it is a *pulsar* (discussed in Unit 18). The Crab is an extended mass of gas that appears filamentous in photographs. It is moving and changing so that photographs taken a few years apart show noticeable differences. The color of the light indicates that much of the gas is hydrogen.

Doppler shifts indicate that the gas in the Crab which is moving directly toward us is moving at about 1,400 km/s. The angular radius of the nebula is increasing about 0.19 second of arc per year, and the present average radius is about 2.5 minutes of arc. If we assume that the expansion is approximately spherical, so that the parts moving across our line of sight and contributing to the 0.19 second of arc per year expansion are also moving at 1,400 km/s, we find that the expansion must have begun about 800 years ago. The supernova in that region was observed just over 800 years ago, and the identification of the Crab nebula with the Guest Star of 1054 seems very reasonable. We believe that a star blew up, sending most of its material out into space to form the Nebula and leaving part of the material in a very dense star which we see now as a pulsar.

Part of the light from the Crab nebula comes from normal energy transitions within atoms, and part comes from a process we have not mentioned before—the *synchrotron process*. Light so produced is synchrotron radiation. According to classical electromagnetic theory, when a charged particle such as an electron is accelerated, it emits electromagnetic radiation. When electrons are driven to very high speeds, approaching the speed of light, and are then bent into circular paths in machines such as cyclotrons or synchrotrons, they radiate because of the centripetal acceleration, and the loss of energy to that radiation is one of the problems with large accelerators. Such radiation loss can be reduced either by making the accelerators straight or by making the circle very large. In the Crab nebula electrons are traveling at speeds very nearly the speed of light, and they are moving through magnetic fields which cause them to take curved paths. If an electron moves directly with or against a magnetic field, the electron is not deflected; if it moves obliquely across the field it takes a helical path. If the path is anything other than a straight line, the electron undergoes acceleration and consequently radiates energy. Part of that radiation is in the radio part of the spectrum, but very-high-energy particles produce radiation in the visible or even shorter parts of the spectrum. One of the clues which tells us that some of the radiation from the Crab nebula is synchrotron radiation is that it is polarized. (See Unit 12 for discussion of polarization of light.)

OTHER VARIABLES

Only two further kinds of variable stars will be discussed, although other types are known and have been grouped into classes.

T Tauri Stars

T Tauri stars are stars of spectral type F, G, K, or M, but they are somewhat more luminous than the main-sequence stars of those types. In addition to the usual absorption spectra, they have emission lines of hydrogen and ionized calcium. They fluctuate by 2 or 3 magnitudes in a completely erratic and unpredictable fashion.

A clue to the nature of T Tauri stars is the fact that they are always found in large gaseous nebulae. Over 200 have been found in the Great Nebula of Orion alone. It is believed that these are young stars, still contracting from the cool gas and dust from which stars are formed, and not yet on the main sequence. Perhaps their energy is still coming from gravitational potential energy released as they contract, and the nuclear processes which maintain mature stars have not yet begun to control their behavior.

Planetary Nebulae

The name *planetary nebulae* is a misnomer, for such stars have
nothing to do with planets. They are very hot, typically
50,000 K, and are surrounded by a thick shell of gas which
glows by fluorescence. The shell is expanding slowly, at speeds
of 20 to 50 km/s. (If that does not seem slow, recall the expansion speeds of the shells around novae and supernovae.) The
best known of the planetary nebulae is the Ring nebula in
Lyra, but about 1,000 are known in the galaxy. They appear to
be stars which slowly release a considerable amount of
mass—perhaps one-tenth the mass of the sun—over a period
of several thousand years. Because the expanding shells must

The planetary nebula NGC 2392 photographed in green light. (*Lick Observatory photograph.*)

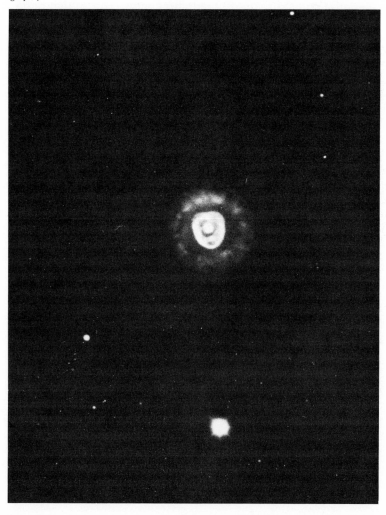

become so thin that they cannot be seen after a few thousand years, and yet we see a large number of planetary nebulae, we assume that the phenomenon by which they form is relatively common among very hot stars. This appears to be one way a massive star can lose mass without undergoing the terrible disruption that marks a supernova.

The shell of a planetary nebula looks in the telescope like a ring because the amount of gas in our line of sight is greater near the edge than in the center. If the gas were really in the form of a ring, however, we could expect to see many planetary nebulae at such an oblique angle that the ring would appear to be a flat ellipse. In fact, all the rings appear to be almost circular, suggesting that the ring of a planetary nebula is the edge of an approximately spherical shell.

QUESTIONS

1 A Type I Cepheid is observed to have a period of 50 days and an apparent magnitude of 10. How far is it from us?

Ans $\approx 10^4$ pc

2 An RR Lyrae star is observed to have an apparent magnitude of 13.5. How far away is the star? Ans $\approx 4,000$ pc

3 Two Cepheids are observed to have periods of 30 days. If one is a Type I and the other Type II, what is the difference in their absolute magnitudes? Which is the more luminous? If both have the same apparent magnitude, which is the more distant? Ans ≈ 1.5 magnitudes

4 From the Wien displacement law, compute the wavelengths at which the peaks of blackbody curves occur for temperatures of 2000 and 2500 K. Sketch the curves and determine where the visible region lies relative to the peaks. Use this information to explain why long-period variables may have a large change in magnitude but only a small change in temperature.

Ans 1,450 nm; 1,160 nm

5 Verify from the data given concerning the Crab nebula that the explosion which created it occurred about 800 years ago.

6 From the data given about the Crab nebula, compute its distance from us. (Combine the speed of expansion with observed angular speed.)

7 Compare the shells of novae, supernovae, and planetary nebulae.

8 Suppose that a supernova became 100 million times as luminous as the sun. What would be its absolute magnitude? From what distance would it appear as a first-magnitude star?

Ans -15; 1,600 pc

12
our galaxy

One of the striking sights of the sky—if one can get away from city lights and smog—is the Milky Way, a band of light stretching across the heavens. Although it had been observed from ancient times, its nature was not known until Galileo turned his telescopes on the sky and observed that the Milky Way is a band of stars so faint or far away, and so close together, that the naked eye does not distinguish them as individual stars. Some of the prominent constellations such as Cygnus, Cassiopeia, and Orion lie in the Milky Way, but they are made up of relatively nearby stars or stars that are so unusually bright that they stand out.

SIZE AND OUR POSITION

In 1750 the English telescope maker, Thomas Wright, suggested that if we live in a flat aggregation of stars, a disk shaped like a grindstone, its appearance should be something like the Milky Way. Looking along the plane of the grindstone, we would see stars to a great depth, and their total effect might be one of faint and diffuse light. But if we looked toward the side, as in Fig. 12.1, we would see fewer stars, and could look out into "empty" space. In the latter part of the eighteenth century, William Herschel undertook to determine systematically whether Wright's suggestion was consistent with observations. Herschel looked at selected areas in the sky and counted the stars within a standard size field. He eventually counted the stars in about 700 fields scattered over the sky; in some fields he found only 1 star, but in others he found as many as 600. On the assumption that the stars are all equally luminous, he estimated the distance to the faintest stars he could see. His conclusion was that Wright had been correct: We live in the center of a grindstone-shaped community of stars. This assemblage of stars is our galaxy—the Milky Way galaxy.

The Milky Way is broken by dark streaks, rifts in which no stars are found. Early astronomers thought that the dark areas were regions without stars, analogous to cracks in a grindstone, but we now know that some light does come from the dark areas, and that they appear dark because clouds of dust obscure

Mosaic of the Milky Way from Sagittarius to Cassiopeia, composed of several wideangle photographs. (*Hale Observatories.*)

most of the stars in them. In fact, we live in a dirty, hazy region of space, and as we look out to great distances, particularly along the plane of the galaxy, we must constantly allow for the dimming and obscuration of our view caused by the clouds of dust that surround us.

Although the general notion formed by Wright and Herschel about the form of the galaxy is correct, they were wrong about one very significant detail. As we look along the plane of the galaxy, the distance to which we can see is determined not by the extent of the stars in that direction but by the dust which blocks the view. Because of the dust, it was plausible to believe that we were located at the center of the disk. A more accurate determination of our position did not come until 1917, when Harlow Shapley became interested in globular clusters, which are bright enough to be seen at great distance.

Shapley counted more than 90 of the clusters, and they were concentrated toward Sagittarius. In some of them he could pick out individual stars, some of which were apparently RR Lyrae variables. Because the absolute luminosity of RR Lyrae variables was known with fair accuracy, he was able to estimate the distance to those globular clusters in which he could see the variables. When he then estimated the luminosity of the total cluster, that estimate could be used to determine the distance to those that were so far away that individual stars in them could not be seen, and Shapley estimated the distance to all the globular clusters he had found. When the clusters were all located in three-dimensional space, they formed an approximately spherical group, which was not centered at the sun. Using the best modern distance measurements instead of the ones used

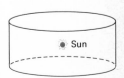

FIGURE 12.1 The grindstone galaxy suggested by Thomas Wright, with the sun near the center. Looking toward the face of the disk, one sees fewer stars than if one looks along the midplane.

by Shapley, astronomers now say that the center of the group of globular clusters appears to be a point in the direction of Sagittarius about 30,000 light years from the sun. Since we can expect globular clusters to move in orbits about the center of mass of the galaxy, and they should be arranged more or less symmetrically about that center, it is clearly implied that the sun is not at the center of the Milky Way. We now believe that our galaxy has a diameter of approximately 100,000 light years and the sun is about 30,000 light years from the center.

CONSTITUENTS OF THE GALAXY

The most obvious constituent of the galaxy is stars—stars alone, double stars, stars in open clusters, stars in globular clusters. But along with the stars the galaxy contains gas and dust, and magnetic fields wind through it. And the structure of the main disk of the Milky Way is probably not so much like a grindstone as it is a pinwheel. We will discuss all these characteristics, and then try to make some guesses about how the galaxy must have begun.

Disk

Most of the stars of the galaxy are found in a rather thin disk, whose diameter is something like 100,000 light years and whose thickness is of the order of 3,000 light years. But the galaxy is really more nearly a sphere with a diameter of about 100,000 light years, and throughout the volume of the sphere outside the disk roam globular clusters and individual stars. The region outside the disk is called the *halo,* and the single stars which move through that region are termed *halo stars.* Both the globular clusters and the halo stars move on orbits about the center of the system, so that during each trip around the center they pass twice through the disk. Within the disk we find galactic clusters, young stars, gas, and dust.

Globular Clusters

About 120 globular clusters associated with our galaxy are known, and many are seen about other galaxies. Only one, M 13 in Hercules, can be seen at all with the unaided eye from the northern hemisphere; it appears as a faint fuzzy patch at the edge of visibility. (M 13 is about 25,000 light years distant, and so to be visible at all it must be quite luminous.)

The number of stars in globular clusters ranges from perhaps 10,000 to 1,000,000, with the typical value probably in the range 100,000 to 300,000. The diameters of the clusters range

The galactic (open) clusters h and χ Persei. (*Lick Observatory photograph.*)

from about 80 light years to 3 times that much. Globular clusters almost certainly are stable configurations of stars. The individual stars within the clusters are moving, so that if we could see the motions speeded up tremendously the clusters might resemble swarms of bees, but few of the members escape from the groups.

There is no evidence that globular clusters contain either gas or dust, and we see no stars in any of the clusters, or elsewhere in the halo, that are identifiable as young. Stars in the globular clusters, like the other stars in the halo, are Population II objects, which means that they have very small quantities of metals.

Galactic Clusters

Galactic clusters are so named because they are found near the plane of the galaxy, but they are also often called *open clusters* because the individual stars within them can usually be resolved. Approximately 800 of them are known. The number of stars within a cluster varies from about 10 to perhaps 1,000, with 100 being a typical value. All those stars lie within a volume with a radius probably less than 10 pc, with 1 to 2 pc being a common radius. Stars in clusters are more closely spaced than other stars, and one open cluster, the Pleiades, has been recognized as an aggregation from very ancient times. Homer mentions the Pleiades, as does the author of the

The globular cluster M 92 (NGC 6341) photographed with the 120-in reflector. (*Lick Observatory photograph.*)

book of Job. What those observers could not know is that instead of the 7 stars traditionally seen in the Pleiades, the cluster contains at least 250 stars. Many of those can be seen with binoculars, and the others are visible through more powerful telescopes. Another galactic cluster which has some stars bright enough to be seen with the naked eye is the Hyades; both it and the Pleiades are located in the constellation Taurus.

Galactic clusters have a property which can sometimes be used to determine whether a particular star is or is not a member of the cluster. Stars in a cluster all move on approximately parallel paths, and if the cluster is near enough that the common velocity of the stars can be calculated, whether or not a star has that velocity can be used to determine if it is a member of the group. For quite distant clusters, velocities cannot be determined because the proper motions are too small to be measured, and it is not always possible to be certain whether a star is a member of the cluster or whether it only appears so because it is in front of or behind the cluster from our viewpoint. Open clusters are not strongly gravitationally bound together, and in time the small differences in the velocities of the stars in them will carry the stars apart from one another so that the clusters will disperse.

Dust

Dust in interstellar apace manifests itself in several ways, but perhaps the most striking are the dark patches visible in the Milky Way. There we see the effect of clouds of dust lying

within a few thousand light years of us. If the clouds are sufficiently far away that a large number of stars are in front of
them, the clouds are not easily seen, but if a cloud is near
enough that few stars are between it and us, its blocking of the
light from the stars behind produces a contrast which makes it
easy to see. Clouds of dust which are sufficiently opaque that
they appear as dark regions in the sky are called *dark nebulae.*

Reflection nebulae Sometimes a cloud of dust surrounding
or lying close to bright stars appears to glow because of the light
it reflects from the stars. The effect is similar to the glow one
sees over a city at night when there is a light fog. The fog particles scatter light from below in all directions so that a faint
luminosity reveals the presence of the fog even though its particles cannot be seen. Dust which is visible because of this scattering mechanism is called a *reflection nebula;* an example is
the dust around Merope, one of the Pleiades.

Obscuration Another way the presence of dust in regions
where it cannot be "seen" manifests itself is through a general
obscuration of distant objects. Distant stars, seen through the
haze, appear fainter than they actually are, so that if we know
their distances, we underestimate their absolute magnitudes, or
if we can estimate their absolute magnitudes, we overestimate
their distances. This is the effect which caused astronomers a
few decades ago to think that the galaxy faded away some
10,000 light years from the sun. The obscuration also shows up
in observations of other galaxies, for we see few if any as we
look near the plane of our galaxy. The reason for the *zone of
avoidance,* as Shapley called it, is that near the plane of the
galaxy dust is dense enough to block our view; as we look more
nearly at right angles to the plane the dust is thin enough for us
to be able to see through it.

Reddening of starlight Another effect by which dust
reveals itself is the reddening of starlight. Many years ago astronomers noticed red-appearing stars which had spectra of the
O or B type, which meant they should be blue. This reddening
effect is caused by dust through which the light passes on its
way to us. When small particles scatter light, they scatter blue
more efficiently than red. This phenomenon is responsible for
our blue sky and red sunsets. Light reaching the earth from directions away from the sun has been scattered by the atmosphere, and the molecules of the atmosphere are so small that
they scatter blue much better than red. Hence our daytime sky
appears blue. In the evening, as the sun is setting, the light from
it passes through a long atmospheric path along which the blue
light has been scattered out to the side (causing the sky to appear blue to persons living to the west of us). When the blue
light is extracted from white light, the remainder appears red,
and hence the sunset is red.

Fortunately, scattering caused by small particles is quite strongly affected by the wavelength of the light, and this fact permits astronomers to determine how great the effect of the dust has been. The continuous radiation from a star is very nearly described by a blackbody curve. If we can see the surface of the star and measure the light being emitted at two frequencies, we can determine which curve describes that star by taking the ratio of the intensity of emission. If the light has been partially scattered and hence reddened, two frequencies are not enough, for one will be affected by the scattering more than the other. If three frequencies are used, however, one can use the fact that all three, if corrected for the scattering, should lie on a blackbody curve and deduce what the temperature of the source is and how much light has been lost by scattering.

Polarization of starlight Another indication of the presence of interstellar particles or grains of solid material is the polarization of starlight. As we discussed in Unit 5, light consists of oscillating electric fields. Any changing electric field is accompanied by a magnetic field, but because the electric field is the part of the wave which interacts with matter most importantly, we usually discuss it and ignore the magnetic field. The direction of the electric field is always perpendicular to the direction of propagation of the light, so that in a pencil of light, the electric field vectors are all in a plane perpendicular to the pencil. In ordinary light those vectors point in all directions within that plane, and the light is said to be unpolarized (Fig. 12.2). If, however, all those vectors are parallel to one line, the light is said to be linearly polarized or plane polarized. Partially polarized light is a mixture of polarized and unpolarized light.

Ordinary, unpolarized light can be polarized completely or partially be several processes. Scattering is one such process, and both the blue light of the sky and the light scattered by reflection nebulae are partially polarized. Light which is scattered to the side by molecules in the atmosphere of the earth is polarized, but light transmitted through the atmosphere is not. Reflection from a surface can produce polarization, and the light reflected from a window, a lake, or a table top is partially polarized. Certain materials, such as calcite and quartz, have the ability to separate unpolarized light into two beams, each totally polarized and with their planes of polarization

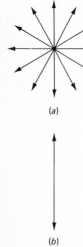

(a)

(b)

FIGURE 12.2 The diagrams show the orientation of the electric field vector in a beam of light coming out of the paper. Part (a) shows the vectors for unpolarized light, with electric fields directed in many directions; part (b) shows the vectors for plane-polarized light, with the electric fields all lying in one plane.

perpendicular to one another. Polaroid is made of a crystalline
material which has the ability to separate the light into two
beams and then to absorb one of those beams, so that only
polarized light is transmitted. In order for a sheet of Polaroid
to polarize light, all the crystals in it must be aligned, for if the
orientations were random, the light transmitted would still
have random planes of polarization, i.e., it would be unpolarized. Light coming great distances through space may be
polarized by interaction with particles in space, but only if the
particles are anisotropic and if they are aligned. Anisotropic
particles might produce partial polarization in either of two
ways: by absorbing the light vibrating in one plane, as Polaroid crystals do, or by preferentially scattering away light
vibrating in one plane, so that what remains to continue
forward is partially polarized.

The light from some stars is partially polarized, and because
the amount of polarization is correlated with the amount of reddening of the light, astronomers believe that the polarization is
produced in space, probably by rod-shaped dust particles
which tend to be aligned parallel to one another. The only explanation we have for such alignment is the presence of a magnetic field, and even a very small magnetic field over large distances could cause the alignment necessary to produce polarization. Polarization, then, is evidence for the existence of dust
in the shape of rods or bars and also for the existence of interstellar magnetic fields. Additional evidence for magnetic fields
comes from observations of the radio waves coming from pulsars.

Description of particles The effects of interstellar dust on
starlight permit us to estimate the density of the dust and the
size of the particles. Interstellar dust is probably not at all like
the dust of a plowed field, and we use the term "dust" to distinguish the material from gas. We know that the particles are
solid, not gas, because they are much larger than gas molecules
and much more efficient at dimming light. The amount of gas
required to produce the same extinction of starlight would have
a mass which is impossibly high. The number of particles in a
cubic mile must be small, however, for the dimming of stars is
inappreciable out to distances of about 100 pc. Throughout
"clear" space the number of particles probably is of the order of
100 per cubic mile, and dark nebulae may have more than that
by at least a factor of 100. Another indication that the particles
are small fragments of solids rather than gas is the way the scattering varies with wavelength. Over the visible range, the dust
dims the light of stars very nearly proportional to $1/\lambda$, where λ
is the wavelength. In other words, light of 400 nm is scattered
from the beam, and thus reduced at the earth, twice as much as

light of 800 nm. When the particles scattering the light are small compared to the wavelength of light, which is the case for the air molecules which scatter the light to make the sky blue, the scattering is proportional to $1/\lambda^4$, and light at 400 nm is scattered 16 times as much as light at 800 nm.

Our best hypothesis is that the dust is composed of solid, elongated particles with dimensions of approximately 10^{-5} cm. They probably are not metallic, and are almost certainly not ice. They might be flakes of graphite (carbon), and they might be graphite flakes coated with ice, but some evidence suggests that they are silicates. Not enough is known about the chemistry of materials under the conditions of empty space to permit us to make very good guesses about how such grains form, and the formation and eventual fate of the particles must be left as unknowns at this time.

Gas

The gas between the stars is quite distinct from the dust, although the two are often found together. Some of the gas glows by fluorescence and can be seen; some of it produces absorption spectral lines in the light of stars; and some emits radio waves which are detected by radio telescopes.

Emission nebulae   Here and there throughout the heavens are great clouds of glowing gas, glowing because very hot stars

An emission nebula (the "Rosette") in Monoceros, photographed in red light with the 48-in telescope. (*Hale Observatories.*)

embedded within them are emitting a flood of short-wavelength radiation that ionizes or excites the gas nearby. A typical example is the Orion nebula. These *emission nebulae* usually appear reddish when they are photographed in color because the most important element present in the gas is hydrogen, and the Balmer alpha line is the most intense in the radiation. If the gas surrounds a very hot star, hydrogen out to some distance is partially ionized in what is called an *H II region;* outside that region the hydrogen is neutral, although it may be excited so that it still glows.

Absorption lines in stellar spectra   In 1904 an absorption line with strange behavior was found in the spectrum of Delta Orionis. The star is a single-lined binary, and therefore all its spectral lines shift back and forth. It was found that one line, at 393.4 nm, did not shift, however. That line is produced by ionized calcium, and the only way to explain its failure to show the doppler shifts shown by the other lines is to assume that the calcium absorbing the light is between us and the star.

Other lines are found in the spectra of spectroscopic binaries and of very hot stars. Sodium, potassium, calcium, titanium, and iron have been detected in the gas of space by this method, as have some simple molecules such as CH and cyanogen. Cool stars have so many absorption lines of their own that the effect of atoms between the stars and us cannot be seen.

21-Centimeter line of hydrogen   For many years radio astronomers have been "seeing" cold hydrogen gas, and in recent years they have also seen a great array of molecules. Neutral, cold hydrogen can be seen by radio telescopes because of a small energy change which can occur within the atom. Both the proton and the electron in the hydrogen atom act somewhat like small bar magnets, but they are restricted to one of two positions relative to one another: either the magnets are parallel to one another or they are antiparallel. In other words, if we simplify the matter and think of them as actual small bar magnets, they can be parallel with their north poles pointing in the same direction, or they can be parallel with their north poles pointing in opposite directions. (The magnetic properties of these particles are associated with their *spin*, a characteristic they have which may be compared to the rotation of a planet on its axis. The restriction on the magnetic alignment is another way of saying that their spin axes must be parallel, and they must either appear to rotate in the same direction or in opposite directions.) The energy of the atom is slightly higher if the magnets are parallel than if they are antiparallel, and if atoms are in the higher-energy state, after a long period they spontaneously flip into the lower-energy state and the energy difference is released as a photon of radiation. The energy is very low, and consequently the wavelength of the radiation is long—21 cm to be exact.

The average time atoms in interstellar space take to drop from the higher of the two energy states to the lower is 11 million years. If the typical atom radiates its energy at 21 cm only after it has been in the upper energy state for about 11 million years, the fact that we detect the line easily in many directions indicates that there are many hydrogen atoms wandering around between the stars. Atoms are carried into the upper energy level by collisions. About every 400 years any given atom will collide with another hydrogen atom, and in the collision they are likely to exchange electrons. The newly acquired electron may be parallel or antiparallel to the proton, and so the effect of the collision may be to make no change in the atom, to raise it to the higher level, or to drop it to the lower level. Any atom in the upper level is much more likely to lose its energy in a collision than to lose it by radiating a 21-cm quantum of energy, but because of the huge number of atoms which are at all times in the upper state, some are always producing the 21-cm radiation.

If the hydrogen is moving toward us or away from us, the 21-cm line is doppler-shifted, and thus we can determine the radial speed of the hydrogen. We can also separate the radiation from several clouds, with different doppler shifts.

Molecules A great number of molecules which emit radiation in the range detected by radio telescopes have been found: hydroxyl (OH), carbon monoxide (CO), cyanogen (CN), carbon monosulfide (CS), silicon monoxide (SiO), water (H_2O), ammonia (NH_3), hydrogen sulfide (H_2S), hydrogen cyanide (HCN), formaldehyde (HCHO), thioformaldehyde (HCHS), methanol (CH_3OH), methylcyanide (CH_3CN), formic acid (HCOOH), formaldimine (CH_2NH), methylacetylene (CH_3C_2H), acetaldehyde (CH_3HCO), ethyl alcohol (CH_3CH_2OH), vinyl cyanide (H_2CCHCN), and formamide (NH_2HCO). The discovery of these molecules has spurred the search for more, and raised two important questions: How are such molecules formed in space? What significance has the presence of such complex molecules in the clouds of matter from which we presume that stars and planets might be formed for estimates of the probability of life elsewhere in the universe? If a planet forms from matter already containing these molecules, perhaps potential life-producing processes begin from a point further along the road than we once thought. Certainly the discovery of organic molecules in clouds hundreds or thousands of light years from the earth proves that the processes that produced organic materials on the earth cannot have been unique to our location or evolution.

The processes by which molecules are formed in space are not known, but it is likely that they are formed in clouds containing dust. Perhaps they are formed by reactions that occur on the surface of dust particles. If dust were not present to shield the molecules from ultraviolet radiation from stars, the molecules would have short lives.

STRUCTURE OF THE GALAXY

As we look out into space beyond the edges of our galaxy, we see other galaxies. Those galaxies take many forms, but most of them are elliptical, irregular, or spiral. Elliptical galaxies are elliptical or spherical clusters of stars with no other visible structure. Irregulars are just what the name implies—chaotic jumbles of stars and dust with no apparent structure. The spirals are giant pinwheels, rotating majestically, which take millions of years to complete one rotation. After viewing the variety of other galaxies, a natural question to ask is: What kind of galaxy is ours?

Spiral Structures

Clearly our galaxy is not elliptical, for most of the stars are compressed into a rather thin disk. The actual extent of the galaxy is almost spherical, but an observer looking at us from another galaxy would not see the halo stars and perhaps not even the globular clusters. Only the disk would be visible. The only remaining question then concerns spiral arms, for we could be a solid disk or even a thin irregular cluster. At least two kinds of evidence indicate that our galaxy is a spiral.

Spiral arms shown by stars The spiral arms of a galaxy stand out not because they are the only places stars are found but because they are outlined by giant blue O and B stars. We can attempt to see spiral structure by trying to locate stars of that type. In our galaxy O and B giants are found which appear to form parts of three approximately parallel lines, presumably spiral arms. We live in one of the arms. It is called the Local arm or Orion arm because the Orion nebula is in it. About 6,000 light years farther from the center of the galaxy we see part of the Perseus arm; some 4,000 or 5,000 light years nearer the center is another string of blue stars, marking the Sagittarius arm.

Spiral arms shown by hydrogen clouds Additional evidence for spiral structure comes from radio telescopic observation of neutral hydrogen by the 21-cm line. As we look with a radio telescope along the plane of the galaxy, perhaps some-what toward the center but not directly toward it, we see evidence of clouds of hydrogen with several different doppler shifts. We can explain the presence of several clouds with different velocities if we assume (1) that the hydrogen is concentrated in the spiral arms, and (2) that the galaxy is rotating, so that the velocity the hydrogen has with respect to us reflects at least partly the differing speeds of a rotating disk of stars. The assumptions are plausible because they are good descriptions of the conditions we see in other galaxies, and this interpretation of the 21-cm radiation tells us that our galaxy is of the spiral type, although many details are not known.

Rotation

Other evidence for the rotation of the galaxy comes from a study of nearby stars. If stars near the sun are moving on approximately Keplerian orbits, we should expect stars nearer the center of the galaxy than we are to be moving faster than we, and stars farther out from the center to be moving more slowly. If we look at many stars in these positions, taking an average speed to eliminate the effect of the individual star's own motion, we find this expectation borne out. Apparently we are all riding a colossal merry-go-round. We are approximately 30,000 light years from the center and we are moving about 250 km/s because of the rotation. We probably make one complete circuit in about 250 million years.

Mass

In the neighborhood of the sun, the density of stars in space is approximately 0.12 star per cubic parsec. Estimating the volume of our disk, we compute the number of stars to be about 2×10^{11}. Since the average mass of all stars is probably about one-half the mass of the sun because of the great number of small red dwarf stars, we estimate that the mass of all the stars is about 10^{11} solar masses. This omits the mass of the interstellar gas, which cannot be negligible.

If we treat the sun as a planet moving in an orbit about a gravitational center, we can ask what the mass at the center must be for a body to take 250 million years to make a circuit of a path having a radius of 30,000 light years. Carrying out the

(a)

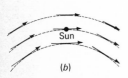

(b)

FIGURE 12.3 Part (a) shows the line of sight of a radio telescope seeing three clouds of hydrogen, each of which has a different radial velocity relative to us because of the rotation of the galaxy. Part (b) shows the relative velocities of stars near us because they are nearer or farther from the center of the galaxy.

computation, we get a mass of a little over 10^{11} solar masses. If we make a more careful computation, taking into account the fact that the mass of the galaxy is not all concentrated at the center, and that it is not arranged in a spherically symmetric distribution, we still find that the mass required to produce the observed motion of the stars on the periphery is of the order of 10^{11} solar masses.

Nucleus

When we observe other galaxies which we suspect are very much like our own—the Andromeda galaxy, for example—we see that they have small, dense nuclei. We cannot see the nucleus of our own galaxy in visible light, but we have reason to believe that a nucleus exists.

Observations by visible light in the few places where we can look through the dust near the plane of the galaxy suggest that near the nucleus there is a bulge of stars perhaps 3,000 light years in radius. Within that central bulge both infrared and radio observations suggest the existence of a nucleus about 30 light years in diameter. Infrared light can penetrate the dust to greater distances than can visible light, and observations in infrared show a small bright spot where we should expect the nucleus to be. Its nature is not known; it may be a dense cluster of bright stars, and part of the infrared radiation may be radiation from dust in the nucleus which is heated by the light from those stars.

The radio source known as Sagittarius A lies in the direction of the nucleus, and the strongest part of the source is thought to be the nucleus itself. In the direction of the nucleus we also find clouds with concentrations of molecules and clouds of hydrogen moving outward at high speed. It may be that the nucleus of our galaxy is the scene of violent activity, as is true of the nuclei of some other galaxies we can see, but we are unsure of its extent and we know nothing of its cause.

POPULATIONS I AND II

The stars and other components of the galaxy can be divided into four classes based on their position: halo, disk, spiral arm, and nucleus. But they can be divided on the basis of other criteria into what are called Population I and Population II objects. Those names were suggested by Walter Baade in 1944 as a result of his study of the stars in the Andromeda galaxy, but it is recognized now that two categories alone are not adequate. Various subdivisions are used, with Population I stars at one end of a continuum and Population II stars at the other. One ordering sometimes used is: extreme Population I, old Population I, young Population II, and extreme Population II. Another is Population I, Disk Population, and Population II. The populations differ in at least four characteristics, and if one understands these differences, the various names are self-explanatory.

Metal Content

The most fundamental difference between the populations is the concentration of metals revealed by their spectra, such as magnesium, calcium, iron, and nickel. Often the concentration of only those elements heavier than helium is considered: Extreme Population I objects are metal-rich, and extreme Population II objects are metal-poor. The concentration of all the elements heavier than helium in Population I objects may be more than 10 times as great as in Population II objects, and the concentrations of individual elements (iron, for example) may be more than 1,000 times as great. The sun is an old Population I star; that means that it does not have as many heavy elements as some stars which we see, but that it has far more than those considered Population II stars.

Location

The second difference, the one which caused the initial division into two categories, is location. The stars and clusters in the halo are Population II objects. All the stars which can be observed in globular clusters are metal-poor, as are the individual halo stars. The stars which occupy the outer parts of the disk are intermediate; they have more metal than the stars in the halo, but not so much as the objects in the thin inner layer of the disk. They are sometimes called Disk Population objects. Objects in the very thin center layer are Population I. Population I objects are particularly concentrated in the spiral arms; they include gaseous nebulae, O and B stars, T Tauri stars, and galactic clusters.

Velocity

The third characteristic which serves to distinguish members of the populations is velocity relative to the sun. This is not an important distinction, but it helps confirm the explanation for

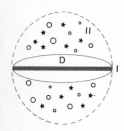

FIGURE 12.4 Cross section of the galaxy. The region labeled I is the thin layer within which are found Population I objects. The spiral arms are within this layer. The region labeled D contains Disk Population stars. The region labeled II contains globular clusters and halo stars—Population II objects.

the origin of the populations. Stars moving at low speed relative to the sun tend to be Population I stars; those moving at high speed tend to be Population II stars. Remember that the sun is moving at a considerable speed around the galaxy, and that stars which have low velocities relative to us are moving about the galaxy at that same high speed. Stars which appear to us to have high speeds may actually be moving around at low speeds. In particular, stars which spend much of their lives in the halo but occasionally pass through the disk will appear to us to be high-velocity stars if we happen to be near them as they are going through the disk approximately at right angles to our path. Since halo stars are Population II stars, it is not surprising that those stars presently near us which have high velocities relative to us should also be Population II.

Age

The fourth distinction between the populations—age—is inferred. We believe that extreme Population II stars are the oldest in the galaxy, and extreme Population I stars are the youngest. If we say that the sun is an old Population I star, we are saying that it is not old enough to be Population II, but it is not one of the youngest in Population I.

EVOLUTION

The observed characteristics of the galaxy can be understood in terms of an evolutionary scenario. We presume that originally the galaxy was a huge, tenuous mass of slowly rotating gas—primarily hydrogen with perhaps some helium—which began gravitational collapse some 10 billion years ago. As the density increased because of the collapse, stars began to form. Those earliest stars formed of only hydrogen and helium, before metals had been created, and they still show few metals in their atmospheres. We do not know if any of the very first generation of stars are visible; in any case, they are extreme Population II. Halo stars and globular clusters date from that earliest period. At the time they formed, the galaxy was approximately spherical, and consequently their orbits take them around the center in directions which may be inclined at large angles to the plane of the disk.

As time went by, the spherical mass of gas collapsed further, and because it had some rotational motion from the first, it rotated faster as its size decreased. The gas, which was free to continue contracting, moved inward toward the central plane and left the halo stars and clusters to pursue their lonely orbits. As the rotational speed increased, gas in the equatorial plane became unable to respond further to the gravitational pull

because of its circular motion, and contraction along the equator ceased. Nothing provided such a limit on collapse in the
other direction, however, and gradually the spherical shape
changed to oblate spheroidal, then to a thick disk, and finally to
a very thin disk. Now all the gas and dust from which new stars
are made is in a very thin layer in the disk, and all the new stars
are in the same region.

Perhaps early in the history of the galaxy giant stars formed
and went through their life cycles quickly, spilling metals
formed within their interiors out into space, so that before the
gas of the galaxy collapsed to a disk it was enriched by mixing
with the metals formed in the first-generation stars. As the disk
became thinner, more stars returned metals they had produced
to the gas of interstellar space, so that the material from which
new stars were forming constantly became richer and richer in
metals. Thus the concentration of metals in a star depends
primarily on when the star formed, so that a star's population
characterization is an indication of its age. Stars such as the
sun were formed from primeval gas and recycled metals.

It appears that eventually our galaxy will have all its material bound up in stars, or at least so much bound up that new
stars cannot form. After that, we can anticipate only a slow
dying. First, the blue and white giants will burn out, not to be
replaced, and later even the yellow and red stars will come to
the ends of their lives.

QUESTIONS

1 Why is it reasonable to suppose that the globular clusters
are arranged rather symmetrically about the center of mass of
the galaxy?

2 If the globular clusters move on highly eccentric orbits about
the center of the galaxy, rising far above the plane of the disk,
where do they spend most of their time? Why?

3 If you were drawing the disk of the galaxy to scale, and it
had a diameter of 10 cm, what would be the thickness?

Ans ≈ 3 mm

4 Describe the distinguishing characteristics of globular clusters and of galactic clusters.

5 On the basis of the discussion in the text of the light-scatter-
ing properties of particles of different sizes, explain why the
sky is more blue when the air is clean than when it is filled
with water or dust particles which are much larger than air
molecules.

6 Suppose that all the lines except one observed in the spectrum of a star have doppler shifts, indicating that the star is
moving away from us at a moderate speed. The one line shows

no doppler displacement. What might you suspect is the reason the one line does not show the same shift?

7 Look at the picture of NGC 4565 (p. 151). Assume that it is like our galaxy, and the diameter of the disk is the same as ours, and measure the photograph to compute the thickness of the disk and the diameter of the nucleus.

8 Assume that the sun can be treated like a planet moving about a mass concentrated at the center of the galaxy. Using the period and the distance given in the text, compute the mass of the galaxy.

9 Explain how the evolutionary description of the galaxy explains the existence and positions of stars of different populations.

13
stellar evolution

"The unchanging stars" have been a symbol much beloved by poets, and to the casual observer the stars do indeed appear to be unchanging. But even slight reflection will convince us that they must change, and, in fact, that all stars have finite lifetimes. Stars shine by converting mass into energy, and eventually that process progresses in each star as far as it can go, until the energy with which to illuminate the sky is no longer available. None of the stars now shining can have been shining forever, and none can last forever. But the stars we see do not come with their ages printed on them, nor do we have records going back millions of years to compare with present observations, and therefore we must use indirect methods as we try to understand how stars change as they are born, reach maturity, become old, and die.

LIFE EXPECTANCIES OF STARS

Stars change very slowly during most of their evolution, and the time we have been watching them is like an instant in the life of a man. Understanding the evolution of stars would be almost impossible if they did not develop in accordance with some relatively simple physical laws. Stars are gas, and the physical laws describing the behavior of spheres of gas are rather simple and are well understood. Stars are held together by gravity; their collapse under gravity is prevented by the development of high temperatures in the cores. Their energy is released by gravitational contraction and by nuclear fusion; all these processes are understood well enough that we can construct models of stars and compute how they should change with time. In the process of trying to understand the evolution of stars, therefore, we construct models and determine how our models age, and then try to find examples among the visible stars corresponding to the different ages and stages of stellar life we have predicted.

This program of action depends on the assumption that we see stars of different ages. We see stars which are hot, blue, class O, or class B. We see red giants, and red dwarfs and white dwarfs. We see stars which are variable, either regularly such as

Cepheids or explosively such as novae and supernovae. Which of those characteristics are functions of age and which have other explanations? Are any of them functions of age? Why should we not suppose that the stars which make up our Milky Way galaxy all formed at about the same time and are hence all about the same age? More evidence will be considered later, but one simple answer to the question can be given now. During an important part of every star's life, it derives its energy from the conversion of hydrogen into helium, as the sun is doing now. Every star begins life with a certain amount of hydrogen, and when the hydrogen (or some fraction of it if the star does not mix its surface hydrogen into the core) is burned up, the star must change in some radical way. A reasonable guess would be that the more hydrogen a star has the longer it can burn hydrogen, but things do not work that way, for the more massive a star is the brighter it is, and the luminosity increases much more rapidly than the mass (see Fig. 10.10). The sun (1 solar mass) is estimated to have a hydrogen-burning lifetime of 10^{10} years. Probably the most massive stars are not more than 50 solar masses, and if a star with 50 times the sun's mass burned hydrogen at the same rate as the sun it would last 50×10^{10} years. But a star 50 times as massive as the sun burns hydrogen nearly 1 million times faster than the sun, and so we might expect its life to be about 500,000 years. (Table 13.1 shows the life expectancies of stars of different masses.) We think that the sun has been shining very much as it is now for about 5 billion years, and clearly the most massive and bright stars we see cannot have been shining at their present rate for anything like that long. They must, therefore, be younger than the sun.

The environment in which these massive, bright stars are found lends support to the belief that they are young, for the hot, blue stars are found in or near clouds of gas and dust—ma-

TABLE 13.1   APPROXIMATE MAIN-SEQUENCE LIFETIMES OF STARS OF DIFFERING MASSES

Spectral type	Mass $(M_\odot = 1)$	Luminosity $(L_\odot = 1)$	Life on main sequence, yr
07.5	30	8×10^4	4×10^6
B0	16	1×10^4	2×10^7
A0	3	60	6×10^8
F0	1.5	6	3×10^9
G2	1	1	1×10^{10}
K0	0.8	0.4	2×10^{10}
M0	0.5	0.06	9×10^{10}
M5	0.2	0.01	2×10^{11}

terial from which we presume stars might be made. At the other end of the evolutionary process are white dwarfs, in which we think no nuclear processes are occurring, and which are dead so far as development is concerned. Evidently they have reached some kind of stellar death. And so we know something about some stars at the ends of their lives. Do all stars begin as blue giants or supergiants? Do they all end as white dwarfs? How do they get from one stage to another?

STAR MODELS

Our knowledge of the evolution of stars has been helped immeasurably by the development of high-speed computers. We set up a model of a star by imagining it divided into thin shells, something like an onion, and then consider the outermost shell first. Energy is not being released within that shell, and all the energy which enters it at the bottom must emerge at the top. The rate at which that energy flows out is the luminosity of the star. But the rate at which energy is transported across the shell depends upon whether the transfer is by radiation or convection, and upon the difference in temperature across the shell and the opacity of the gas of which the shell is made. The opacity, in turn, depends upon the composition of the shell, particularly the concentrations of elements heavier than helium. Further, the pressure upward at the bottom of the shell is equal to the weight downward of the shell. When assumptions are made about the composition of the gas, the equations which relate all these variables can be solved, and the shell's temperature, density, and pressure at its lower surface can be computed.

Continuing with the analysis of our model, we then consider the next shell downward. The same process is repeated, but the conditions at the upper surface of this shell must be the same as those at the bottom of the first shell. The pressure at its inner surface must be just enough to support the weight of the layers above. The process is continued downward toward the center of the star until the region is reached where nuclear reactions are releasing energy. Below that point the energy flowing out of a shell upward must be equal to that flowing in from the bottom plus that released within the shell. In this region the compositions of the shells begin to change because of the effect of the nuclear reactions which have occurred in the past. When we finally reach the center, then, we know the conditions there (temperature, pressure, density, composition), and how much energy is being released within the star.

Such step-by-step calculations were tedious and slow before

the advent of electronic computers; now they can be done rapidly. It should be noted that the process is not so simple as this description makes it sound. If we begin with a set of assumptions about the luminosity and composition, and find, after working our way to the center of the star, that the energy being released is too much or too little to produce the luminosity required at the outside, we must work back through the chain of computations with appropriate modifications until everything is consistent.

After a satisfactory model of a star has been developed, it can be allowed to evolve. If the original assumption was that the star began as almost pure hydrogen with trace amounts of other materials, we can recompute the model for the condition which will exist when part of the hydrogen in the core has been converted into helium. As time passes, the amount of helium in the core increases, and hence models computed for later and later stages must be different from the original.

In considering the results of computer-aided studies of models, we must remember that details of the models depend upon assumptions about composition and opacities, both of which are known poorly. Also, the models neglect mass loss from the stars during their lives, as well as the effects of such factors as magnetic fields and rotation. Different sets of assumptions and different computer programs have been used to follow the evolution of stars like the sun or stars having masses somewhat higher than the sun's, and the results are similar, although not identical. One computation may predict that a certain phase of a star's life will require 5×10^8 years and another program may give 6×10^8 years for the same phase, but the difference is not significant.

EVOLUTION OF STARS OF 1 SOLAR MASS

Let us consider first the probable evolution of a star the mass of the sun. Exact details of its development depend upon the concentrations of such elements as lithium, carbon, and oxygen, and when we construct a model, we make assumptions about what those concentrations are likely to have been. Different stars of the same mass may have had slightly different concentrations of those elements, and therefore their evolution may be described more or less well by the model.

Protostar

Here and there in space we see huge clouds of gas and dust, and we believe that all stars form within such clouds. Imagine a spherical region within such a cloud—a region large enough that the entire solar system could be placed within it. The density of gas within the cloud is very low, but because of random movements of the gas, the density of matter within our imaginary sphere fluctuates. Now suppose that at some time the density of gas within the sphere is considerably above average, so that more mass than normal is within the sphere. This mass exerts a gravitational attraction upon the gas at the edge of the sphere, and if that attraction is great enough, more gas flows into the volume than is leaving it by the normal random movement of gas molecules. As the net inflow continues, the gas within the volume begins to contract because of gravitation, and when it has progressed to the point that the random movements of the parent cloud will not cause it to disperse, it may be said to have begun to form a *protostar*. It is not yet visible, however, for the gas is almost if not entirely transparent, and the collapsing material has not separated itself enough from the larger cloud to appear different.

If the protostar has the same mass as the sun, it becomes opaque when it has a radius somewhat larger than the radius of the orbit of Pluto and when the temperature is a few kelvins. Dark spots, called *globules*, can be seen in some photographs of interstellar clouds, and they are believed to be protostars in this early stage of development. It may take a million years from the time a protostar separates from the parent cloud until it becomes opaque, but after it becomes opaque, it collapses fast enough, perhaps 100 m/s at the outer edge, that it goes through the next stage of its development in a few years.

Onto the H-R Diagram

When the radius of the collapsing cloud has decreased to 1 AU, the central temperature is of the order of 2000 K and the protostar is beginning to radiate enough energy to become visible. It is very faint and very red, but its luminosity is increasing rapidly. Within months the internal pressure has become great enough to halt the rapid collapse of the outer layers, and the luminosity has become several hundred times that of the sun. The surface temperature is not higher than that of the sun, and it may be somewhat lower, but the star has a radius of about 0.25 AU and so is very bright. During the final months of its collapse, the star has temperature and luminosity which place it on the H-R diagram, and its motion can be represented as in Fig. 13.1.

On the Main Sequence

After the luminosity reaches its peak, the star undergoes internal adjustments. A core develops in which convection is not

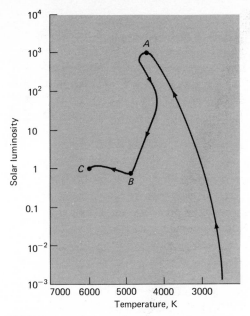

FIGURE 13.1 Path of a star of 1 solar mass to the main sequence. Point A is reached in months. The time from A to B is of the order of 10 million years, and from B to C is another 30 million years. At C the star is on the main sequence.

the main means of transferring energy, and a slow period of maturation begins. During that time the luminosity declines slowly so that the point on the H-R diagram which represents the star moves generally downward. The stages the star passes through on its way to the main sequence and the approximate times are given in Fig. 13.1. When a star of the sun's mass reaches approximately the position on the H-R diagram now occupied by the sun, it is said to have reached the main sequence.

A star having the mass of the sun spends its mature middle years on the main sequence. With time it may become slightly more luminous, but no significant changes occur within it for about 10 billion years. During that time the energy of the star is produced by the burning of hydrogen to helium within the core, but a star the mass of the sun does not mix material from the outer layers into the core, so that the core gradually becomes depleted of hydrogen and enriched with helium. At the end of the 10 billion year period of stability, the core will include about 10 percent of the mass of the star. At this point, the core will be so nearly pure helium that hydrogen burning will virtually cease within it, although some hydrogen burning continues in a thin shell outward from the core.

Red Giant Stage

As hydrogen burning decreases within the core, less energy is released, and the tendency of the core is to cool. When the temperature begins to fall, the pressure drops greatly, and the weight of the layers above compresses the core. The compression raises the temperature, so that instead of cooling as hydrogen burning begins to stop, the core actually becomes hotter. The hydrogen burning taking place in the shell around the outside of the core continues, but as the core shrinks, the shell becomes thinner. It may seem that the higher temperature should cause the core to expand to its original size, but that does not happen. If the core expanded, its temperature would drop and the pressure would again go down. What happens is that as the core becomes more compressed and dense, the force on its outer layers becomes ever greater. The system acts as if all the core's mass were at the center, and the force on the outer layers of the core increases as $1/r^2$ as it shrinks. Thus ever higher temperatures are required to produce the pressure which can support its own weight plus the weight of the outer layers of the star.

The extra radiation streaming out from the collapsing core pushes on the outer layers of the star so that they begin to expand. The first effect is to make the star more luminous, but in time, after perhaps 2 billion years, the outer layers of the star expand to such a size that their temperature begins to drop, and while the core becomes hotter and hotter, the outer layers become cooler and redder. On the H-R diagram, the point representing the star begins to move to the right; it now moves rather rapidly, and the almost horizontal rightward movement occupies only about 1 billion years. Then the star moves upward rapidly to become a red giant, the upward movement into the red giant region requiring only a few million years.

By the time a star becomes a red giant, its radius is more than 50 times that of the sun. Its surface temperature has dropped enough that it has changed from yellow to red, but the greater surface area causes the luminosity to be about 1,000 times that of the sun. The outer layers of a red giant are so tenuous that they would behave as a good vacuum if they could be transported to the earth.

Helium Burning

By the time the star becomes a red giant, its core has a temperature of about 100 million K. At such a temperature helium nuclei can combine to form beryllium 8 by the reaction

$$^4_2\text{He} + {}^4_2\text{He} \rightarrow {}^8_4\text{Be} + \gamma$$

The beryllium-8 nucleus is very unstable and almost immediately disintegrates again into two helium nuclei, but after the
temperature has become high enough that the nuclei can be
formed, some are present at all times. Occasionally before a be-
ryllium-8 nucleus splits back into helium nuclei it is struck by a
third helium nucleus, which combines with the beryllium to
form carbon 12 ($^{12}_{6}C$). The process by which three helium nuclei
(alpha particles) are changed to one carbon nucleus is called
the *triple-alpha process*. The carbon 12 is stable, and in the
core the star begins to convert helium to carbon.

When helium burning begins, the point on the H-R diagram
representing the star begins to move down and to the left again,
back toward the main-sequence line, as the outer temperatures
increase but the luminosity decreases (see Fig. 13.2). Some
computations indicate that in stars of about 1 solar mass the
helium burning is explosive, sometimes beginning and blowing
the core of the star apart within a matter of hours. That rapid
burning is called the *helium flash*. The effect of disrupting the
helium core is to reduce the rate of hydrogen burning in the
shell around the core so that the conditions which produced
the expansion to red giant stage are partially reversed and the
star almost retraces its path as it moves downward. Very possi-

FIGURE 13.2 The path of a star of 1 solar mass from
the main sequence into the red giant region. The movement downward from the peak is initiated by the
helium flash.

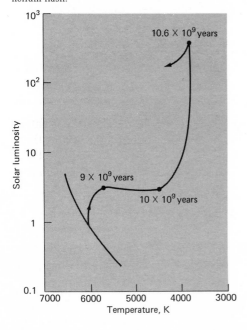

bly a new core forms as the star moves up and to the right on the diagram again, and the process may be repeated. The repetition cannot be exact, however, for the inner regions of the star still have the elements made during the first helium burning, and their presence provides new initial conditions for later stages of development.

White Dwarf

In order to trace the evolution of a star of 1 solar mass beyond the helium-burning stage, the computations for the model must be made for very small increments of time, and the computer time required is so great that detailed calculations have not been carried out. It is believed that the core eventually uses up all its helium, leaving only heavier elements. Even before that condition has been reached, however, electron degeneracy becomes important in the core, and the pressure arising from the exclusion principle (to be discussed in the next section) supplies about half the pressure required to support the outer parts of the star. It is not known whether the path of the star is a steady progression or whether it moves back and forth across the H-R diagram, but we do know that the star eventually becomes a white dwarf, located in the lower left region on the H-R diagram. When the star has reached that position, most of its hydrogen and helium have been converted into heavier elements, and the nuclear fires in its interior have gone out. The star may still be quite hot, but no energy is being released within it and all its radiation comes from its store of thermal energy. As that store is exhausted, the star becomes a *black dwarf*. The final cooling is so slow, however, that it is possible that no star within our galaxy has had time to become a black dwarf. A star of 1 solar mass would become a white dwarf approximately as large as the earth, and no further changes would occur in its size as it slowly cooled. The density of a white dwarf is such that one cubic inch of it, if brought to the earth, would weigh 10 tons.

Electron degeneracy The structure of a white dwarf or of the core of a massive star in which degeneracy forces are becoming important is quite different from that of an ordinary star. In the ordinary star, the force which prevents collapse under the weight of the outer layers of the star is the pressure of hot gas. Under the conditions existing throughout most of the volume of most stars, the electrons and nuclei of the elements present move at high speeds because of the temperature, and their collisions with one another produce the force which prevents their being compressed into an excessively small volume.

As the material in a stellar core is crowded into a smaller and smaller volume after the exhaustion of hydrogen, the interaction of electrons with one another becomes the most important interaction, and eventually it stops the compression. The entire core of the star becomes like one giant atom, and the same restriction which prohibits any two electrons within an atom having identical descriptions prevents the electrons within a small volume from having identical energies and momentums. The necessity that electrons be different from one another in energy and momentum (the *exclusion principle*) produces the condition known as degeneracy. If the mass of the star is not so great that its weight overwhelms the repulsion forces produced by degeneracy, those forces stop the contraction of the star. In a white dwarf, the forces countering the weight of the outer parts of the star are those arising from electron degeneracy.

The electrons in a white dwarf cannot lose energy, because if an electron lost energy it would become just like another electron which already has that lower energy, and that duplication does not occur. But the nuclei of the atoms, which are crowded in among the electrons, can lose energy, and it is their energy which is radiated away from a white dwarf.

Effect on Earth of Solar Evolution

The effects on earth of the changes in the sun as it ends its main-sequence life will be drastic. When the sun begins its march toward red giant status, its luminosity will increase greatly, and any life on earth will find the new conditions increasingly difficult. By the time the sun is a red giant it will have expanded enough to reach almost to the orbit of Mercury, and in fact it may have engulfed that planet. By then it will be sending the earth about 1,000 times as much radiation as we now receive, and temperatures on the earth will have climbed to far above the boiling point of water. (Of course, the inhabitants of earth will have a grace period of hundreds of thousands of years after the sun begins its rise up the H-R diagram toward the red giant region during which they can make plans and preparations to leave the earth.) Later, when the sun has shrunk to the white dwarf stage, the earth will be very cold.

STARS WITH GREATER MASS THAN SUN

We have an obvious interest in studying the evolution of stars of the mass of the sun, for our future is tied to their development. The evolution of stars somewhat less massive than the sun (0.7 solar mass, for example) or slightly more massive (say 1.5 solar masses) has been studied in detail; it is not signifi-

cantly different from the evolution of stars like the sun. More
massive stars have convective cores during the time they are on
tha main sequence, and more of their energy is generated by the
CNO cycle than in a star like the sun, but the only important
difference in the history of the star which can be observed from
the outside is the rapidity with which it moves through the
stages of its life. Instead of 10 billion years on the main
sequence, a star of 1.5 solar masses spends about 2 billion
years, and all its other stages of development are similarly shortened. A star of less than 1 solar mass takes longer than the sun
to pass through the same stages of life.

Stars with much less mass than the sun have not been studied with the same care, for we have little chance of comparing
the predictions of computations with observations. Stars of
very low mass evolve so slowly that there may be none visible
which have left the main sequence. Very massive stars, on the
other hand, are of great interest because they evolve so rapidly
that we can hope to see them in all stages of development. Our
best chance to check theory against observation is by studying
those stars of 3 or more solar masses.

The position at which a new star eventually reaches the
main-sequence line depends almost entirely upon its mass. The
more massive the star, the higher on the line it comes to rest,
and the shorter the time it rests there before moving upward

FIGURE 13.3 Path of a massive star after it leaves the
main sequence.

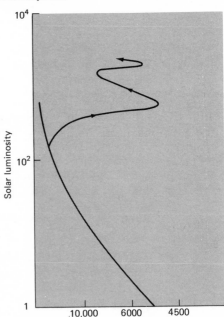

and to the right into the red giant region. A star of 5 solar masses, for example, is on the main sequence for only about 100 million years before beginning to move upward. The sequence of events in its core and the hydrogen-burning shell around the core are similar to those of the star of 1 solar mass except that everything happens much faster, and the eventual outcome may be different. The star of 5 solar masses may go through several stages of burning light elements to heavier ones as it uses up first hydrogen, then helium, then carbon, etc. The ignition of each stage causes a drift to the left in the diagram, and the star seems to move back and forth several times, as shown in Fig. 13.3. During those back-and-forth movements, the star passes through the region on the H-R diagram occupied by Cepheid variables, and it seems likely that Cepheid variables are stars passing through a phase of instability; it is likely that such a phase lasts a small part of the star's life. The later stages of the life of a star this massive have not been followed carefully with models, and we have many questions about the demise of massive stars for which we do not have satisfactory answers. Such answers as we do have will be discussed in a later section.

COMPARISON OF MODELS WITH OBSERVATIONS

We need to describe the comparisons of computer models with observations in some detail, since they provide the evidence which causes us to believe that the models have enough resemblance to reality to make them worth discussing.

Stellar Clusters and Associations

Much of the support for our hypotheses concerning the evolution of stars comes from the study of stellar clusters and associations. Galactic clusters and globular clusters were described in the previous unit; *stellar associations* are another kind of grouping.

Galactic clusters are of value in testing our models of evolving stars because they all formed from a common cloud of material, in about the same region of space, at about the same time. We may be reasonably confident that they all began with very nearly the same composition and that they are nearly the same age; differences we see must be the results of different rates of evolving. Globular clusters appear to have formed early in the history of the galaxy, and all the stars in one of the clusters must be approximately the same age. They are all very old stars.

Stellar associations are groups containing one type of star in greater than average numbers. The most important types of associations are the O associations, made up of O and B stars, and the T Tauri associations, made up of T Tauri stars. T Tauri stars are variable stars which are believed to be very young, perhaps not yet evolved onto the main sequence. They are always found associated with clouds of gas and dust, and they may also be associated with O and B stars. The existence of such associations can be explained by the assumption that a high density of T Tauri stars, and O and B stars, is found in regions where large numbers of stars are being born. By the time those stars have moved far from the places of their birth, they will have evolved past the point that they are O or B type, for all O and B stars are massive, and have rather short lives on the main sequence.

The Pleiades

To begin to see how study of stellar clusters is related to the making of computations describing the evolution of stars, let us consider the Pleiades cluster. It is near enough that the stars within it can be identified with reasonable certainty, and it contains enough stars to give us a reasonable sample. Two features of the H-R diagram of the Pleiades (Fig. 13.4) are important to notice: The number of A and B stars in the Pleiades is higher than would be true in a sample of the same number of stars taken in the volume immediately surrounding the sun, and the cluster contains no giants, supergiants, or white dwarfs. Possibly we would not *see* white dwarfs, but the ab-

FIGURE 13.4 Approximate H-R diagram for the Pleiades, an open cluster. Note the large number of very bright stars and the absence of red giants.

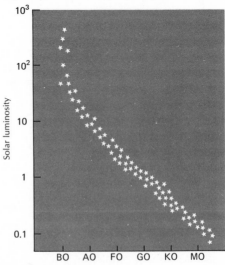

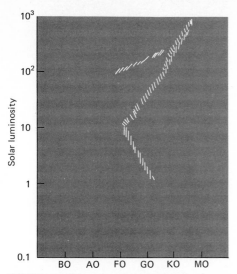

FIGURE 13.5 H-R diagram for M 3, a globular cluster.
Note the absence of main-sequence stars brighter and
hotter than F type and the presence of many red giants.
The horizontal branch is also noticeable.

sence of visible giants indicates that there are none in the cluster. How can we interpret this description of the characteristics
of the stars in the Pleiades?

The absence of red giants or white dwarfs suggests that the
cluster is not old enough for those stars to have developed. The
presence of an abnormally high number of A and B stars also
suggests that the cluster is young, for these stars remain on the
main sequence for short times. The dust which surrounds some
of the bright stars in the Pleiades and appears as reflection
nebulae lends support to the idea that the stars may be young.
The cluster is not so young, however, that stars of low mass
have not had time to evolve to the main sequence; K- and
M-type stars appear to have reached their main-sequence positions. It seems reasonable to conclude, therefore, that the
Pleiades is a young cluster, but not an infant group.

Globular Clusters

To see what the effects of age are, let us consider a globular
cluster. The H-R diagram for the globular cluster M 3 is shown
in Fig. 13.5. The differences between this diagram and the one
for the Pleiades are striking. The main sequence for M 3 terminates at approximately F-type stars; O, B, and A stars are missing. The red giant region is heavily populated. All the clues visible on this diagram suggest that the cluster is old. We may
presume that at one time its main sequence was completely

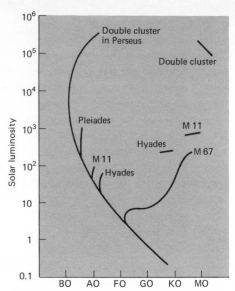

FIGURE 13.6 Composite H-R diagram for several clusters. The Double Cluster in Perseus is the youngest; M 67 is the oldest.

populated, but the O stars used up their hydrogen and moved to the right to become red giants. A short time later the B stars went the same route, and still later the A stars evolved off the main sequence. All those stars now appear as red giants or they are crossing the diagram to the left, headed toward the white dwarf region. (They are said to be in the *horizontal branch*.) Perhaps the cluster contains white dwarfs, but we cannot see them. We presume that if we look at this cluster at a still later time, the F stars will have left the main sequence and the turn-off point will have dropped lower.

Figure 13.6 shows a composite diagram for many clusters, and it can be seen that they have quite different turn-off points. This difference is interpreted as the effect of differing ages—the younger clusters having higher turn-off points from the main sequence—and it is consistent with our models for the development of stars. On the basis of the models and the time we predict stars should require to evolve off the main sequence, we can estimate ages for the various clusters.

DEATHS OF STARS

We have considered the birth and maturity of stars; now we must give some attention to their old age and deaths, considering in more detail the changes in internal structure which occur as stars evolve off the main sequence.

Very Low Mass

Stars with mass below about 0.08 solar mass probably never develop temperatures in their cores high enough to cause hydrogen burning. As they condense from the primeval cloud, their internal temperatures rise and they become luminous, but they never reach the main sequence because they never ignite the hydrogen-to-helium process. Instead, the interior of the star becomes more and more dense, until finally electron degeneracy prevents further collapse. Until the collapse is stopped, the star is changing gravitational energy to heat and radiation. After the collapse ceases, no more energy is released, and the star is in the white dwarf stage.

Stars of 1 Solar Mass

The end of a star of 1 solar mass has been discussed. After the hydrogen in the core has been exhausted, the core shrinks and helium burns to carbon, while hydrogen continues to burn in a shell outside the core.

When the helium in the core has all been burned to carbon, helium burning may continue in a shell outside the core, just as hydrogen burning may continue in a shell higher up in the star. The next important set of nuclear reactions burn carbon to form heavier elements, but they require a temperature of about 200 million K, and it is unlikely that a star the mass of the sun will reach such a temperature. It may undergo explosive burning, which will eject part of the mass to form something like a planetary nebula, but whether or not that happens, most of the star's mass eventually becomes a white dwarf. In that state, it finds final rest, cooling slowly.

Stars Greater Than 1 Solar Mass

A star with a mass greater than 1.2 solar masses (a value known as the *Chandrasekhar limit*) cannot move across the H-R diagram and downward to the white dwarf stage, a possible course of development for less massive stars, and the fate of stars with greater mass than 1.2 holds particular interest.

Planetary nebulae During all their lives stars lose mass, both by converting it to energy and by "evaporation" of surface material, but the rate of loss for a star like the sun is quite low. Some stars are losing mass at a great rate but are not exploding; these are the stars forming *planetary nebulae*. The stars at the center of nebulae are hot—perhaps hotter than any of the stars on the main sequence. When they are plotted on the H-R diagram they fall to the left of the main-sequence line but above the white dwarf region. There is reason to suppose that these stars have exhausted the nuclear burning possible within them and that radiation pressure or turbulent motions within the stars are forcing their outer layers out into space. We see the shells because the ejected matter is still near enough to the parent star to fluoresce under the intense illumination from the central star. We presume that by the time the process of mass loss has ended, the star has brought its mass below the Chandrasekhar limit and is ready to become a white dwarf.

Supernovae If the mass of a star is sufficiently low, degeneracy of the electrons in the core can prevent the contraction that would follow the completion of one phase of nuclear burning, so that the core never becomes hot enough to ignite the next stage. Thus the core may be left all helium or all carbon, and the star will end its life with shell burning the only source of energy. But if the star is so massive that degeneracy of electrons cannot prevent the contraction, hydrogen burning is followed by helium burning, which is followed by carbon burning, and so on until the core is composed of the elements of the iron group—iron, cobalt, and nickel. At this point the generation of energy by the formation of ever-heavier nuclei stops, for in order to form the elements lying above iron in the periodic table, energy must be added to the system to make the reaction go. What happens to stars beyond this point is not known with certainty, except that some of them become supernovae. Two different mechanisms have been proposed as triggers of supernova explosions: Both provide explanations accounting for a sudden cooling of the core. One suggests that as the core temperatures reach the range of a few billion degrees, nuclear reactions which release neutrinos become common. The neutrinos can escape from the star almost as if the overlying layers were not present, and so they carry away energy at a rate which photons cannot match. In fact, they carry away energy so rapidly that the core is suddenly cooled below the point at which its pressure can support the layers above, and it collapses. The collapse causes temperatures to shoot up abruptly, and the result is an explosion which thrusts most of the mass of the star out into space, leaving a small, very dense core. The other suggestion is that as temperatures reach the few billion kelvin range, when the core is almost pure iron, the energy of photons becomes great enough to begin to break up the iron nuclei—a process called *photodisintegration*. The effect is the same as energy loss by neutrinos, for the nuclei which are broken absorb energy, cooling the core and precipitating collapse and reexplosion.

The star which is left, at least in some cases, is called a *neutron star*. Its further collapse is prevented by degeneracy of neutrons rather than electrons. In the moment of collapse, electrons are forced to combine with protons, forming neutrons, and the star which remains is almost like one gigantic atomic nucleus. It has no energy sources within it, and it is merely cooling, but further collapse is prevented by the degeneracy forces acting on the neutrons which are its most plentiful constituent. What we know about neutron stars has been learned by studying pulsars, and they will be discussed further in connection with pulsars.

The time scale for supernova collapse and explosion is startling after one has been considering changes measured in millions and billions of years. After the core of a star cools enough that it cannot sustain the weight of the atmosphere above it, a cooling process that may take a few seconds or minutes, the collapse of the core occupies a few seconds. More time is required for the shock wave to propagate outward to the outer layers of the star, and the star's brilliance will rise more slowly yet because of the time required for the outer layers to expand, but the damage to the core which makes all the later changes inevitable occurs in seconds.

Just as the Chandrasekhar limit gives a mass above which a star cannot be a white dwarf, there is a limit, as yet unnamed, above which a star cannot be a neutron star. Computations of that limit are not as easy as for the case of electron degeneracy, but we currently think it lies between 2 and 3 solar masses. It seems possible that the collapsed core of a very massive star might be above that limit, in which case it could not form a neutron star. The collapse would be unchecked, and the center of the star would form what is known as a *black hole*. This possibility will be considered later.

NUCLEOSYNTHESIS

Now that we have some understanding of the processes which occur within stars, we can consider the origin of the elements of which we are made. As we shall suggest in the sections on cosmology, the universe in which the first stars began to burn probably consisted of hydrogen and helium; if any other elements were present, they were in insignificant amounts. All the elements we know heavier than helium have been formed in stars.

We have already suggested how elements up to the iron group are formed in the nuclei of stars, and we have mentioned some of the ways in which those elements may be brought out of the star and thrust out into space to form the raw material for a next-generation star. But how do we explain the existence of elements heavier than iron? What has been the genesis of gold and silver and mercury and uranium? Probably all those elements are formed during the few minutes when a supernova is blowing itself apart. During that tremendous blast, neutrons are released in great numbers, and the neutrons can combine with those heavy elements already formed within the core to produce the elements we know—up to uranium. Subsequent radioactive decay of unstable species of nuclei produces the variety of isotopes we find.

Origin of Elements of Earth

Nucleosynthesis, the process by which the elements are formed, is a topic of tremendous interest now, and views on it are changing rapidly. Some recent work suggests that the relative abundances of the elements found in the solar system, which is the only place we can make careful measurements, can be best explained by the assumption that all the elements heavier than helium were formed during the explosions of supernovae, and that the elements formed slowly within stellar interiors have not made a significant contribution.

The conclusions we must draw from these considerations is that the sun is a second- or third-generation star, and that it and the planets of the solar system are made from recycled materials. Early in the morning of the life of our galaxy stars must have formed, stars unlike the ones we have been describing because they had no heavy elements to add to the opacity of the outer layers. Those stars went through life cycles, and in their deaths they returned to the primeval gas of the galaxy some of the elements they had formed or that were formed in their moments of death. That material enriched the turbulent clouds of gas with small amounts of oxygen, carbon, nitrogen, iron, gold, etc. From such an enriched cloud—a mixture of the debris from the deaths of first-generation stars and unprocessed hydrogen and helium—the sun and its planets formed.

Population I and II Objects

The reason for the correlation between position within the galaxy and metal content discussed in connection with Population I and Population II stars now becomes clear. Globular clusters formed when the material of the galaxy was still approximately spherical, and when it was mostly hydrogen and helium. The stars in those clusters may be producing heavier elements within their cores, but their atmospheres reflect more nearly the original material from which they formed, and they are metal-poor. As the galaxy evolved, becoming more flattened, the process of enriching the interstellar material with metallic debris from spent stars continued, and therefore the later an object formed, the more metals it had. But the later it formed, the nearer the disk it probably was. And now, the most metal-rich stars, the extreme Population I stars, are in the thinnest part of the disk.

QUESTIONS

1 Consider a star of 7 solar masses. Use the mass-luminosity diagram given in Unit 10 to estimate its luminosity, and then estimate its life on the main sequence.

2 What kind of argument can you give that a star whose luminosity is 10,000 times that of the sun probably is younger than the sun?

3 The more massive a protostar is, the more rapidly it goes through contraction to the point that it appears on the H-R diagram. Why should that be so?

4 Describe the appearance of a star of 1 solar mass at the time it reaches its maximum luminosity before moving onto the main sequence.

5 What property of a star determines its position on the main-sequence line?

6 If, in becoming a red giant, a star increases in radius by 50 times and in luminosity by 1,000 times, by what factor does its surface temperature change?   *Ans* ≈ 1.3

7 If a white dwarf became cold and black, would its electrons still be moving?

8 Why is the turn-off point O in the H-R diagram for a cluster an indication of the age of the cluster?

9 Why are clusters studied in connection with stellar evolution?

10 What is the Chandrasekhar limit?

11 What is the probable origin of the gold found on the earth?

14
the universe of galaxies

Probably every grade school child now learns that the Milky Way galaxy, in which the sun and the earth are located, is only one of many such galaxies, each comprising many stars. But that information was acquired by astronomers very slowly, and the fact that the Milky Way does not constitute all the visible universe was not fully realized until the early 1920s.

NATURE OF THE NEBULAE

The earliest observations of the heavens with telescopes revealed nebulous patches of light which did not appear to be stars; these were named *nebulae*. We now know that the nebulae include stellar clusters, clouds of luminous gas, clouds of dust reflecting star light, and galaxies, some of which are relatively near to us and some of which are at the limits of vision with telescopes.

As early as 1750 Thomas Wright suggested that the Milky Way is an aggregation of stars of finite extent, shaped something like a grindstone, and at about the same time Immanuel Kant suggested that the nebulae were other *"island universes,"* but there was little evidence to support either speculation.

In 1845 Lord Rosse, who built a 72-in telescope in Ireland, first observed spiral structure in nebulae, and some 10 years later he published a drawing of the Whirlpool galaxy which caused a stir in astronomical circles. Not until 1889 was the spiral structure of a galaxy shown on a photograph.

Spectroscopic Evidence

The spectroscope was first used to obtain a spectrum of a nebula by Sir William Huggins in 1864. He reported that the spectroscope had ended the controversy—the light from nebulae consists of bright lines, indicating that the nebulae are clouds of tenuous gas, not aggregations of stars. Later, however, he found that his conclusion applied to only some of the nebulae; others give the type of spectrum produced by stars.

262   In 1899 a spectrum of the Andromeda galaxy was pho-

tographed; the spectrum indicated that the nebula consists of
stars similar to the sun.

Distance Evidence

The critical question to answer in determining whether the
nebulae are parts of the Milky Way or are external systems was
one of distances—the extent of the Milky Way and the distances to the nebulae. For a number of reasons, some of them
quite reasonable, the distances to the nebulae which we call
galaxies were consistently underestimated, and they were
usually placed near enough to us so that astronomers assumed
they must be part of our system. And of course some of the
nebulae—the objects we call emission or reflection nebulae,
and the stellar clusters—are indeed part of the Milky Way system.

In 1924 Edwin Hubble first reported the use of Cepheid variables in determining the distances to other galaxies, and although we know now that the distances he obtained were too
small, they clearly placed those objects outside the Milky Way.
It was at last clear that the Milky Way and the other galaxies
are indeed island universes as Kant had suggested, but finding
the proof for that guess had required more than 170 years of
work.

DISTANCE MEASUREMENTS

Those island universes are widely dispersed. We find fewer
than 20 within 2.5 million light years, but thousands within 50
million light years, and we see no end as we look out into
space. Counting the galaxies visible with a telescope like the
200-in would be a tedius and probably worthless task, but from
samples of parts of the sky in which the visible galaxies have
been counted it has been estimated that something of the order
of 1 billion galaxies must be within reach of the 200-in telescope. The Great Galaxy in the constellation Andromeda (M 31)
is about 2.5 million light years distant.

Distances to relatively nearby galaxies can be computed by
many methods, and agreement among the methods gives astronomers confidence in the results. For more distant galaxies,
methods calibrated from the nearer sample are used.

Cepheids

For galaxies near enough that Cepheid variables can be recognized, their period-magnitude relation can be used to determine the absolute magnitudes of the stars, and the combination
of that with the apparent magnitudes yields the distances.

Novae

The novae in our own galaxy appear to have an average peak absolute magnitude of about −7.5, and we assume that the novae in other galaxies are about the same. Particularly in a galaxy where many novae have been seen, as in M 31, the average value of maximum brightness can be used with some confidence.

Globular Clusters

About 120 globular clusters are known associated with our galaxy, and they appear to have an average absolute magnitude of about −7.5. On the assumption that the globular clusters in nearby galaxies are like those in our own galaxy, we can compute the distances to the galaxies.

Blue Supergiants

The brightest stars seen, blue supergiants, appear to have about the same intrinsic luminosity wherever they are found, and if such stars can be seen in distant galaxies, they can be used to determine the distance. Such stars have an absolute magnitude of about −9 and can be seen to about 80 million light years.

H II Regions

As we pointed out in Unit 12, some extremely hot stars are surrounded by regions of ionized hydrogen called *H II regions*. The hydrogen glows by fluorescence so that such a region can be seen for a very great distance, although from a great distance it may be hard to distinguish from a very bright star. The diameters of H II regions depend upon the temperatures of the stars emitting the ionizing radiation and the density of the gas around the stars, but they may be several tens of light years. The diameters of H II regions can be correlated with the type of galaxies in which they are seen, and those diameters provide a means of extending distance measurements beyond the point at which individual stars can be seen.

Brightest Galaxy in a Cluster

Finally, galaxies often occur in clusters, and if the brightest galaxy of each of many clusters is studied, it is found that they all have about the same luminosity. And so for distances so great that individual stars and even H II regions cannot be seen, the luminosities of galaxies themselves become a tool to estimate distances. Galaxies are not all alike, and even the brightest are not all alike, and measurements obtained in this way are not very precise. They may even be wrong by considerable factors, but we have no other evidence to use, and even poor estimates are better than none.

Redshift

The spectra of distant galaxies show a systematic shift to the red, which seems to be proportional to their distances from us, and the magnitude of redshift can be used to compute the distance to the source. (Redshift is discussed in detail in Unit 15.)

TYPES OF GALAXIES

The impression one forms from seeing published pictures of galaxies is that they are all majestic spirals. The giant spirals are eminently photogenic, and they are readily recognizable objects. Furthermore, they are examples of the type structure we believe we live in. The fact is, however, that not all galaxies are spirals. Of the 1,000 brightest galaxies we can see, only about 77 percent are spirals; the remainder are irregulars (3 percent) or ellipticals (20 percent). One reason these are not usually offered when someone wants a beautiful picture of a galaxy is that an irregular galaxy may look something like an out-of-focus photograph of a bank of lights, and an elliptical may resemble a badly overexposed photograph of a street lamp. At best, if an elliptical looks like an astronomical object, it cannot be easily distinguished from a globular cluster.

Most of the brightest galaxies we see are spirals, because their average luminosity appears to be several times that of other types, but we must not jump to the conclusion that most galaxies are spirals. The very bright members of the population of galaxies can be seen for great distances, whereas the faint members may be overlooked. The sample of 1,000 brightest galaxies is biased just as a sample of the 1,000 brightest stars would be. From a consideration of the brightest stars one would conclude that the most common stars are blue, very luminous giants, when in fact the most common stars are faint, red dwarfs. But to form that conclusion, we must make a detailed census of a region of space near enough that we can count most of the stars in it, so that we can be confident our sample fairly represents the population. Our best indication of the actual numbers of the different types of galaxies comes from an examination of those galaxies lying near us, the so-called *Local Group*. Within that small sample, about 60 percent are ellipticals and only about 20 percent are spirals.

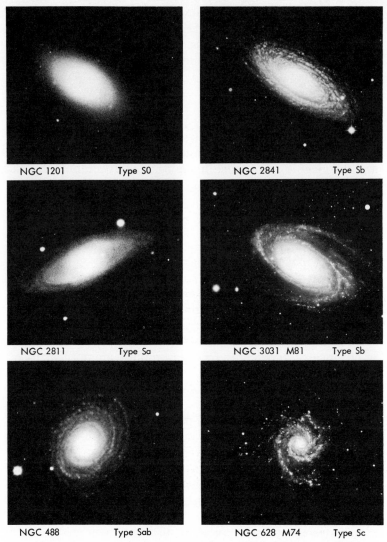

NGC 1201   Type S0

NGC 2841   Type Sb

NGC 2811   Type Sa

NGC 3031 M81   Type Sb

NGC 488   Type Sab

NGC 628 M74   Type Sc

Six spiral galaxies with their classifications. The S indicates that they are spirals, and the 0, a, b, and c indicate how tightly the arms appear to be wound. (*Hale Observatories*.)

Irregulars

Irregular galaxies contain dust, gas, and bright stars, and they appear to be places where stars are still being formed. But mixed in with Population I objects like new stars and gas clouds are such Population II objects as globular clusters. The separation of the two populations into different parts of the galaxy does not appear to have occurred in irregular galaxies as it has in spirals. In fact, the irregulars seem to have no organization or structure; they are a jumbled mixture of stars, gas, etc.

The Clouds of Magellan, two satellite galaxies of the Milky
Way, are irregulars.

267

THE UNIVERSE
OF GALAXIES

Ellipticals

Elliptical galaxies seem to contain little if any gas and dust, and
their stars appear to be almost entirely Population II objects.
Ellipticals are classified on the basis of the eccentricity of the
profile they show to us. Some look round; we cannot tell
whether such a galaxy is spherical or whether it is shaped like a
football and we are viewing it from one end. But because their
profile is round, we say these galaxies have an eccentricity of
zero and call them globular. Others appear to be highly flattened, like a football viewed from the side; they have eccentricities greater than zero. One can make a statistical analysis
designed to find what shapes the population of galaxies must
have to produce the number of profiles we actually see in a random sample. For example, if they were all spherical, we should
see none flattened, but if they all were highly elongated, we
should see fewer with a circular profile than we actually observe. Evidently they actually range from spherical to quite
elongated forms.

In the typical photograph of an elliptical galaxy, most of the
center of the picture is solidly overexposed, and at the edges
the density of stars thins so that individual stars can be seen.

An elliptical galaxy, M 87 (NGC 4486). This long exposure
shows globular clusters in the outer part of the galaxy.
(*Lick Observatory photograph.*)

Actually there is no definite edge, for if longer exposures are taken, the halo of stars which can be seen becomes larger. Evidently at the outer part of the galaxy the density of stars falls off gradually, and the number of edge stars seen, and hence the apparent diameter of the galaxy, depends upon the photographic exposure time. The appearance of a solid wall of light in the center, is somewhat misleading, for the centers of those galaxies are not packed solid with stars. In fact, a rifle bullet or a spaceship or even a star could pass straight through the center of such a galaxy with little probability of hitting a star. The appearance of solid light is caused by the diffraction of light by our collecting instruments plus the scattering of light in the atmosphere, two effects which make the images of individual stars spread from their theoretical point size to overlapping disks.

Some elliptical galaxies, however, known as *dwarf ellipticals*, are so sparsely filled with stars that they do not appear to us like a solid core of stars. In fact, they are transparent, and we can see more distant objects through them. This points up the range of size of elliptical galaxies. At the low end, they may be as small as 1 million solar masses, but at the upper end they may be as huge as 10^{13} solar masses, or 100 times the size of the Milky Way.

Spirals

Spiral galaxies are classified on the basis of how tightly the spiral arms are wound, and on whether the spiral arms appear to begin from a nucleus or from a bar passing across the galaxy. About one-fifth of the spiral galaxies are of the barred type. All spiral galaxies appear to be rotating, with the spiral arms trailing.

Spiral galaxies are mixtures of Population I and Population II objects. Our own galaxy and M 31 are probably somewhat larger than the average, but in other respects they are typical of spirals. All those spirals which are near enough for such details to be visible have globular clusters, halos, open clusters in the spiral arms, nuclei, and disks, in addition to the spiral arms. The spiral arms are outlined, as in our own galaxy, by young, exceptionally bright stars, open clusters, and H II regions; between the arms are Disk Population stars. More than 200 globular clusters have been counted in the halo of M 31, and globular clusters are visible about many other galaxies whose distances from us are not too great for such small objects to be seen. Spiral galaxies appear to range in diameter from approximately 20,000 to over 100,000 light years; both M 31 and our own

The barred spiral NGC 3992. Note that the spiral arms appear to begin
at the ends of a bar through the nucleus. (*Lick Observatory photograph.*)

galaxy are in the vicinity of 100,000 light years, at the upper
end of the range.

THE LOCAL GROUP

As we begin to think of the universe as made up not of planets
or even of stars, but of galaxies, we find that we have several
near neighbors. Within about 3 million light years of our galaxy
are 16 or perhaps 18 other galaxies; those and the Milky Way
comprise the Local Group. Until a few years ago it was believed
that there were 17 galaxies in the Local Group, but now it is
suspected that Maffei 1 and Maffei 2, discovered in 1969, also
belong to the group. Both these galaxies lie in the constellation
Perseus, however, and we view them through such a haze of
dust that their distances have not been determined with certainty.

Within the Local Group, M 31 is the most luminous, with an
absolute visual magnitude of about −21, and probably the largest. The next brightest is our own galaxy. Third in the list is a

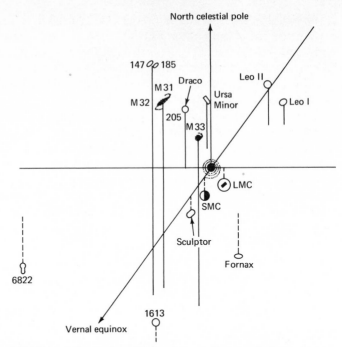

FIGURE 14.1 The members of the Local Group. Our galaxy is shown at the center; it and M 31 are the largest members of the cluster. SMC and LMC are the Small and Large Magellanic clouds.

spiral known as M 33, and then come the Large and Small Magellanic clouds. The first three in terms of luminosity are spirals, and the next two are irregulars. Leaving Maffei 1 and 2 out of consideration, 10 of the remaining 17 in the Local Group are dwarf ellipticals, 4 are dwarf irregulars, and 3 are spirals. The Local Group does not contain a giant elliptical such as many galactic clusters have. The diameter of M 31 appears to be about 150,000 light years; the diameter of our galaxy is about 100,000 light years; and the diameter of M 33 is 40,000 light years. The diameters of the smallest dwarf galaxies in the group are somewhat over 3,000 light years. It is possible that other galaxies which are part of the Local Group are hidden from us by the dust in the disk of the Milky Way.

MASSES OF GALAXIES

Rotation Curve

The most accurate method of computing the mass of a galaxy is applicable only to spirals which are viewed somewhat obliquely. If the rotation velocity at different distances from the nucleus can be determined from measurement of doppler

shifts, then the velocity of rotation can be plotted as a function
of distance from the center. Such measurements have been made both by taking many spectra at different positions, and also by setting the slit of a spectrograph across the galaxy so that the entire galaxy is observed at once. When the latter method is used, the spectral lines are slanted, and the amount of their displacement from the normal position at points corresponding to different parts of the galaxy gives the rotational velocity. If the rotation curve can be obtained, one can estimate the mass of the galaxy very much as we estimated the mass of our own galaxy through computations using the sun's speed around and distance from the center. Essentially the method is to equate the centripetal force required to hold stars on circular paths at their distances from the center to the gravitational force produced by the stars nearer the center. From such measurements we conclude that spirals have masses lying between 10^9 and 2×10^{11} solar masses.

Pairs

Some galaxies occur as pairs which appear to be rotating about one another. We cannot wait to observe a rotation, as we can with double stars, and so we cannot obtain the kind of information which makes a clear mass determination possible. In particular, we have no way of knowing the angle at which we are observing the rotation, so that the velocities we obtain from doppler measurements are lower limits of the actual velocities. We can obtain therefore only a lower limit on the sum of the masses of a single pair. If many pairs can be studied, however, we can use statistical methods to determine the probable range of masses they possess.

Virial Theorem

A second statistical method involves the study of galaxies in clusters. If we assume that a cluster is gravitationally bound, then the velocities of the individual members relative to the center of mass can be related to the total mass of the cluster by a mathematical relation known as the *virial theorem*. The velocities are observed (the velocities in one direction, our line of

FIGURE 14.2 When the image of a galaxy seen almost edge-on is placed on the slit of a spectrograph, each line of the spectrum is slanted because of the doppler effect. If the velocity of each part of the galaxy is not proportional to its distance from the center, the spectral lines are not straight, for the displacement to the right or left of each segment of the spectral line is proportional to the velocity of that part of the galaxy which produced the segment.

A portion of the Coma cluster of galaxies. This cluster lies at a distance of about 300 million light years and has a diameter of at least 10 million light years. We can observe more than 1,000 galaxies in the cluster and we suppose that most of the members are too faint to be seen, so that the actual number may be many tens of thousands. (*Lick Observatory photograph.*)

sight, can be observed, and the true velocities are statistically inferred), and from that the total mass can be calculated. The method is also dependent upon the assumption that the clusters have been in existence long enough that the kinetic energies of the members are randomly distributed.

Clusters range in size from one with four members to those with thousands of galaxies. (The one with four members is misnamed Stephan's Quintet—one "member" is not part of the cluster!) One of the largest, the Coma cluster, is more than 150 million light years from us and has a diameter of between 20 and 25 million light years. The Coma cluster has been studied extensively, and its mass estimated by use of the virial theorem.

The mass of the cluster estimated in that way is far greater than the mass estimated by counting the members and assuming that they have masses like others of their type studied at closer range. Another way of saying the same thing is that we do not see enough mass in the Coma cluster to hold it together; the use of the virial theorem may not be legitimate because the cluster may be in the process of flying apart. Attempts to detect intergalactic gas by the x-radiation it should emit have been unsuccessful, however, and we have no satisfactory explanation for the discrepancy between the results obtained by the two different methods.

EVOLUTION OF GALAXIES

Our understanding of stellar evolution has provided a framework for understanding many seemingly unrelated facts about stellar objects. If we had an equivalent understanding of the processes of galactic evolution, we might be able to relate observations of galaxies which now are isolated facts. Unfortunately, we do not. The existence of three distinctly different types of galaxies led early in the study of these objects to two theories about their evolutionary significance, but we now know that both theories were wrong. Hubble thought that all galaxies began as elliptical aggregations, with time they flattened to form spirals, and finally they tended to fall into disorganization and become irregulars. Shapley took exactly the opposite view, that is, that the first form to take shape is the irregular, and that it changes into a spiral, which finally winds its arms up so tightly that it becomes an elliptical. Hubble's view can be eliminated by consideration of the composition of the types of galaxies. Irregulars and spirals have young stars and the material to make more stars; ellipticals seem to be all old stars, and are devoid of the dust and gas required to form more stars. Evidently ellipticals do not change into the other forms. Shapley's view is consistent with the composition of the galaxies, for one could reasonably conclude that as spirals age and all their Population I materials disappear, they will change into galaxies containing only Population II objects, and that description fits the ellipticals. For dynamic reasons, however, we do not believe that spirals change into ellipticals.

Thus our understanding of the origin and development of galaxies is still in a primitive state. We lack any explanation of all the observations which has the compelling simplicity and consistency of the theory of stellar evolution. What *can* be said, therefore, is tentative, and it contains more questions than answers.

All galaxies appear to contain old material. The Clouds of Magellan, both irregular galaxies, contain globular clusters and RR Lyrae stars, both of which are presumed to be old Population II objects. Irregulars and spirals appear to be mixtures of materials of all ages, but the oldest objects in them are probably as old as the material in ellipticals. Perhaps, then, we should assume that all galaxies are approximately the same age and look for characteristics within them which might explain why they have evolved differently or at different rates. Surely we can say that ellipticals have evolved more rapidly than other types; if one adopts as an indication of the degree to which evolution has run its course the proportion of the mass of the galaxy bound into stars, ellipticals are far along the evolutionary track. Neither irregulars nor spirals have progressed so far, and at the rate at which star formation seems to be occurring, they will be producing new stars for a long time to come.

What can explain the differences among the types? We have no satisfactory explanation, and this is a research topic of great interest now. Since the rate at which material is incorporated into stars probably depends strongly upon the density of the gas which is to form the star, it may be that for some reason the gas which has formed elliptical galaxies reached higher densities at the time star formation was beginning than did the gas which was to form spirals. Perhaps the greater rotation rate of the spiral materials prevented their collapsing to the densities reached by the ellipticals. We simply do not know.

Problem of Spiral Structure

Another intriguing puzzle about galaxies involves the nature of the spiral arms. The stars near the nucleus of the galaxy move faster than those farther out (approximately as predicted by Kepler's third law), and therefore the arms should wind themselves up. The outer ends should trail farther and farther behind, until instead of majestic pinwheels the galaxies should appear as tightly wound concentric circles. And the rate at which such winding up should occur is great enough that all spirals should by now have lost their spiral shape. Clearly something occurs to reconstruct the spirals or to maintain them. The best explanation we have now takes into account the fact that spirals are outlined by very bright blue supergiants and by H II regions, both of which may be the result of a transient increase in the density of gas within the plane of the galaxy. If a wave of some kind travels out through the galaxy, compressing gas so that it produces a new generation of hot, blue stars, then those stars will outline the path of that wave. If the wave travels in a spiral form, the path will be a spiral arm. As time passes, stars will move out of the arms, at the same time they are evolving past the point at which they can serve as the beacons for spiral arms. As one arm loses its brilliance, other beacon stars will be formed by new waves propagating themselves outward. In this view, the spiral arm is a temporary structure, but one which constantly renews itself, changing gas in the plane of the galaxy into new type O and B stars. However, the origin and even the existence of such a wave is uncertain. A barred spiral galaxy has two spiral arms which originate at the ends of a bar passing through the nucleus of the galaxy; the cause of the bar structure is not understood.

ACTIVE GALAXIES

As we have said, most of the galaxies we can see appear to be systems of many stars in which the stars are quietly or explosively being born and dying. The energy being radiated by the galaxies is the energy released by stars, and although we do not understand the processes by which the galaxies acquired their forms or how they evolved, there is reason to suppose that those things can be explained by the use of familiar principles and laws.

But certain galaxies are unusual in some way, and, in particular, some appear to be releasing energy or to have released energy at a rate we cannot explain. We shall lump all the different kinds of unusual galaxies under the heading *active galaxies*, and consider three different types: radio galaxies, galaxies with explosive cores, and galaxies with abnormally bright nuclei.

Radio Galaxies

In 1944 Grote Reber, one of the pioneers of radio astronomy, observed that his radio telescope was detecting some type of discrete source in the constellation Cygnus. Since it was the first radio source found in the constellation it was given the name Cygnus A, and in 1954 it was identified as a galaxy. Cygnus A is the second brightest radio source we see in the sky.

Early radio telescopes could tell little more than that the sky is not uniformly bright in the radio region; they could reveal the presence of discrete sources but could not locate them at all accurately. As soon as the science progressed to the point that the locations of radio sources could be determined with some precision, catalogs began to be prepared listing and giving the positions of radio sources. The most famous of those catalogs is the series produced at Cambridge: The 1C catalog appeared in 1950, and it was followed by 2C, 3C, etc.; most recently the 6C has been published. An object designated 3C 295 is number 295 in the third Cambridge radio catalog. As such catalogs became available, optical astronomers began trying to identify the radio sources with visible sources. One of the earliest identifications made was Taurus A, which is the Crab nebula. The Crab is part of our galaxy, the remnant of a supernova explosion some 900 years ago, but most of the strong radio sources are not within our galaxy. In fact, many of them are galaxies themselves. Two galaxies early identified with radio sources are NGC 4486, known as radio source Virgo A, and NGC 5128, also known as Centaurus A. Of the radio galaxies known, about 30 percent are in clusters of galaxies, and each is optically the brightest in the cluster.

Exploding Galaxies

Most of the radio galaxies show a phenomenon called *radio doubling;* the radio signals come not from the galaxy which is visible on photographs but from two outer regions, one on either side of it. The appearance suggests that some kind of explosion within the galaxy has ejected material in opposite directions and that the radio emission comes from that material, which is flying away from the parent galaxy. The ejected material is not visible, and so we assume that it is gas rather than stars. Perhaps the gas is hot enough that it has many free electrons, and is also carrying with it magnetic fields. The electrons moving at high speed in magnetic fields produce the radio emission. Some sources, like Centaurus A, exhibit a multiple structure suggesting that explosions ejecting material have occurred repeatedly.

The most spectacular explosion known is in 3C 231, also known as M 82. About 1 million years ago a huge explosion in the core of that galaxy expelled over 5 million solar masses of hydrogen, which is now seen moving above and below the plane of the galaxy. The kinetic energy of the moving gas is about 10^{48} J.

Energy Problem in Active Galaxies

The energies and powers involved in discussions of galaxies are so large that we can make sense of the numbers involved only by making comparisons. For example, both the Milky Way

FIGURE 14.3 Radio doubling. Radio signals come from the regions labeled *R* on either side of the optical galaxy. Radio galaxies often show this configuration, with the radio signals coming from regions separated from the galaxy seen by an optical telescope; no visible light comes from the *R* regions.

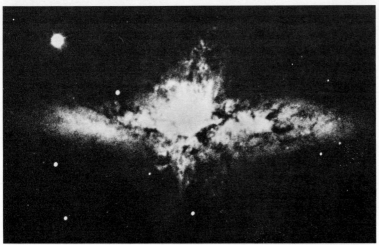

The galaxy M 82 (NGC 3034). This galaxy seems to have suffered an explosion in the distant past which expelled hydrogen above and below the plane of the galaxy. This photograph, taken in the red light of hydrogen with the 200-in telescope, shows faint filaments extending 10,000 light years on each side of the center of the galaxy. Photographs in blue light show filaments extending 25,000 light years outward. (*Hole Observatories.*)

and the Andromeda galaxy appear to radiate in the radio region at a rate of about 10^{31} J/s (10^{31} W). Is that a large amount of radiation? Recognizing that 10^{31} is a large number does little to answer the question, but if we know the power being radiated by those galaxies in the visible region we can make some comparison. Their visible radiation occurs at a rate of about 10^{37} W. Therefore the visible radiation is greater by a factor of 10^6 than the radio emission, and consequently the radio output plays a small part in the total expenditure of energy by the galaxy.

By contrast we consider Cygnus A, whose optical output is comparable to that of the Milky Way or Andromeda but whose radio output is 10^{38} W. It radiates 10 times as much energy in the radio region as in the visible, and its radio emission is 10 million times that of the radio-quiet galaxies.

Now let us take another look at that 10^{48} J estimated for the kinetic energy of the hydrogen being expelled from M 82. We can get some notion of the meaning of that number by considering the energy that would be released if the entire mass of the sun were converted to energy. The sun has a mass of 2×10^{30} kg, and the speed of light is 3×10^8 m/s; when those numbers are substituted into $E = mc^2$, the equation yields 1.8×10^{47} J. That is, if all the mass of the sun were converted into energy, it would be 1.8×10^{47} J. Hence the kinetic energy of the hydrogen in M 82 represents the total rest mass energy of 5 solar masses. (Notice that this is not the energy released

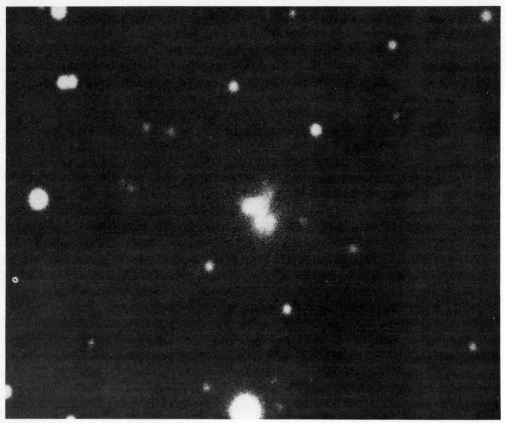

The Cygnus A source of radio noise, photographed with the 200-in telescope. (*Hale Observatories.*)

by five stars like the sun. At the present time the sun is converting 0.7 percent of its mass into energy, for as hydrogen disappears, helium is formed, and most of the mass remains. In the life of the sun, it will convert less than 1 percent of its mass to energy.)

The energy problem can be pointed up more clearly by considering Hercules A. Like other radio galaxies, Hercules A appears to produce radio emission by the synchrotron process, which means that it has large numbers of electrons with high energies moving through magnetic fields. Energy is stored in magnetic fields, and the electrons have energy, and one can estimate how much energy must be in each to produce the radiation actually seen. The results of the estimates are that the total energy stored in the regions in Hercules A that produce radio waves must be at least 10^{53} J, which is 100 thousand times greater than the kinetic energy of M 82. (We have not discussed the magnetic field and electron energies in M 82.) If a single explosion drove the radio-source regions out of Hercules A, it gave the two regions an amount of energy at least equal to the total masses of about 1 million suns. Stars usually convert only about 1 percent of their mass into energy, and so we can say that the energy stored in the radio-emitting regions of Hercules A is equivalent to the available energy of 10^8 stars like the sun.

What could release such stupendous energies in such a short time? Nuclear fusion, which fuels the stars, is a slow and sedate process compared to what must cause the explosions of galactic cores. A number of suggestions have been offered, but none seems entirely adequate. Among those things which have been considered are the following:

Matter-antimatter annihilation: Normal matter consists of atoms with positive nuclei and negative electrons; antimatter consists of negative nuclei and positive electrons. If matter and antimatter meet, they disappear in a flash of radiation—changed totally into energy. The concept of antimatter is discussed more fully in a later unit.

Catastrophic gravitational collapse: If matter undergoes total collapse to form a black hole (discussed later), a significant part of its mass energy might conceivably be released, although the mechanism for such release is not known.

Rapid collisions in closely packed stars: If in a galactic core the density of stars becomes so great that collisions become common, vast amounts of energy undoubtedly are released; whether the amounts needed to explain strong radio sources would be produced is doubtful.

Chain reaction of supernova explosions: This is similar to the preceding suggestion, except that it assumes that many stars are near the supernova point and when one explodes, it triggers others in a chain reaction.

Seyfert Galaxies

We have considered two phenomena which are evidence of unusual activity in the nuclei of some galaxies: radio emission and the ejection of gas. Other galaxies have another indication of strange behavior—very bright nuclei. We will consider particularly Seyfert galaxies, although N-type galaxies are similar.

In 1943 Carl Seyfert noticed some spiral galaxies with unusually bright nuclei. When these objects are photographed with short exposures, their nuclei show up as if they were stars; longer exposures reveal the full structure of the galaxies. About a dozen are known, which represents 1 to 2 percent of the spirals we can see well, and it may be that the Seyfert stage is a stage in the life of all spiral galaxies.

The stars in the outer parts of the galaxies are normal yellow or reddish stars, but the nuclei of Seyfert galaxies are very blue. They emit more ultraviolet than can be explained by what we know of the stars in the galaxy, and some of the light from the nucleus is polarized, suggesting that it is synchrotron radiation. Several of the Seyferts also radiate very strongly in the infrared part of the spectrum, perhaps because the nuclei contain dust which is heated. The most important characteristic of the Seyfert galaxies, other than the small nuclei, is the broad emission lines in the spectra of the nuclei. Normal galaxies show primarily absorption lines, because those are produced by stars. When emission lines are seen, they are narrow and sharp, produced by H II regions where interstellar gas is excited by the radiation from hot stars. Astronomers usually interpret broad lines as an indication that the emitting gas is in violent motion. That is, some of it is moving toward us and hence yields radiation doppler-shifted to the blue, but at the same time some is moving away and produces radiation doppler-shifted to the red. If that interpretation is made of the hydrogen-emission lines seen in some of the Seyfert galaxies, the gas must be in random motion at speeds of several thousands of kilometers per second. At such speeds the gas would soon escape from the nucleus of the galaxy, and so we must suppose that some unknown mechanism exists to hold the gas in the nucleus, that the gas is constantly replenished, or that the outburst is of short duration.

Another characteristic of Seyfert nuclei is that their intensity of output varies strongly, and at least two have been observed to change greatly in a time of a few months. The significance of this type observation will be discussed when we consider QSOs, for they also show short time variations. At least two of the Seyferts are also radio sources, and the radio output fluctuates widely.

One Seyfert galaxy studied with particular care is NGC 4151, which has been observed with Stratoscope II, a 36-in telescope carried by a balloon to about 80,000 ft. NGC 4151 is about 30 million light years away, and its nucleus is no more than 12 light years across. There are reasons to believe that that nucleus contains about 10 billion stars, which would make stellar collisions frequent enough to explain part, although not all, of the energy output of the nucleus.

Because of some resemblances between Seyfert galaxies and QSOs, Seyferts have been studied with particular interest since QSOs were discovered. So far, however, we have more questions than answers, and Seyfert galaxies, like radio galaxies and galaxies with exploding cores, are places where nature is releasing energy on a scale which baffles us and stupefies the imagination.

QUESTIONS

1 How do the spectra of gaseous nebulae and stars differ so that one can be sure that the spectrum of the Andromeda nebula indicates that it is made up of stars?

2 Suppose that exactly 20 galaxies are nearer than 2.5 million light years from us. If the density of galaxies in space is uniform, and those 20 represent an accurate sample, how many should lie within 50 million light years?   *Ans* 1.6×10^5

3 Suppose that a nova with a peak apparent magnitude of 17.5 is observed in another galaxy. About how far away is the galaxy?   *Ans* 10^6 pc

4 How is it possible for an emission nebula (H II region) to be brighter than the star which is supplying the energy to make it glow?

5 How can we be virtually certain that elliptical galaxies do not evolve into spirals?

6 Sketch a spiral galaxy, and look at one arm. Assume that the stars at each point on the arm obey Kepler's third law and compute the relative periods of stars half way out along the arm and at the end of the arm. Draw the arm as it will look when the stars half way out have completed one revolution.

7 The kinetic energy of moving mass is computed from $KE = \frac{1}{2} mv^2$. Show that if the kinetic energy of 5 million solar masses is 10^{48} J, the speed must be approximately 450 km/s.

8 Explain the steps by which distance measurements are extended from the nearest stars to the most distant galaxies.

15
life in the universe

Life in the universe—other than on the earth—is a subject which has enormous interest for us, perhaps because we do not like to think that we are alone in the vast universe. We have absolutely no evidence that life exists anywhere else, and any discussion of this subject must begin with the probabilities that the conditions in which life might be possible exist elsewhere and then concern itself with the probabilities that life would arise if the conditions were right.

We shall not try to define life; that is a task for the biologist. In this discussion, however, we shall focus our attention on the conditions for life of the sort we know, based on oxidation reactions, requiring temperatures slightly above the freezing point of water. Hal Clement has written a science fiction novel called *Iceworld* about creatures who live on a planet near a blue star, a planet which is so hot that sulfur is a vapor, and the creatures breathe sulfur vapor. Such odd metabolism may be possible, but for the present we will bypass that kind of consideration.

The reasons for considering the possibility that life, and more particularly intelligent life, exists elsewhere in the universe are two: the intrinsic interest of the subject, and the fact that for the first time in the history of man we may have the capability of communicating with intelligent creatures on planets about other stars. Within recent years a great amount of effort has gone into considerations of the ways in which signals might efficiently be sent to civilizations about other stars or ways in which we might recognize radio messages such civilizations may be sending. One serious attempt has been made to hear radio communications from planets which may be circling two nearby stars.

LIFE IN THE SOLAR SYSTEM

Before turning our attention from the solar system to consider planets out in the greater void, we should give brief consideration to the other planets of our own system on which life might be found. Mercury is unlikely because of the rather high temperature and the small amount of atmosphere. The surface of Venus is much too hot, but the suggestion has been made that life might arise in Venus's clouds and that living organisms

may be floating in the atmosphere. Mars is still a possibility, although the lack of water and atmosphere seem to preclude any
high form of life. Mars may have low forms of life, and the only
way to determine whether or not it does is to land craft on the
surface to sample the surface material. Even if the reports from
unmanned craft are negative, it will not definitely settle the
matter, for the area sampled will be miniscule. Jupiter appears
a poor place to look for life, except that the atmospheric constituents of Jupiter (methane and ammonia) are similar to those
hypothesized to have existed on the earth when life began here.
The top of Jupiter's atmosphere is cold, but the atmosphere
may have warmer layers deeper down, for heat is flowing out
from the planet. Any life which may exist within the dense atmosphere and under the high pressures of Jupiter would appear strange to us, but we cannot be sure that none exists.

It seems, therefore, that the probability of finding life anywhere in the solar system except on the earth is small, with the
possible exception of Mars, and the probability of finding intelligent life elsewhere in the solar system is nil.

Consideration of the probability of life outside the solar system involves first estimating the probability that habitable
planets exist elsewhere, and then the probability that life will
have arisen on such planets if they exist. We will begin by discussing the estimate of the probability that planets habitable by
man exist; this probability is designated by those who make
these computations as P_{HP}—the probability of habitable
planets. We will follow particularly the arguments of Shlovskii
and Sagan and of Dole.

PROBABILITY OF HABITABLE PLANETS

We know of dark companions of a few nearby stars, but we cannot be sure that any are of such small mass that they can
reasonably be called planets. Probably the star which has been
studied most carefully in search of planets is Barnard's Star.
Because it is near us and has the largest proper motion of any
star seen, deviations in its motion because of planets might be
visible. Peter van de Kamp of Swarthmore College has studied
photographs of Barnard's Star taken over a period of more than
50 years, and he has concluded that the star has one, or more
likely two, planets of approximately the mass of Jupiter. However, in late 1973 George Gatewood of the University of Pitts-
burgh and Heinrich Eichhorn of the University of South Florida
reported that they had analyzed other photographs taken over
the same period without finding any wobble in the path of the

star which would indicate the presence of planets. Whether or not Barnard's Star has planets, it is clear that information about other planets must be mostly indirect. Perhaps if we had large telescopes on the moon, outside the atmosphere of the earth, we would be able to see planets about the nearer stars, but under no conceivable conditions would we be able to see planets at great distances. Recall the analogy in which the sun was represented by a basketball and the earth by an appleseed 100 ft away; the stars, represented by basketballs, would be separated by approximately 5,000 mi. Space is mostly empty, and the distances are huge; those appleseeds would be hard to see even with large telescopes!

Origin of the Solar System

Crucial to our guesses about the probability of finding planets about other stars is our understanding of the origin of our own planetary system. If it came into existence through a freakish accident, then it may be the only such system in the universe. On the other hand, if the processes which brought it into existence were common as stars formed, similar systems may be frequent companions of stars.

Any theory purporting to explain the origin of the planets about the sun must explain most of the significant regularities in the system, the most important of which are as follows.

1 All the planets move almost in a plane (the plane of the ecliptic); they do not move with odd orientations.
2 All the planets revolve about the sun in the same direction; there is no retrograde motion.
3 All the planets rotate on their axes in the same direction with two exceptions: Venus rotates the wrong direction, but slowly, and Uranus is tipped almost 90° to the plane of the ecliptic.

One might hope that a theory of origin would also explain the change from terrestrial to Jovian types and the variations in mass from low to high and back to low, as well as certain other characteristics, but the three characteristics listed above *must* be explained. Many explanations have been offered, but we will consider only three: the *random-capture* hypothesis, the *encounter* hypothesis, and the *condensation* hypothesis.

Random-capture hypothesis This hypothesis assumes that roaming about in space are cold bodies the size of planets, and at times in the past the sun has come close to such bodies, which have been captured to form the planets we know today. This hypothesis has two defects which are so overwhelming that it can be abandoned immediately. One is that bodies wandering in space which found themselves attracted to the sun as it passed by would not be captured unless they could lose energy as they came near. After one was captured it might help take energy from another to permit its capture, but the first could not be captured unless it could pass through an atmosphere or in some other way lose energy. The second defect of the explanation is that it is highly unlikely that nine bodies captured in such a random way would all have orbits lying almost in the same plane. One should expect rather that the orbits would look somewhat like the stylized diagrams of the atom in which electron paths are shown with many inclinations.

Encounter hypothesis The encounter hypothesis, associated with the names Jeans, Chamberlain, and Moulton, enjoyed great popularity in the early part of this century. It proposes that a star passed near the sun, and its gravitational attraction caused a great arm of matter to be lifted from the sun and strung out into space. As the star passed, the matter lifted out from the sun was pulled so that it began to move around the sun, and subsequently the planets condensed from it. If this is the way the planets of our system formed, then they may well be the only planets in the galaxy or the universe, for such near collisions of stars are very rare events everywhere except in the dense nuclei of galaxies.

The explanation has so many problems, however, that it has been abandoned. Two reasons for discarding it will be mentioned. One is that if another star came near enough the sun to pull out the amount of material required to make the planets, it would be so close that it could not give that material the *angular momentum* which the planets have, and the angular momentum distribution within the solar system is one of the characteristics which must be satisfactorily explained. Angular momentum is one of the quantities which can be computed which remains constant in an isolated physical system. A precise definition of it is difficult to give without using more mathematics than is desirable here, but almost everyone has some intuitive feel for it. Angular momentum depends upon the speed of rotation and upon the distribution of mass around the axis of rotation. Other things being equal, the farther the mass is from the axis of rotation, the greater the angular momentum. If no outside forces tending to change the rotation act on a body, its angular momentum remains constant. That is why a figure skater begins a spin with arms outstretched and then wraps them about his body; as the arms are drawn in, the rotational speed increases so that the angular momentum remains the same. A diver doing a flip goes off the board with her body straight, then doubles up so that her rotational speed increases, and finally straightens out so that her rotational speed returns to its original low value before she hits the water. During the maneuvers, the angular momentum remains the same. A cloud of gas rotating slowly in space which begins to contract because of mutual gravitational attraction of its parts rotates faster and faster as its size decreases because its angular momentum remains constant, and as the mass moves nearer the axis, the speed of rotation must increase.

The solar system has angular momentum arising primarily from two motions—the rotation of the sun on its axis, and the revolution of the planets about the sun. (The rotations of the planets contribute a small amount more.) The interesting thing about the system is that the planetary revolutions contribute 90 percent of the angular momentum! The sun rotates rather slowly, and even though it is very massive, its angular momentum is far less than that of the planets. At the equator the sun's surface moves about 2 km/s. If all the planets were dropped into the sun with their present angular momentum, its mass would be increased negligibly, but its speed of rotation would increase to 100 km/s at the equator.

The theory of the dragging into space of a filament from the sun cannot explain how that filament could have been given enough angular momentum (rotational speed about the sun) to account for the angular momentum possessed by the planets now.

The second problem with the encounter theory is that the gas pulled into space should disperse, not form into planets. Because the planets have relatively little of the hydrogen and helium which are the principal constituents of the sun, the mass of material originally drawn out must have been enormous, so that when the light gases evaporated the observed mass of heavier elements was left. But such a large mass could not have come from the atmosphere of the sun alone—it must have come from deep within the sun. The result is that the temperature of the gas would have been very high, and, because of this high temperature, the gases would have dispersed before they could have condensed into planets.

Condensation hypothesis   The third explanation of the origin of the planets is that they condensed from the same mass of gas which gave rise to the sun, and that they were formed at about the same time as the sun. If this theory is correct, then we may reasonably expect that ours was not a unique experience, and that planets may have formed about many stars.

The first attempt to work out this explanation in a detailed way was made by Kant and by Laplace. In general terms, they assumed that as a cloud of gas condensed to become the sun, part of it formed into a disk because of its rotation, and in the disk condensations formed which became the planets. The explanation fell into disfavor when theoreticians discovered, late in the nineteenth century, that it could not explain the distribution of angular momentum in the system. So long as we count on gravitational forces alone to cause things to happen, we conclude that the sun should have most of the angular momentum, and any satisfactory theory of origin must include some mechanism by which angular momentum was transferred from the sun to the planets. Within recent years such a mechanism has been proposed, and a modern version of the Kant-Laplace nebular hypothesis is connected with such names as Hoyle and Kuiper. In general outline, the explanation goes as follows.

The solar system began as a huge cloud of gas and dust, which began to contract because of its gravitational attraction. It had a small rotational velocity, and as it contracted, the rotational rate increased to conserve angular momentum. Eventually the central condensation, the protosun, was surrounded by a disk of matter which was rotating so fast that it could not contract any more. The rotation did not inhibit contraction that made the disk thinner, but the material of the disk was unable to fall into the protosun because of its high angular speed. The sun was also rotating at a high rate of speed. These two parts—the protosun and the disk—were joined by a magnetic field, originating in the sun and extending outward. Since the gas in the ring was partially ionized, it interacted with the magnetic field. Magnetic fields act at times like rubber bands. One can speak of magnetic field lines as if they had physical existence, and in this case the easiest way to visualize what perhaps happened is to think of magnetic field lines almost like bicycle spokes radiating from the sun and linking the sun to the disk rotating about it. The effect of the magnetic lines was to pull back on the sun and forward on the disk, so that the sun's rotation was slowed and the disk's rotation was increased. But if the disk tried to increase its rotational speed, it expanded, so that it drifted out from the protosun. Because of eddies and turbulence in the disk, in time huge protoplanets condensed within it, each having a large amount of hydrogen and helium. Still later the increasing heat of the sun drove away all the remaining material of the disk and most of the material of the protoplanets which had formed. The earth probably lost more than 99 percent of its material, and in its initial state, after the remnants of the disk were blown out of the system, its surface was bare, with no atmosphere. The present atmosphere formed later as volcanic action brought to the surface gases originally trapped within this planet.

This explanation can account for the angular momentum distribution, for it assumes that much of the angular momentum of the sun was transferred to the planets by means of the magnetic field. It accounts for the fact that the planes of the orbits are all about the same, for all the planets formed from the same disk. It explains the common directions of rotation on their axes by supposing that when the protoplanets were huge, they were tidally coupled to the sun, always turning the same face to the sun, but when they lost mass they broke loose and continued to turn in the same direction. And one can make conjectures about the variations in mass by making guesses about the variations in the thickness of the ring.

This account does not answer all the questions we can ask about the formation of the planets about the sun, but it seems plausible to most astrophysicists working in the field, and it is almost universally accepted as being the most likely explanation. It has the happy characteristic, from the standpoint of our present discussion, of offering reasonable expectation that the formation of planets is a likely accompaniment of the formation of stars.

Rotation Rates of Stars

Another observation which perhaps tells us something about the frequency of the occurrence of planetary systems deals with the rotation rates of stars. The rate at which a star rotates on its axis can sometimes be measured by the doppler broadening of its spectral lines. One side of the star is approaching us while the other side is receding from us, and consequently the light from the entire disk of the star is a mixture of light from the center which displays no doppler effect and light from the limbs which is shifted toward the red from one edge and toward the blue from the other edge. The result is that a given spectral line is wider than it should be if the star were not rotating.

The rotation rates observed seem to be a strong function of the stellar type: O, A, and B stars have high rotation rates, up to more than 300 km/s at the equator; and cool stars, F, G, K, and M types, have low rotation rates. The sun, for example, a G2 star, has a rotation rate of 2 km/s. The change from high speed to low speed does not occur gradually, however, but it changes rather abruptly within the F types. F-type stars are subdivided into nine subclasses, designated F0, F1, F2, . . . , F9. Between F0 and F2 types the rotation rate drops drastically.

If the sun had all the angular momentum of the solar system, it would have an equatorial velocity of about 100 km/s, about like an F0 star. We have hypothesized that its rotation rate is low because that angular momentum has been transferred to planets, and it is tempting to suggest that all the stars with low

rotation rates, that is all the stars cooler than F2, have planets to
which angular momentum has been transferred. Let us examine
this suggestion more closely.

It seems odd that planets would form with all stars which were destined to be cool (of low or medium mass) and never with stars of higher mass which were to become O, B, or A types. Is there, perhaps, a difference between the stars which would cause the transfer of angular momentum to be more effective in one type than in another? As a matter of fact, there is such a difference. If the angular momentum is transferred by magnetic field lines, acting somewhat like stretched elastic bands, those lines must be attached to a layer of the star which is highly ionized and therefore a good conductor of electricity. A cool star has a rather deep atmosphere of neutral hydrogen, and the magnetic field lines pass through it and are attached to a deeper layer, where the ionization is almost complete. The atmosphere has enough ionization, however, that it cannot move past the lines without some effect, and therefore the magnetic field acts upon a thick layer of the star's atmosphere and hence can be very effective in slowing its rotation rate.

In the hot star, the ionization layer is near the top of the atmosphere, and consequently the magnetic field is far less effective in braking the rotation of the star because it interacts with a relatively small part of the star's mass.

If the magnetic field is to slow the star's rotation, it must transfer the angular momentum to some other matter. In the case of the sun, we have assumed that the angular momentum was transferred to the planets or to the disk of material which was to become the planets. If such a disk of material is orbiting the star, it seems likely that a similar transfer can take place. One cannot eliminate as a possibility, however, the transfer of angular momentum to a stellar wind flowing out from the star. The sun is emitting particles which leave it at high speed and flow past the earth; this is called the *solar wind*. Since the particles are often ionized, the wind interacts with the magnetic field of the sun. The solar wind is too low in intensity to make a significant change in the sun's angular momentum, but it may not have always been so low, and the stellar wind about another star might be sufficiently high that it would be able to carry away enough angular momentum to account for the low rotation rates of cool stars.

The above analysis of the rotation rates of stars suggests that all or most stars of class F2 and cooler have planets, but it does not prove it.

Probably between 30 and 50 percent of the stars are members of binary pairs or more complicated systems of three, four, or more stars. Perhaps the processes which lead to star formation always produce more than one body, and if the bodies formed are large enough to be stars the result is a binary pair or multiple system, but if the bodies are not that large, they are planets. This would suggest that at least all single stars are accompanied by planets and that binary systems may have planets.

Conditions for Habitability

The consideration of rotation rates has focused attention on stars cooler than F2, and now another consideration reinforces the interest in those stars. If life as we know it is to have developed, the star about which a planet moves must have been stable for a long time; perhaps as much as 3 billion years is required for life to begin and develop to the point it has reached on the earth. The earth probably began without an atmosphere. The original atmosphere came from volcanic activity, but it has been converted over a very long time into an atmosphere capable of supporting our type of life through the action of ultraviolet light on water vapor and the action of plants. If a planet is to develop a breathable atmosphere and some form of life, its star must remain on the main sequence for a long time; that consideration eliminates planets circling O-, A-, and B-type stars because their sojourn on the main sequence is too short. We seem to be forced to conclude that *only* stars F2 and cooler can be the source of energy for planets on which life has a chance to develop.

There is a region about a star within which a planet is likely to have the temperatures which make life possible. That zone has been called the *ecosphere,* and one must estimate the probability that if planets exist about a star, at least one will lie within the boundaries of the ecosphere.

Dole* has summarized the characteristics which planets must have to be habitable for man. The most important are these:

1 The mass must be greater than 0.4 and less than 2.35 earth masses. The lower limit is set by the requirement that the planet be able to hold an atmosphere for the time required for life to develop. The upper limit is set by the belief that man would find uninhabitable a planet with a g at its surface more than 1.5 times the value of g at the surface of the earth. (A planet with composition similar to that of the earth with such a value of g would have a mass approximately 2.35 the mass of the earth.) That upper limit restric-

*Stephen H. Dole. "Habitable Planets for Man," 2d ed., p. 103. American Elsevier Publishing Company, Inc., New York, 1970.

tion does not necessarily apply to life which might arise on the planet, however, for evolutionary processes surely could produce life adapted to a greater value of g.

2 The period of rotation on its axis must be less than about 96 h. The reason for this limit is that a reasonable rotation rate will ensure acceptable temperature extremes. If the day is too long, the daylit side will become too hot and the night side will become too cold. Man probably would not find habitable a planet which always turned the same face toward its primary. Whether life could arise on planets we would find unsuitable in this respect is debatable, but the restriction on permissible temperature variation seems more likely to be valid for all life than does the restriction on the value of g.

3 The eccentricity of the orbit must be less than about 0.2. The reason for this is again an acceptable range of temperatures, but this time during the course of a "year." If the eccentricity is too high, the planet will come so close to the star at one time that its temperature will go very high, and then will go so far away that the temperature will fall drastically.

4 Finally, the mass of the star must lie between 1.43 and 0.72 solar masses. The upper limit is set by the requirement that the star spend a sufficiently long time on the main sequence that life can develop; a mass of 1.43 solar masses corresponds to an F2-type star. The lower limit comes from a more obscure argument; that is, if the star is smaller than 0.72 solar mass (type K1) a planet would have to be so close to the star to get enough heat to support life that tidal forces would cause the planet to always turn the same face to the star.

Dole concludes that our galaxy contains about 600 million habitable planets, that it is probable that one habitable planet lies within about 27 light years of us, and that 50 habitable planets probably lie within 100 light years of us.

PROBABILITY OF LIFE

The problems now change from physical and astronomical considerations to biological. If habitable planets exist, what is the probability that life has arisen on some of them? Unfortunately, we know of life in only one place, and consequently our basis for estimating probabilities of its development is very small.

Origin of Life

The explanation of life which is most commonly accepted postulates that life began from simple molecules which were produced in the atmosphere, perhaps by the action of lightning or of ultraviolet radiation from the sun. From the time simple organic molecules arose until life appeared, and then until man appeared, the process was one of random rearrangements and combinations of the materials available, and it can reasonably be inferred that what happened once could happen again. Estimating the probability that the same sort of thing would happen again within a period of 3 billion years, however, is a matter of making a wild guess.

Organic Molecules in Space

Within recent years the probability that life has arisen on other planets has seemed to increase because of the discovery of organic molecules in space. Almost 30 molecules have been found in gas clouds in space, some of them organic molecules which might be precursors of amino acids and other molecules essential to life. Since those molecules probably exist in clouds which will eventually form stars, and perhaps planets, one may conclude that a new planet may not begin with a totally barren surface, on which organic molecules must be formed from the basic elements by and action of lightning or some other intense energy source. Perhaps enough molecular contaminants are present from the first to provide a starting point for organic evolution.

This kind of argument has been strengthened by the discovery of organic molecules, particularly amino acids, in meteorites. Almost certainly these molecules are abiotic in origin, but at some time in the past they have formed in the rather inhospitable environment of space. Probably the meteorites in which we have found such amino acids are fragments from the planetoid belt, between Mars and Jupiter, and they bring with them indications of processes which occurred in that region when planets were forming. The fact that they contain amino acids which are essential for life strongly suggests that the surface of the earth may have had more building materials for life from the first than has usually been assumed.

Clearly the more evidence of the precursors of life we find in space, where there is no life, the more likely it seems that under favorable conditions on a planet, life may have actually developed. Some astronomers and physicists seem to be quite optimistic about the development of life when the conditions are right, to the extent of assuming that every planet which is habitable is inhabited, or to stating flatly that somewhere out in space intelligent creatures are watching us, trying to detect evidence of life here. Biologists tend to be more cautious about predicting life. They are more impressed than physical scientists with the many improbable events which must happen in

the course of billions of years of evolution in order to produce
man, and they are less sure that life is the inevitable result of favorable conditions. On the matter of intelligence, there is even
more divergence, with the biologists again pointing to the
many ways evolution could go, even if life exists, without ar-
riving at what we would call intelligence. Persons exasperated
by the blunderings of mankind have sometimes even
suggested that intelligence has not appeared on the earth!

POSSIBILITY OF COMMUNICATION

Is there any reason to care about whether intelligent life exists
elsewhere in the universe other than idle curiosity and the fact
that it is fun to think about such possibilities?

Radio Communication

There is a reason, and it is that we now have the capability of
attempting radio communication with other civilizations at a
state of development comparable to our own.

In the autumn of 1960, an effort called Project Ozma, named
for the queen in the Wizard of Oz books, began under the direction of Frank Drake. The 85-ft radio telescope at Green Bank,
West Virginia, was used, and for a total time of about 200 h it
looked (or listened) for radio indications of intelligence from
possible planets about two stars, Epsilon Eridani and Tau Ceti.
The stars were chosen because they seemed likely candidates
for habitable planets, and because they are relatively close,
each about 11 light years from the earth.

At what frequency should one listen for possible radio trans-
missions from another planet? If we hope to eavesdrop on a
conversation or a radio broadcast for local consumption, there
is no way of guessing a frequency to try, but if we assume that
the natives of that planet are trying to signal to us or other life
in the universe, then we can use clever arguments to choose a
frequency (or wavelength). Drake used the 21-cm wavelength
we have met before for the Project Ozma search. The argument
that leads to that choice goes something like this.

If a civilization is trying to signal to others which may be out
among the stars, it will have to choose one or a few wavelengths
in order to keep the power requirements reasonable. It will try
to find a wavelength for which the recipients are likely to have
receivers—preferably very sensitive detectors. Any civilization
which is sufficiently advanced to be able to produce such
radio equipment will have discovered the 21-cm line of hydrogen and is likely to have built radio telescopes designed to detect that wavelength efficiently because of its importance to as-

tronomers. Therefore if one were attempting to choose a wavelength for which receivers probably would be available, the 21-cm wavelength would be a natural choice. And so, supposing that any other intelligent creatures would reason as we do, Drake listened at 21 cm. He found no indication of artificial radio signals, but after all, 200 h is not long to listen.

This sort of search presupposes more than that habitable planets exist, that life exists, and that intelligence exists. It presupposes that at the present time a civilization exists at the state which makes interstellar communication possible, and that may be an unlikely condition. How long have human beings been in a state one would call "intelligent?" If one is willing to grant that we are in that state now, then we have been so for thousands or even millions of years. But how long have we had instruments capable of listening to planets 11 light years away? The answer is: only a few years. And how long will this civilization remain at a level high enough to mount such an effort? That is not known, and estimates of the probable life expectancy of our civilization fluctuate with optimism or pessimism about arms reduction and control. In any case, the history of the world suggests that a civilization capable of carrying on interstellar conversations may not have an infinite lifetime. If the inhabitants of a planet circling one of our neighboring stars sent signals to the earth during the time of Pericles, or during the reign of Elizabeth I, or even during the Second World War, they must have concluded either that life does not exist here or that intelligence has not arisen in this planetary system.

If everyone in the universe listens, but no one sends messages, communication will be slow. And so work has been done to plan strategies by which signals could be sent toward stars likely to be supporting life in order to maximize the probability of making contact with that life for a fixed amount of investment of time and money. Besides identifying the stars which are the best candidates for investigation, such strategies must consider the frequencies at which messages should be broadcast, the length of broadcasts, the frequency with which signals should be beamed at each star, and the schedule of listening for response.

Interstellar Travel

Communication by radio seems to be the most efficient way of making contact across interstellar distances, but can we eliminate from consideration actual travel? Most people know about techniques which appear in science fiction writing, such as those involving space warps, travel through "hyperspace," and others. What are the *realistic* chances of interstellar travel?

They are very slight. Despite science fiction, so far as we know, nothing can travel faster than light. Even if a ship could be driven to almost the speed of light, it would require about 100 years, measured in our frame of reference, to travel to a star 100 light years distant, and it would require the same time to return home. There is no way known to overcome that limitation.

There is a way to overcome half the limitation, however. The 200 years travel time will be real for persons on the earth who sent the ship away, but the time may be less for the crew of the ship. If, for example, the ship travels to a star 100 light years from the earth, makes a short stop, and then returns to the earth, and if most of that trip can be made at 99 percent the speed of light, the crew will have aged only 28 years, 14 years for each leg of the trip. (When they are traveling at 99 percent the speed of light, they will see the distance they must travel to be 14 light years.) Thus it is quite possible that crew members could travel to a planet 100 light years away and return, but they would return to the happy embraces of their great-great-great-grandchildren.

This prediction of the special theory of relativity is very well established, and there can be no doubt now of its accuracy. It may sound as fanciful as the space warp of some popular science fiction, but it is a cold fact. It has been abundantly tested, although in not so dramatic a way as by an actual trip to another star. In spite of the solidity of this prediction, however, interstellar travel would be extremely difficult, for the energy required to drive a ship to 99 percent the speed of light is quite considerable, and no conceivable source of such energy is known. We must regretfully conclude, therefore, that for the present space travel to other inhabited planetary systems is out of our reach.

Nonscientists often do not understand why scientists dismiss notions of travel by jumps through "hyperspace" or by use of a space warp, and they may think that the scientists are being unscientifically dogmatic in asserting that such matters do not deserve study. A brief explanation of the attitude of most scientists toward such topics may be useful.

Any method of travel at speeds greater than the speed of light would have to be based upon a science which we do not have. Such scientific knowledge may be found, and such a science may be developed, but we cannot now see the roots of it. One cannot say that it will not happen, for several times in the past abrupt changes in the direction of scientific development have occurred, and theories have been developed or facts have been found which were not extrapolations of previously known science. At the present time, however, we can see no way to extrapolate from present scientific knowledge to any theory leading to a method of traveling above the speed of light. If someone should offer 1 billion dollars for the development of a method of traveling faster than light, no physicist in the world would have any idea in which direction to begin working. So long as that condition exists, scientists remain skeptical about the possibility of developing that kind of travel.

One other method of searching for evidence of life elsewhere in the universe should be mentioned. Some people look at legends in our own past for indications that the earth has been visited in the past by creatures from superior civilizations—civilizations which have learned to travel the distances between the stars. It has been suggested, for example, that the wheels Ezekiel saw in the air were spaceships, and that what appear to be abrupt upward steps in the culture of ancient peoples are indications of education by creatures from other planets. No such speculations have been proved, however, and probably none can be proved.

Finally, we turn to a consideration of the form life might take if it arose on another planet. The argument that similar needs will produce similar form has even been taken to the extent of suggesting that any intelligent creature anywhere will have two eyes. (It is not clear why two eyes are superior to three equally spaced around the head, if indeed the head is to be a universal structure. Perhaps on another planet nature has filled in that regrettable blind spot we humans have in the backs of our heads.) Arguments insisting that we have the only reasonable form sound very much like Pangloss, however, and they are not accepted by everyone. Loren Eiseley, in *The Immense Journey*, has this to say about the probability that on another planet evolution has taken the same route as on the earth and has produced creatures just like us.

> In a universe whose size is beyond human imagining, where our world floats like a dust mote in the void of night, men have grown inconceivably lonely. We scan the time scale and the mechanisms of life itself for portents and signs of the invisible. As the only thinking mammals on the planet—perhaps the only thinking animals in the entire sidereal universe—the burden of consciousness has grown heavy upon us. We watch the stars, but the signs are uncertain. We uncover the bones of the past and seek for our origins. There is a path there, but it appears to wander. The vagaries of the road may have a meaning, however; it is thus we torture ourselves.
>
> Lights come and go in the night sky. Men, troubled at last by the things they build, may toss in their sleep and dream bad dreams, or lie awake while the meteors whisper greenly overhead. But nowhere in all space or on a thousand worlds will there be men to

share our loneliness. There may be wisdom; there may be power; somewhere across space great instruments, handled by strange, manipulative organs, may stare vainly at our floating cloud wrack, their owners yearning as we yearn. Nevertheless, in the nature of life and in the principles of evolution we have had our answer. Of men elsewhere, and beyond, there will be none forever.[1]

QUESTIONS

1 If a body of planetary size were wandering in space and came near the sun, on what sort of path would it be likely to move as it passed near the sun?

2 Why does the condensation hypothesis for the formation of the solar system require the presence of a magnetic field?

3 Summarize the three arguments given in the chapter for suspecting that other stars have planetary systems.

4 If life arose on a planet with a surface gravity considerably greater than that of the earth, how might the structures of living things differ from those of earth?

5 Why is it suggested that high forms of life are not likely to have arisen on planets circling stars with masses twice the sun's mass?

6 Verify that if planets in our neighborhood are assumed uniformly distributed in space, if one is to be expected within 27 light years of us, 50 are to be expected within 100 light years.

7 On the same assumption as the previous question, how many habitable planets should be expected within a distance of 200 light years? *Ans* 400

8 If radio signals were sent to Tau Ceti, how long should one wait to begin listening for a reply? *Ans* 22 years

9 If one could travel 80 percent of the speed of light, how long, measured in the earth frame of reference, would a ship require to travel to Tau Ceti? *Ans* 14 years

---

[1]Loren Eiseley, "The Immense Journey," p. 161, Random House, New York, 1957.

the old
the new
and the
unknown

three

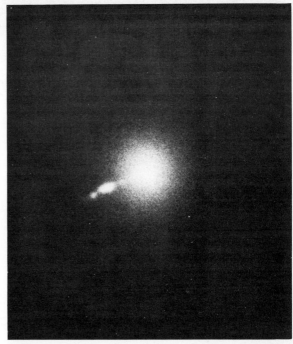

2

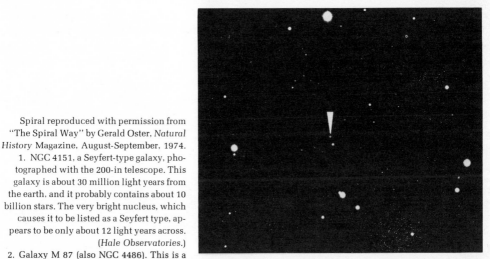

Spiral reproduced with permission from "The Spiral Way" by Gerald Oster, *Natural History* Magazine, August-September, 1974.

1. NGC 4151, a Seyfert-type galaxy, photographed with the 200-in telescope. This galaxy is about 30 million light years from the earth, and it probably contains about 10 billion stars. The very bright nucleus, which causes it to be listed as a Seyfert type, appears to be only about 12 light years across. (*Hale Observatories.*)

2. Galaxy M 87 (also NGC 4486). This is a radio source known as Virgo A. Note the jet of luminous material projecting from one side. The light from the jet is partially polarized, indicating that it is produced by the synchrotron process. (*Lick Observatory photograph.*)

3. Photograph (200-in telescope) and spectrum of 3C 295, a galaxy which is a strong source of radio noise. It probably is more than 2,500 million pc distant. (*Hale Observatories.*)

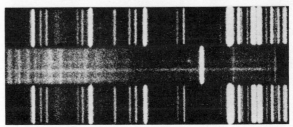

3

In this section we deal with new discoveries—most of them made since 1960—of things which are old, some as old as the universe itself. We will consider how the structure of the universe has changed and try to extrapolate back to the beginning.

Cosmology, the study of the structure of the universe, and cosmogony, the study of the origin of the universe, are not new

topics; questions about the history of the universe and its large-scale structure have always fascinated many people. But cosmology has been a field of study with so many questions and so few facts that it has bordered on pure speculation. Within the past half century, however, and even more within the past quarter century, some facts have become available, and although the questions still

RELATION BETWEEN RED-SHIFT AND DISTANCE FOR EXTRAGALACTIC NEBULAE

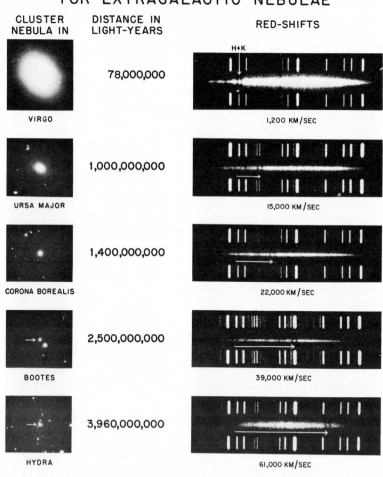

| CLUSTER NEBULA IN | DISTANCE IN LIGHT-YEARS | RED-SHIFTS |

VIRGO — 78,000,000 — H+K — 1,200 KM/SEC

URSA MAJOR — 1,000,000,000 — 15,000 KM/SEC

CORONA BOREALIS — 1,400,000,000 — 22,000 KM/SEC

BOOTES — 2,500,000,000 — 39,000 KM/SEC

HYDRA — 3,960,000,000 — 61,000 KM/SEC

Red-shifts are expressed as velocities, $c\,d\lambda/\lambda$. Arrows indicate shift for calcium lines H and K. One light-year equals about 9.5 trillion kilometers, or 9.5×10^{12} kilometers.

Distances are based on an expansion rate of 50 km/sec per million parsecs.

4. (Hale Observatories.)

outnumber the facts, enough progress has been made to engender optimism and excitement. The most significant change is not that answers to old questions have been found, for most of the answers are still ambiguous, but rather that for the first time in human history the experimental procedures which may produce the answers to our questions can be described, and the equipment with which to make the important measurements seems to be within our technical capabilities. Possibly the answers to our questions about the fundamental nature of the physical universe will prove as elusive as the end of the rainbow, always receding as we attempt to approach it, but at least within the past two decades the rainbow has become much brighter, and we are encouraged to hope that

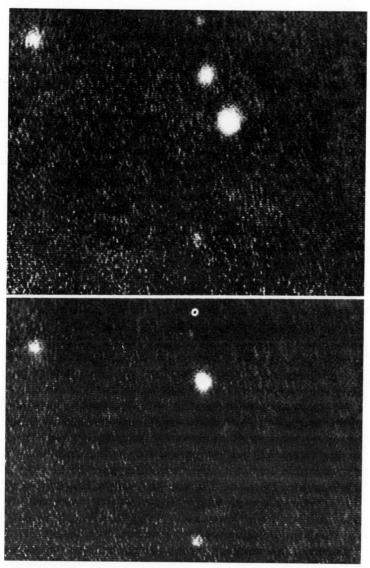

5. Television photographs of the pulsar in the Crab nebula, taken at minimum and maximum light. Notice that at minimum light one star (the pulsar) disappears entirely. (*Lick Observatory photograph.*)

we may reach at least part of the gold which is supposed to lie at its end.

Quasars, pulsars, and black holes regularly make the front pages of newspapers, and they have become common topics for the science sections of newsmagazines. Within the scientific community excitement has run high over these same discoveries—excitement that has expressed itself in conferences, symposiums, and special sessions of professional meetings. The educated reader should know enough of the background of these discoveries to understand what he reads in the popular press, and this section is intended to provide that knowledge. More important, however, the educated reader should know *why* these discoveries are significant. They are more than mere curiosities. They pose challenges to some of the basic concepts of the physical scien-

Quasi-stellar Radio Sources

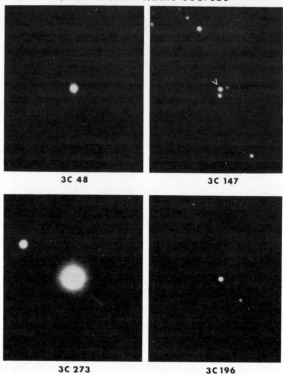

3C 48

3C 147

3C 273

3C 196

6

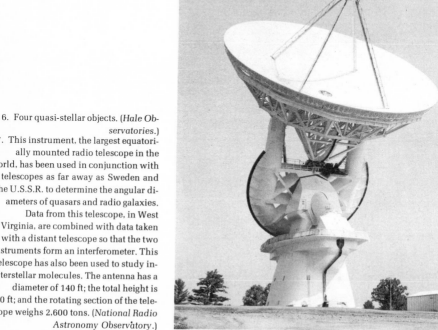

6. Four quasi-stellar objects. (*Hale Observatories.*)

7. This instrument, the largest equatorially mounted radio telescope in the world, has been used in conjunction with telescopes as far away as Sweden and the U.S.S.R. to determine the angular diameters of quasars and radio galaxies. Data from this telescope, in West Virginia, are combined with data taken with a distant telescope so that the two instruments form an interferometer. This telescope has also been used to study interstellar molecules. The antenna has a diameter of 140 ft; the total height is 200 ft; and the rotating section of the telescope weighs 2,600 tons. (*National Radio Astronomy Observatory.*)

7

ces, and they suggest that nature may have important processes of energy transformation that we do not yet understand. The problems raised by these discoveries have stimulated reflection and research on the most fundamental problems of physical science, and for that reason knowledge about them should be part of the equipment of every educated mind.

8. The curvature (zero, positive, or negative) of a surface could be determined by measurements made on the surface of the sum of the interior angles of a triangle or of the ratio of the circumference of a circle to its radius. Analogous tests could indicate the curvature of the three-dimensional space of our universe.

9. A black hole is a region where the curvature of space-time has become so great that nothing can escape from it. The fall into a black hole is a one-way trip, and objects which fall in are destroyed by tidal forces within the hole.

10. The universe is expanding, which means that distant galaxies are receding from us and from each other.

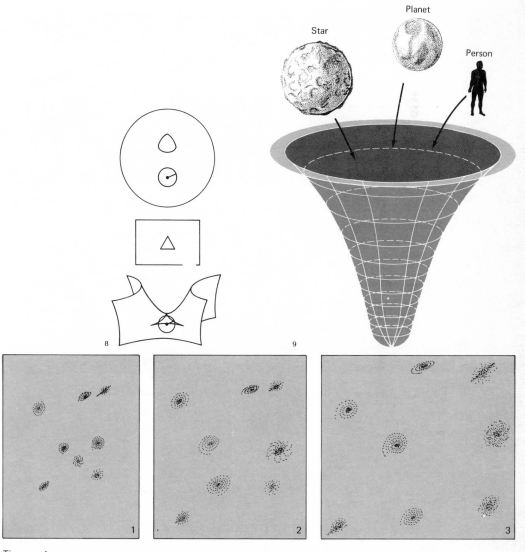

Star

Planet

Person

8

9

1

2

3

Time ⟶

16
the cosmological problem–a first look

In order for the significance of some of the measurements and controversies to be discussed in the following units to be apparent, we must have some grasp of notions basic to cosmology. This unit will present those notions, and detailed study of cosmological models will be presented in a later unit.

EXPANSION OF THE UNIVERSE

The most fundamental idea underlying all cosmological speculation is that the universe is expanding. We can give two arguments to support the belief, and we begin with Hubble's law.

Hubble's Law

As we have seen, from the early days of telescopic observation, nebulae which could not be resolved into individual stars were observed. As late as 1921, however, prominent astronomers rejected Immanuel Kant's idea that some of the nebulae were "island universes," and continued to maintain that the Milky Way galaxy comprised all the universe we can see. One of the men who at that time championed the belief that the spiral nebulae are other galaxies of stars suggested that we might be able to see 10 or even 100 million light years into space! Such distances, which then seemed to tax the imagination, are now commonplace; galaxies 10 million light years away are near neighbors.

In 1924 Edwin Hubble established clearly that some of the spirals contain Cepheid variables, and his estimates of their distances put them far outside the Milky Way. To estimate distances to galaxies so far away that their Cepheids could not be seen, Hubble assumed that the brightest stars seen in all galaxies are about the same, and from galaxies showing both superbright stars and Cepheids he determined the magnitudes of the bright stars. Then to be able to discuss galaxies so distant that no stars can be resolved, he determined that the absolute luminosities of galaxies do not vary tremendously, and so the luminosity of an entire galaxy can be used as a basis for determining its distance. His distances were all too small, but they were sufficient to permit the next step in the story.

When the light from a galaxy is passed into a spectrograph, certain lines which many stars show prominently can be seen, and a doppler shift can be observed if the galaxy is approaching or receding from us. The effect of the motion of stars within the galaxy is to broaden the lines. As early as 1912 the doppler shift of the center of the Andromeda galaxy was measured, and the galaxy was found to be approaching us at a speed of about 200 km/s. Later, when the rotation of our galaxy was recognized, our motion around the edge of the giant merry-go-round was found to be carrying us toward Andromeda at a speed to account for half that motion; the galaxy is actually approaching our own galaxy at a speed of only about 100 km/s. Before 1920 the speeds of several galaxies had been measured, and although some were approaching, most were moving away from us.

After the demonstration that the galaxies are stellar systems entirely outside our own, Hubble began comparing the doppler shifts shown by galaxies with his distance measurements, and by 1929 he could say that outside the Local Group and out to 6 million light years, the velocity of a galaxy is proportional to its distance. Within the Local Group, the 17 or 18 galaxies with which our own is associated and which are our neighbors, velocities are rather random; some of our neighbors are approaching us and some are leaving us, but all are traveling at low velocities. At greater distances, however, all galaxies are leaving us, and the velocities become very large. At the limit of Hubble's measurements, the velocities were approaching 1,000 km/s.

Cosmological redshift We are making an assumption which should be pointed out and discussed. What one actually observes is a displacement of spectral lines toward the red end of the spectrum; that displacement is called the *cosmological redshift*. The assumption is that the redshift is a doppler effect, and that we can compute a velocity from measurements of the redshift.

From the time of the discovery of the cosmological redshift to the present some people have argued that it is not a doppler effect—that some effect other than relative velocity causes the shift in spectral lines. Most of the explanations have been of the "tired-light" variety—suggestions that as light travels great distances for great times, it becomes "tired" and loses part of its energy. If a photon lost part of its energy, its wavelength would appear greater, and it would appear in the spectrum displaced toward the red. No one, however, has been able to propose any mechanism that would cause light to lose energy. and there is no other evidence that such a thing happens. In the absence of a complete theory that might explain the tiring of light, most astronomers and astrophysicists prefer to interpret the redshift as a doppler effect and to believe that it indicates the velocity of the source relative to us. (The assumption that the doppler effect is the only explanation for redshifts is being challenged now by Professor Halton Arp on different grounds; his evidence will be considered in the chapter on quasi-stellar objects.)

Hubble's constant If we believe that the velocities of recession of distant galaxies are proportional to their distances, we can express that fact in an equation as

$$V = Hr$$

where V = velocity
r = distance
H = Hubble's constant

Because we may want to admit the possibility that H may have changed and may be changing at all times, we should write the equation

$$V = H_0 r$$

where H_0 is the value of the Hubble constant now. When Hubble first announced the equation, he thought that the constant was 500 km/(s)(Mpc). In other words, he thought that for every million parsecs (megaparsec) one goes out, the speed of galaxies increases about 500 km/s. Since 1929 many measurements of that constant have been made, and its value has steadily come down. The most recent measurement, given by Allan Sandage, is 55 km/(s)(Mpc).

That changing value for H_0 does not mean that H is changing; if it changes at all, the effects could be seen only over millions of years. The changing value reflects improvements in the measurements of the distances to galaxies. The redshift is rather easily measured, so that V in the equation can be computed easily. Measurements of r are not easy, however, and as methods have been improved the estimates for the distances to galaxies far away have increased. A report which Sandage published in 1971 in *Carnegie Institution Yearbook 70* mentions eight steps, beginning with trigonometric parallax measurements and ending with computations of the distances of galaxies several hundred megaparsecs distant, at which distance the velocities are 22,000 km/s. The plot of velocity versus distance is shown in Fig. 16.1.

How are we to interpret the observation that galaxies appear to be rushing away from us at great speed in every direction? Does it imply that we are at the center of the universe and that we have such an acute case of moral leprosy that everything else in creation is trying to get away from us? The answer is "no." We are not at the center of the universe, and in fact we have no reason to suppose that our position is in any way unique. If we were situated in any of those other galaxies we would also observe everything running away from us.

Raisin-bread model The easiest way to explain how the universe appears to be acting is in terms of the *raisin-bread model*. Imagine that you make a loaf of raisin-bread. The dough is prepared with raisins embedded in it and set to rise. On one of those raisins is a bacterium. It is an unusual bacterium, for it can see other raisins around it out to a considerable distance, but its view does not reach to the edge of the loaf. So far as it can tell, raisins are dotted throughout space as far as it can see. An hour later the bacterium still sees raisins around it, but during that time the dough has doubled in size, and so the raisin that was originally 2 in away is now 4 in; and the one that was 3 in is now 6 in. Since the raisins have made their respective moves in the same time, their speeds have been proportional to their distances from the bacterium. The one that was 1 in away has moved at the rate of 1 in/h; the one that was 3 in away has moved at a speed of 3 in/h. The bacterium is not justified in concluding that it is at the exact center of the loaf, for it would have made the same observations no matter where it was located so long as it was not at the surface.

Let us imagine a movie made of the motion of the galaxies over eons. If the movie is run forward, they will all appear to be moving away from us, but if the movie is run backward, they will all appear to be moving back toward us. As the film runs backward, however, we notice a curious thing: a galaxy which is 100 million light years away is moving back with a particular speed, and we can compute the time when it will reach us. But a galaxy 200 million light years away is moving twice as fast, and therefore it will reach us at the same time. In fact, Hubble's law, that the velocity is proportional to the distance, implies that all the galaxies will reach us at the same time. When we imagine the record of the universe running in the forward direction, then, we conclude that all the galaxies must have started from some one spot originally; this is the basis of the *big-bang* model of the universe.

FIGURE 16.1 A plot of the Hubble relation: velocities of recession of distant galaxies are proportional to their distances from us.

Hubble time If we want to know how long ago the expansion began, we can compute the time easily from $V = Hr$. If any galaxy which is now a distance r from us has reached that point moving at a speed V, the time it has required is

$$t = \frac{r}{V} = \frac{1}{H} = 18 \text{ billion years}$$

That time, 18 billion years, is called the *Hubble time*. It probably is not the actual time since expansion began, for the Hubble constant almost certainly has been decreasing since the beginning because the mutual gravitational attraction of all parts of the universe is slowing the expansion. Estimations of the exact time since the beginning of the universe depend upon the model of the universe one wishes to use. Reasonable estimates range from 11 to 16 billion years.

Olbers's Paradox

A second kind of argument, known as *Olbers's paradox*, can give indirect evidence for the expansion of the universe. Olbers's paradox, like Bode's law, seems to be a case of inappropriate naming, for Olbers mentioned that the problem was discussed by Edmund Halley, and Halley was not the first to point it out. Furthermore Olbers, whose paper on the subject was published in 1826, seems to have had no notion that he was treating a profound cosmological problem; he was simply trying to estimate the dimming of starlight by interstellar matter.

Regardless of the history of the reasoning, the argument does tell us something about the universe at large. Let us consider the following question: Why is the sky dark at night? One's first answer probably is: Because the sun is hidden by the earth. And of course that is part of the answer. But consider the following sort of reasoning. Let us assume that the stars are spread rather uniformly throughout space. (This is the kind of assumption made by Olbers; we will take advantage of our superior knowledge and replace stars by galaxies, so that we imagine galaxies spread uniformly throughout space. But since the galaxies are made of stars, we can still talk about the stars.) Let us suppose that the density of stars is given by N per unit volume, for example, N per cubic parsec. Now if a large sphere is drawn about the earth, and then a thin shell is drawn about that, as shown in Fig. 16.2, where the radius of the sphere is R and the thickness of the shell is ΔR, the volume of the shell is approximately $4\pi R^2 \, \Delta R$. If there are N stars per unit volume throughout space, the shell contains $4\pi N R^2 \, \Delta R$ stars. But the light we receive from a star varies as $1/R^2$, and so the effect of those stars at the earth is proportional to

$$\frac{4\pi R^2 N \; \Delta R}{R^2} = 4\pi N \; \Delta R$$

The important thing to notice about that result is that the light we receive from the stars in the shell does not depend upon the radius of the shell, for the number of stars in the shell is proportional to the square of its radius, but the effect of each star is proportional to the reciprocal of the square of that radius, and the two effects cancel. Now if we regard all space as being divided into many thin shells, each one should contribute the same amount of light to the earth as any other, and the amount of light the earth should receive from the stars, even at night, would seem to be infinite. That result is not quite correct, for eventually if the stars out to greater and greater distances are considered they begin to lie one behind the other, and so actually the night sky should look like a vault completely covered by stars. The same thing should be true of the daytime sky, and the earth's temperature should be about 10,000 K.

Obviously the earth is not at a temperature of 10,000 degrees, and the nighttime sky is dark; therein lies the paradox. The mathematical part of the argument—the manipulation of 4π's, R's, and N's—is correct, and the explanation of the paradox is not to be sought in some unfair trick with the mathematics.

The suggestion Olbers made to explain the darkness of the night sky is that material in space between the stars blocks part of the light from us. He was right in that dust does dim the stars, and his estimate of the amount of dimming was close to the value we accept now, but that is not a satisfactory explanation of the paradox. If light is pouring in on us at the rate our computation indicates, dust clouds may stop some of it, but they can stop it only by absorbing it. But dust which absorbs light must become warm, and as the dust becomes warm it begins to radiate light again. If that were the only thing between us and that blazing vault of fire, we could expect only a short respite before the dust became luminous, radiating at the same temperature as the stars.

We made an assumption about the extent of the stars, and perhaps that assumption is wrong. In saying that the light

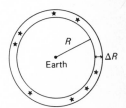

FIGURE 16.2 A thin shell of radius R and thickness ΔR, drawn about the earth, has volume $4\pi R^2 \Delta R$. The number of stars within the shell is presumed to be proportional to its volume.

would become infinite or even that eventually the effect is the same as if every point in the sky is filled by a star, we are assuming that the stars (or galaxies) extend out to infinity; more precisely, we are assuming that the stars extend far enough to provide the very large number of stars we need for this dire result. Perhaps the universe of stars is limited, and somewhere out there the stars come to an end—abruptly or gradually. If that is the case, the shells we described can be extended out only so far, and we need not worry about the sum of an unlimited number of shells. Whether such a limit to the visible universe exists we do not know, but we have not seen it. So far as our best telescopes can reach, galaxies seem to continue at about the concentration we know them at shorter distances, and quasars even seem to increase at very great distances. This does not seem like the solution to Olbers's paradox.

The problem as stated contains another assumption, so well hidden that probably the reader has not thought of it as an assumption. That is that the stars (or galaxies) are stationary. For if they are not stationary, then their motion weakens the light we receive from them. Hubble's law implies that there does exist an horizon of sorts for the visible universe, for galaxies sufficiently far away are moving at the speed of light, and their light can never reach us. For galaxies much closer than that, however, their motion has two effects on the light we receive. First, the light is shifted to the red, and hence each photon has less energy. Second, because the photons which the source emits in some period of time are stretched out so that we receive them over a longer time period, the intensity of the source is decreased. Both these effects combine to make the light from very distant galaxies much less effective in brightening our night sky than the original statement of the problem implied, and possibly this is why the sky is dark. This argument was accepted for many years.

In 1974, however, E. R. Harrison pointed out that the usual discussion of this paradox neglects any consideration of the time required for the night sky to become bright. According to Harrison's computation, if the night sky is to be as bright as the usual discussions of Olbers's paradox propose, it must simultaneously receive light from stars near us and from those out to a distance of about 10^{24} light years. But if light from a region 10^{24} light years deep is required, two conditions must be met: the universe must be at least 10^{24} years old, and stars must radiate energy for times at least as great as 10^{24} years. Whether the universe is as old as 10^{24} years is a matter we must consider, but we already know that individual stars do not remain in a highly luminous state for 10^{24} years. The last point, about the lifetimes of stars, can be put in another way: If all the known matter in the universe were changed into radiation, it would fall far short of being enough radiation to fill all space at the level of intensity Olbers's paradox predicts.

It appears, then, that the night sky should be bright only if three conditions are satisfied:

1 The universe is not expanding, or the expansion is quite slow.
2 The universe has been in existence for at least 10^{24} years.
3 Enough matter is present and being cycled through stars to keep stars shining throughout space for a time of the order of 10^{24} years.

If any one of the three conditions is not satisfied, the night sky should be dark. It is probable that none of the three is satisfied. The short lifetimes of stars probably is most important in producing the dark sky, and the expansion of the universe may be least important. Our best evidence now indicates that the age of the universe is far less than 10^{24} years, and that fact would guarantee a dark night sky even if the other two conditions could be satisfied. Olbers's paradox thus becomes an argument for *either* the expansion of the universe or some limits on its age, but in any case it implies something about the universe which is of importance to the cosmologist.

THE COSMOLOGIST'S PROBLEMS

The task of the cosmologist is to construct models of the universe and then compare the models with the reality which can be seen, hoping that one of the models will have characteristics which match closely the real world. The view we have of the universe from one spot in it and one time does not permit us to answer all the questions we may have by direct observation, and so we are forced to construct models in an effort to reconstruct the past and predict the future. In the construction of such models the cosmologist confronts fundamental problems which he cannot solve but cannot avoid.

Extrapolation of Physical Laws

We think that we understand rather well the physical laws which govern the earth, and we have good reason to believe that those same laws describe the solar system and the stars near us. In order to construct a model of the universe we must use laws to describe the behavior of matter over vast distances and times, and we have no choice but to use the laws we have discovered by experiments and observations in our small corner of space during a time period of a few years or centuries.

How do we know that the same laws of nature operate over distances equal to the diameter of the universe and over times equal to the lifetime of the universe? The answer, of course, is that we do not, but we have no indication that the laws are changed at all at the greatest distances and times we can observe, and in any case, we have no other laws to use. We are forced to use those we know.

Completeness of Laws

A related problem concerns the possibility that we know only approximations to the correct physical laws, and that terms which are so small that they have escaped our notice should be added to the equations expressing the laws. An example of a problem of this type is the *cosmological constant*, a constant which can be placed into the starting equation in general relativity, which has an effect on cosmological models. Probably the constant is zero, but its effect, if it is not zero, is so small that no conceivable experiment on such a small scale as the solar system will reveal its value. If the cosmological constant is not zero, its effects can be seen only on a scale comparable to the size of the universe itself.

Uniqueness of the Universe

Scientists like to have many examples of objects they are studying so that the essential characteristics can be separated from the incidental, individual variations. The cosmologist is denied this aid to his work—he knows only one universe. We have no way of knowing what characteristics of our universe are consequences of fundamental laws and hence would be found in all universes which might possibly exist and which characteristics are accidents of our particular universe. Scientists would like to focus attention on the former and neglect the latter, if they only knew which was which!

Multiplicity of Models

General relativity theory seems to be the only physical theory which can be the basis of models of the universe, but one can construct many different models using general relativity. We might wish for a theory which would permit only one model, so that we could be sure of the nature of our world at distances and times we cannot examine, but our only satisfactory basic theory does not yield a single model, and we are forced to make many models which might match the real world and then to look for similarities and differences.

All cosmological models can be divided into two classes: *evolutionary* models and *steady-state* models.

Evolutionary Models

All evolutionary models are based upon the *cosmological principle*, which is the assumption that we are not at a special position in the universe. Having taken note of the experience of our ancestors, who erroneously thought first that the earth, then the sun, and still later our galaxy was the center of the universe, the modern cosmologist has elevated into a principle the lesson to be drawn from that series of delusions of grandeur, and refuses to consider any model which would give the earth, the sun, or our galaxy a favored position. The cosmological principle is the assumption that any observers similar to us anywhere else in the universe would observe the same large-scale structure of the universe that we see. Such observers might see differences in their planetary systems and their galaxies, but those would be local effects. They should see the same expanding universe we see, with the same value of Hubble's constant. Like us, they should see all distant galaxies receding from them, and should see the same density of galaxies in space that we see.

With the assumption of the cosmological principle to help interpret observations and with the general theory of relativity to predict how matter will behave, one can construct models of the universe which describe an expansion from a dense, hot state to the present large volume, low density, and low temperature. Many models can be fit to the data we have now, and they make differing predictions for the future development of the universe; these will be discussed in more detail later. The feature they have in common, however, is that they predict evolution—a change in the appearance and condition of the universe with time.

Steady-state Models

In the late 1940s three men, Hoyle, Bondi, and Gold, proposed an entirely different sort of model—the *steady-state model*. Although this is actually not a single model but a class of models, the various examples differ among themselves less than the evolutionary models. One of the reasons for introducing the steady-state model was that in the forties evolutionary models and work on stellar evolution seemed to be in serious conflict. The value accepted for Hubble's constant was so large that it implied that the universe had not existed long enough for a star like the sun to evolve to its present form. This apparent conflict has disappeared as later, and presumably more accurate, measurements of Hubble's constant have lowered its value. The fundamental idea of the steady-state models is that the cosmological principle should be expanded into something the authors of the idea called the *perfect cosmological principle*. (Perhaps more modest authors would have been content with something like the extended cosmological principle.) The perfect cosmological principle is the assumption that not only are we not at a special position in space, but we are also not at a special position in time. Not only could an observer go to any other position in the universe and see the same conditions around him, but the same observer could come back at any time in the future (or past) and also see the same conditions. If we could return to observe the universe 10 billion years hence, we would see the same large-scale conditions we see—the same expansion with the same Hubble's constant, the same density of galaxies within view, etc. In other words, nothing changes with time on the large scale.

One consequence of that assumption comes to mind immediately. If the galaxies at great distances are moving away very rapidly, other galaxies must form to take their places so that the total number of galaxies within view remains the same and the distribution with distance remains the same. The steady-state theory requires that matter come into existence steadily at just the rate required to equal the depletion it would otherwise suffer from the expansion. This aspect of the theory is termed *continuous creation*, and it is a prediction which has aroused skepticism but which cannot be checked experimentally. The predicted rate of creation is approximately one hydrogen atom in each cubic meter every 10 billion years. If we could observe a cube 1 km on a side, we should expect one hydrogen atom to appear within the cube every 10 years, on the average. Clearly we are unable to detect such a small rate of creation of matter even if it is real.

The steady-state theory makes other predictions which can be checked, and some of those will be discussed in the following chapters. One prediction, which is inherent in the nature of the theory, is that we shall see no evidence of aging of the universe. In other words, as we look back in time, we shall see the same kinds of objects and the same density of objects at all times. To take one example, the number of quasars visible in the sky should have been the same at all times in the past and should remain the same for all the future.

Perhaps the anticipation of revelations to come in succeeding chapters will not be lessened greatly to mention that the steady-state theory is now almost entirely abandoned. It has run into such serious problems that most cosmologists no longer consider it a possible description of our universe. Attention now is focused almost entirely upon choosing the one of many evolutionary models which best fits our data about the real world.

QUESTIONS

1 If the center of the Andromeda galaxy is approaching us at 200 km/s, at what wavelength is a line in its spectrum seen if the line observed from a laboratory source has a wavelength of 600 nm?   *Ans*   599.6 nm

2 What redshift corresponds to a velocity of 22,000 km/s?

Ans   0.07

3 Verify the statement that if the Hubble constant is 55 km/(s)(Mpc), $1/H$ is 18 billion years (1 pc = 3.1×10^{13} km; 1 year = 3.16×10^7 s).

4 What kinds of observations might be made to attempt to see evidence of changes in the universe in the past to support or refute the steady-state theory?

5 Distinguish between the cosmological principle and the perfect cosmological principle.

17
quasi–stellar objects

Although radio galaxies had been found earlier, the discovery which most clearly ushered in the new era of astronomical findings and excitement was *quasars*, also known as *quasi-stellar objects* (QSOs for short).

DISCOVERY

The story begins in 1960, although at that time no one knew that a remarkable discovery had been made. By then the positions of some radio sources listed in the 3C catalog had been determined sufficiently accurately to encourage optical astronomers to search for visible objects where the radio telescopes were seeing sources. One of the sources thus studied was 3C 48, and at the apparent position of that radio source the optical telescope revealed what appeared to be a star. It had two unusual characteristics, however: its brightness varied irregularly on a time scale of 1 day, and its spectrum showed broad emission lines. Stars normally have absorption lines, and emission lines are most commonly associated with nebulae which are fluorescing, but the emission lines from such nebulae are sharp, not broad. Furthermore, the emission lines were not at the wavelengths of any known lines, and therefore they could not be identified. Not much excitement was generated by the observations.

In 1963 another radio source was located with unusual precision, and again it seemed to be a star. The source 3C 273 happens to lie at such a position in the sky that the moon passes across it. The moon acts like a knife edge cutting off the

FIGURE 17.1 Intensity of radio signal received from a point source as the moon moves across the line of sight.

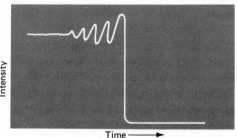

radiation from the source, and as the moon moves across the line of sight, the radio signals from 3C 273 drop, then rise somewhat, and finally drop to zero. What is observed is a diffraction pattern familiar from optics. An analysis of the diffraction pattern, plus the knowledge of the position of the moon, permitted the Australian radio astronomers making the observation to determine the position of 3C 273 with very high precision. It also permitted them to recognize that the radio waves come not from just one source, but from two quite close together.

Recognition of Quasars

When the region in which 3C 273 lies was searched on photographs made with optical telescopes, it appeared that the radio waves come from what seems to be a thirteenth-magnitude star. The star was abnormally blue, and its spectral lines were not recognized as anything familiar. Soon, however, Maarten Schmidt realized that some of the lines looked like the hydrogen spectrum, but they were shifted to the red, and he discovered that if he assumed a redshift of $\Delta\lambda/\lambda = 0.158$, the lines could be explained easily; they were indeed the familiar Balmer lines of hydrogen.

Redshifts are often designated by z, which is defined to be $\Delta\lambda/\lambda$. If the redshift is 0.158, a line which is normally at 400 nm, for example, is shifted an amount $\Delta\lambda = 0.158 \times 400$ nm $= 63.2$ nm, so that the wavelength at which the line actually is observed is 463.2 nm. A redshift of that size indicates that the source is receding from the earth at 16 percent the speed of light, and hence it is a reasonable guess that it is not a star relatively near to us. If we measure its distance by assuming that the redshift arises from the expansion of the universe and hence that the Hubble relation applies to it, we compute a distance for the object of 500 Mpc. But if the object is 500 Mpc away and still appears as a thirteenth magnitude object, it must be approximately 100 times as bright as the brightest galaxy known. That did generate excitement! (If we used the latest value of the Hubble constant, the distance would be more nearly 1,000 Mpc.)

With that clue from the work on 3C 273, Greenstein and Matthews reexamined the spectrum of 3C 48 and found that it could be interpreted if it had a redshift of $z = 0.367$. That clearly implied it was even farther away than 3C 273, and was also a remarkably bright object. Whatever these things were, they were neither stars nor ordinary galaxies. But since they looked on photographic plates like stars, they were designated quasi-stellar radio sources, soon shortened to quasars.

A search for more quasars was in order, but trying to obtain accurate locations of radio sources in the radio catalogs was difficult, and so Allan Sandage hit upon another technique. The quasars which had been found were abnormally blue, and they had high intensity in the near ultraviolet. Sandage photographed regions of the sky through a blue filter, then shifted the telescope slightly and photographed the same region of the sky through an ultraviolet filter. Objects which are very blue (have an ultraviolet excess, to give the technical term) made darker images on the negative through the UV filter, and stars which have normal colors made darker images through the blue filter. A comparison of the two images of each object photographed permitted rapid selection for further study of those that are unusually blue.

Nonradio Sources

The immediate discovery was that there are objects which are like quasars except that they are not radio sources. Various names were given to those objects: interlopers, blue stellar objects (BSOs), quasi-stellar galaxies, etc. Eventually Margaret and Geoffrey Burbidge suggested the name quasi-stellar object for all quasar-like objects whether they are radio sources or not, and that name has become common.

LARGE REDSHIFTS

As more and more QSOs were found, the redshifts which had to be assumed in order to make sense of their spectra increased; in June 1973, z values had reached 3.53. Fortunately, objects with a range of redshifts were found so that spectroscopists could work up gradually to the very large redshifts, and they could be reasonably sure of their identification of the lines which were seen.

We must stop for a moment and consider how one is to interpret a redshift so large as 3.53. For small speeds and small redshifts, we have used the approximate relation

$$z = \frac{\Delta\lambda}{\lambda} \approx \frac{v}{c}$$

where v is the speed of the source and c is the speed of light. If we try to use that relation with $\Delta\lambda/\lambda = 3.53$, we find that the source of the light is leaving us at 3.53 times the speed of light. But we know from relativity that nothing can move at greater than the speed of light, and therefore something must be wrong with our calculation. The problem is that the relation $\Delta\lambda/\lambda \approx v/c$ was derived by assuming that v is very much less than c, which means that z, which is $\Delta\lambda/\lambda$, is very much less than 1. For $z = 0.158$, the condition is satisfied rather well; for $z = 0.367$ it is barely satisfied, but for $z = 3.53$ it certainly is not satisfied at all.

To interpret z's of that size we must go back to the correct
expression given by special relativity. So long as we are considering motion only directly toward us or away from us, with
that speed given by v, the proper relation to use is

$$1 + z = \sqrt{\frac{c + v}{c - v}}$$

The quantity $1 + z$ has a simple meaning:

$$1 + z = 1 + \frac{\Delta\lambda}{\lambda} = \frac{\lambda + \Delta\lambda}{\lambda} = \frac{\lambda_0}{\lambda_L}$$

where λ_0 is the wavelength observed and λ_L is the wavelength
measured in the laboratory.

Suppose that a spectral line which has a wavelength of
300 nm when measured in the laboratory is observed in the
spectrum of an astronomical object to have a wavelength of
600 nm. The change in wavelength $\Delta\lambda = 300$ nm, and $z = \Delta\lambda/\lambda$
$= 300$ nm/300 nm $= 1$. But $1 + z = 2$, and this is the ratio of the
observed wavelength to the laboratory wavelength,
600 nm/300 nm.

Now let us compute the speed required to produce a doppler
shift of 3.53. That is z, and so $1 + z = 4.53$ and

$$4.53 = \sqrt{\frac{c + v}{c - v}}$$

Square both sides of the equation to get

$$20.5 = \frac{c + v}{c - v}$$

from which $v = 0.95c$.

CHARACTERISTICS OF QSOs

How does an astronomer know when he has found a QSO?
What are the identifying characteristics? They are not absolutely certain, and in some cases there is a question about
whether certain objects should be classed as QSOs or something else. All the QSOs share two characteristics: they show
no particular structure and look like stars when viewed
through the telescope; and they have large redshifts. In addition, those with redshifts less than about 3.0 are blue; they may
or may not be radio sources. More than 200 objects are known
which meet the criteria.

Variation in Brightness

One of the important characteristics of QSOs is that many of
them vary randomly in brightness. Both the radio sources and
the optical objects have been found to vary. The quasar 3C 273

is bright enough that it has been recorded on photographs many times during the past 80 years, and these old photographs have now been studied in an attempt to find some periodicity in the variations of the quasar. No periodicity has been found; for the entire 80 years 3C 273 has varied in brightness, but the variations appear to be completely random. Another of the quasars, 3C 446, has varied in brightness by a factor of 2 within a day. This variation will be discussed below as a clue to the size of the source.

Multiple Redshifts

In 1966 absorption lines were discovered in the spectra of some of the QSOs. Stars and galaxies normally have absorption lines, but the spectra of these objects which had been observed up until this time were only emission lines. Now, however, many are known which have both emission and absorption lines. The emission and absorption lines do not have the same redshifts, however. The emission lines always have the larger redshift, and the absorption lines may have one or several smaller redshifts. In at least one object the absorption lines seem to fall into six classes, each with its own redshift. The arrangements of the emission and absorption lines suggest that the source of the radiation has emission lines superimposed on a continuous spectrum. It is moving away from us at a very high speed. Between us and the object are clouds of gas which absorb light from the continuous spectrum; the clouds are receding from us, but at speeds less than that of the source itself. Each cloud, therefore, absorbs light to produce an absorption spectrum at a redshift which indicates its speed, and if we see evidence of six different speeds, there must be at least six clouds. We do not know, however, whether the clouds are associated with the

FIGURE 17.2 Expanding shells of gas around a small source could produce absorption lines with differing doppler shifts. Emission lines come from the source and have the largest redshift.

Observer

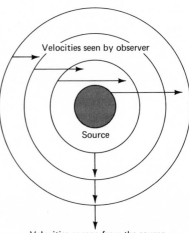

Velocities seen by observer

Source

Velocities as seen from the source

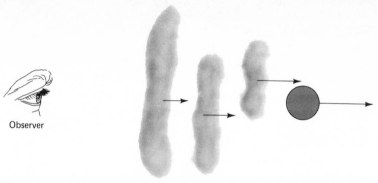

Observer

FIGURE 17.3 Clouds of gas between us and the QSO, each moving at the speed appropriate for its distance from us, could produce absorption lines with differing redshifts.

primary source or whether they are far from it and independent of it. Perhaps the source has shed shells of matter, somewhat like a star sheds mass to become a planetary nebula, and the matter which has been released is expanding about the source. Then the speed a shell has with respect to us is the speed of the source less its expansion speed, and therefore the absorption spectrum produced in the shell has a smaller redshift than the primary source. On the other hand, the light, in its long journey through space, may pass through clouds of gas which are not so far from us, and therefore are not moving away so fast as the QSO, and they may produce the absorption lines. We do not know which of these two pictures is the better description of reality.

Multiple Sources

In 1971 another startling observation related to one QSO was announced: It appeared to contain two sources which were separating at approximately 10 times the speed of light! The radio source 3C 279 was observed by two radio telescopes separated by some distance, and from the interference between the radio signals reaching them it was deduced that 3C 279 consisted of two sources, with a certain angular separation. On the assumption that the object is at the distance which one could compute from its redshift and the Hubble law, the actual separation of the two sources could be computed. A few months later the observation was repeated, and the sources were found to have separated. But when their new separation was computed, it appeared that they must be separating at approximately 10 times the speed of light. Since then the same sort of apparent velocity within other QSOs has been observed.

Physicists and astronomers are extremely reluctant to believe that anything is moving at a speed greater than c, for special relativity, a theory which has been verified in many ways countless times, assures us that such speeds cannot be attained. Various explanations have been offered for the observation. One of the simplest is that perhaps 3C 279 is not so far away as we have supposed. This possibility will be discussed in more detail later, but if the sources were relatively close, the observed angular separation would imply smaller actual separations and therefore lesser velocities. Another possible explanation is that the source may have three or more small bright spots, and their relative brightnesses are changing. (This is sometimes called the *Christmas tree model* because the bright spots might be compared to blinking lights on a Christmas tree.) The radio telescope might "see" several bright spots as two sources, and if those in the center of the line faded, the telescope would seem to see the two move apart. Other explanations have been proposed, but no way exists now to decide which is most likely.

Association with Galaxies

We will mention two other observations, although discussion of their significance will be deferred for a short time. In 1971 James Gunn reported finding a quasar which appeared to be associated with a cluster of galaxies, and it had the same redshift as the brightest of the galaxies. He argued that the quasar must be at the same distance as the galaxies and that its redshift must be as reliable an indication of its distance as the redshift is of the distance to the galaxies. Since that time he has found other examples of the same kind of association.

In the same year, the Burbidges reported finding four cases of quasars which appeared to be associated with galaxies in which the quasar and the galaxy have different redshifts. Since that time Halton Arp has extended this argument by finding at least one QSO which appears to be tied to a galaxy by a bridge of luminous matter, but the QSO and the galaxy have different redshifts.

IMPORTANCE OF QSOs

Since their discovery, QSOs have inspired many papers in scientific journals, have been the subject of scientific conferences, and have provided subjects for numerous popular articles. They have become such common topics of conversation that the name quasar has been adopted as a trade name. What is it about QSOs that has aroused such interest? Why are they important? We will suggest three reasons that they interest scientists, and then we will discuss some problems connected with the study of QSOs.

Nature Unknown

One reason for great interest in QSOs is that they are a class of object we do not understand. We do not know what they are or how they release their energy, and what we do know about them does not fit any model we can construct. Before 1960 astronomers believed they knew all the important kinds of objects and were well on their way to understanding them. There are stars, and they can be divided into many classes. No one could claim that we understand everything that can be known about stars, but the questions that should be asked and the kind of information needed to answer those questions seemed within reach. There are galaxies, and again we have many questions about galaxies, but prior to 1960 we could believe that more research of the kind that had been pursued in previous years would steadily increase understanding of them. Then along came quasars. They appear to be very small and very far away but fantastically bright. They cannot be stars, and they do not seem to be galaxies. Precisely what they are is not known yet, and hence they pose a challenge to the astronomer or astrophysicist.

Oldest Objects Seen

A second reason for great interest in QSOs is that if they are as far away as they seem to be, they are the most distant objects we see. And, being the most distant, they are also the oldest. Perhaps they existed before galaxies formed, or perhaps they are the first stages of developing galaxies; in any case, they are of interest because of their distance and antiquity.

Possible Redshift Anomalies

The third reason for investigating QSOs as thoroughly as possible is that the redshifts of some of them present anomalies which may indicate the operation of an unknown physical law. If, as certain investigators think, some of the QSOs have redshifts produced neither by a doppler effect nor by any other familiar cause, they may lead us to the discovery of a physical principle thus far unsuspected. That, of course, would be exciting and momentous.

PROBLEMS POSED BY QSOs

Energy Source

The first major problem connected with QSOs has to do with their size, distance, and brightness. Rapid variability of the light output indicates that the objects are small; the redshifts imply that they are at great distances; and their brightnesses indicate that they release energy at a tremendous rate—perhaps at a rate 100 times that of a giant galaxy. The problem is: What kind of object is small, perhaps not much larger than the solar system, but brighter than 100 billion stars? The answer is not known.

Now let us take those points one by one and consider them in more detail. Rapid variability indicates that the objects are small because a light source cannot change its intensity greatly in a smaller time than light requires to cross it. Suppose, for example, that a signal telling all the stars to become brighter left the center of a galaxy. That signal could not travel faster than the speed of light, and so the entire galaxy would not suddenly get brighter, but a wave of brightening would spread out from the center. If such a message were sent by the center of our galaxy, the sun would not know of it for 30,000 years. If one star flares up brightly and it triggers another star to flare in the same way, the second star cannot know that anything has happened until light has had time to go from the first star to it. Furthermore, it is extremely unlikely that in anything like a galaxy, with millions or billions of stars, random fluctuations of the stars would cause the large variations in brightness (by a factor of 2 or more) shown by some of the QSOs. Since QSOs often vary in their light output in months, and in at least one case the variation is within a period of a day, they must be not more than a few light months in diameter, and perhaps they are only light hours or days. Certainly they are very small indeed compared to galaxies.

The only way we have to estimate the distance to the QSOs is by interpreting their redshifts with the Hubble law, so that we say that their distances are given by

$$r = \frac{v}{H}$$

Distances computed in that way place most of the QSOs farther away than any galaxies that can be studied. Their distances are measured in billions of light years, the most distant seen being perhaps 10 billion light years from us.

After the apparent brightness of the object is known from observation and its distance is computed from the Hubble law, its intrinsic luminosity can be computed. Clearly, objects farther away than the faintest galaxies we can see must be very luminous to appear bright enough to be mistaken for stars, and in fact QSOs appear to be as much as 100 times as luminous as giant galaxies.

Now we return to the big question: What process can release energy equal to the output of hundreds of billions of stars, all within a volume that normally would be expected to contain only one star? Many answers have been suggested, but we do not know which, if any of them, is right. Nuclear fusion, the process that provides energy for the stars, seems unable to release energy at the required rate. The annihilation of matter and antimatter has been suggested; gravitational collapse of large masses has been proposed; and the violent eruption of matter into our universe from some other part of the universe or another universe (in a "white hole") has been suggested. Perhaps the correct answer is more mundane than these wild and imaginative suggestions, but so far mundane explanations have seemed inadequate.

Local or Cosmological Distances?

We have assumed that QSOs have large redshifts because they are at great distances and are being carried out by the expansion of the universe. We have said that their distances could be computed by using the Hubble law, so that if the apparent velocity of the object as determined from its redshift is 5×10^4 km/s, its distance is $v/H = (5 \times 10^4 \text{ km/s})/[50 \text{ km/(s)(Mpc)}] = 10^3 \text{ Mpc}$, or 1 billion pc. QSOs at such distances are said to be at cosmological distances.

One could explain the redshifts, however, by assuming that the objects are relatively near but are moving at the speeds indicated by the doppler effect. That explanation is called the *local hypothesis*. It has one great advantage—it helps solve the problem of fantastic energy release. If the objects are not far away, then they need not be extremely luminous, and the explanation of their energy source may not require strange and wonderful processes.

The local hypothesis was suggested very soon after quasars were discovered, and the suggestion was that they are small objects ejected from the nucleus of our galaxy or of other galaxies. We have good reason to believe that the nuclei of some galaxies, probably including our own, undergo explosive eruptions, and it is not unreasonable to suppose that objects with masses equal to many solar masses might be shot out at very high speeds. It is unlikely, however, that all the quasars produced in that way should have come from our own galaxy, and if other galaxies are ejecting such objects, some of them should be coming toward us. Consequently a search was carried out for quasarlike objects with blue shifts. The search was unsuccessful, and no QSOs are known which have doppler shifts indicating that they are approaching us.

So long as only two or three quasars, with speeds a few percent the speed of light, were known, one could argue that they had been propelled out from the center of our galaxy. As the number known has climbed above 200, however, and the apparent speeds have reached 95 percent the speed of light, that argument has become ridiculous. Both the probability that we should be the only source of QSOs and the probability that the amount of energy required to give so many objects such high speeds could be released in the core of one galaxy seem vanishingly small. If in offering a solution to the problem of energy release within the QSOs astronomers must explain how many large masses were accelerated to a high precentage of the speed of light, they have solved one problem only by creating another even more severe.

Upper Limit on Redshifts—A Nonproblem

For the first 10 years that the existence of QSOs was known, an important question was: Why do none of them have redshifts larger than 3.0? In April 1973, however, a QSO was found with a redshift of 3.4 and in June 1973 a redshift of 3.53 was reported. At least one reason for the delay in finding objects with such large redshifts is that one of the early adopted criteria for recognizing QSOs is that they are blue when viewed optically, and if the redshift is above 3, they are no longer blue. The objects with redshifts larger than 3 have been found because they are radio active—genuine quasars.

Origin of Redshifts

A fourth problem associated with QSOs is the origin of the redshift. The simplest explanation is that it is a cosmological effect arising from the expansion of the universe, indicating that the objects are at very great distances. That explanation carries with it the difficulties of explaining the great luminosity of what must be rather small objects. The next simplest explanation is that the redshift is a genuine doppler effect, but that the objects are relatively near and are moving with great speeds. We have already discussed this local hypothesis and pointed out that it seems inconsistent with other observations about QSOs, particularly with the complete absence of such objects with blue shifts.

A third possible origin for a redshift is a gravitational field. According to the general theory of relativity, light which originates in a strong gravitational field appears to be shifted to the red if it is observed in a region where the field is weaker. The effect has been observed on the earth, on the sun, and on white dwarfs. Light coming from the sun is shifted slightly to the red because the gravitational field on the sun at its point of origin is stronger than the field at the point of observation on the earth. The gravitational field of a white dwarf is considerably greater than that of the sun, and consequently the redshift of light from a white dwarf is larger than from the sun. Both these produce such small redshifts that their detection is difficult. The effect has even been seen on the earth when light (actually gamma rays) moves upward a few feet, so that the gravitational field is less at the upper elevation than at the lower.

Even though a gravitational field undoubtedly can produce a redshift, it seems an unlikely explanation for the shifts seen in QSOs. In order to produce a redshift of the order of 3, the gravitational field would have to be so intense that the matter of the object would be compressed to high density. But the spectra of QSOs are the spectra of gaseous atmospheres, and, moreover, gases at moderate pressures. It seems inconceivable that gas at the density required to produce such spectral lines could exist on the surface of a body with a gravitational field great enough to produce the observed redshifts.

A few observers have presented evidence which they think indicates that some of the redshifts are produced by an effect which we do not understand at all—some new law of nature. The evidence consists of pictures of QSOs and galaxies in such proximity as to suggest that they are all part of the same cluster, but with the QSO and the surrounding galaxies having quite different redshifts. Margaret and Geoffrey Burbidge have pointed out cases like this, where the clustering seems to imply that objects which must be at about the same distance have very different redshifts.

The man who has been most closely associated with this kind of argument is Halton Arp, of the California Institute of Technology and the Palomar Observatory. Arp has published pictures of pairs of galaxies which appear to be connected by a bridge of gas or stars, but which have different redshifts. If the pairs really are joined, they must be at about the same distance and probably have somewhat comparable speeds, and therefore the difference in redshifts cannot be explained by either the cosmological effect (Hubble law) or by simple doppler shifts. With those two possible causes eliminated, one is virtually forced to postulate a new law of physics. Not all astronomers and astrophysicists are convinced of the reality of Arp's observations, however, and it has been suggested that the apparent bridges of luminous material joining the galaxies are not real, but are produced by a blurring in the emulsion produced by two faint images close together. Other observers have been unable to reproduce some of Arp's observations about the bridges.

On the other side of the argument, James Gunn, also at Palomar, has found QSOs associated with clusters of galaxies where all of them have the same redshift. Since the redshift of the galaxies is usually unchallenged as an indication of their distance, it is reasonable to suppose that the shift of the QSO is

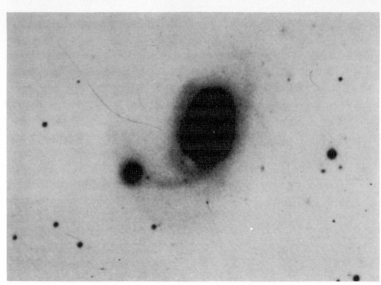

The galaxy NGC 7603 appears to be connected to a smaller companion by luminous filament. The red shift of NGC 7603 corresponds to a velocity of recession of 8,800 km/s, and
the redshift of the companion corresponds to a velocity of 16,900 km/s. The main galaxy is
of the Seyfert type. The photograph is a 3-h exposure with the 200-in telescope. (*Courtesy
of Halton Arp.*)

also indicative of its distance. This observation tends to support the interpretation of all redshifts as being cosmological in
nature. Such observations of QSOs with galaxies can be made
only for those QSOs having the smallest redshifts, for most of
them have redshifts far beyond the maximum seen for any
galaxy.

Change with Time

In 1970 Maarten Schmidt reported a study of a small region of
the sky in which he identified 23 QSOs, all of which he classified by luminosity and redshift. If the distances are computed
from the redshifts, then the numbers lying at differing distances
appear to increase much faster than the cube of the distance,
implying that they were more plentiful in the distant past,
perhaps 5 to 8 billion years ago. This finding is incompatible
with the steady-state theory of the universe.

Relation to Galaxies

Comparison of QSOs with the nuclei of Seyfert and N-type
galaxies has suggested that all three are closely related, and in
fact, that QSOs may be galaxies with very bright nuclei which
are so distant that only the nuclei can be seen. In a few cases objects first identified as QSOs have been found later to have
galactic structure surrounding the bright source.

The argument that QSOs are abnormally bright galactic cores has received support from the study of BL Lacertae and other objects similar to it, all of which seem to be giant galaxies with quasars at their centers. BL Lacertae is a radio source whose radio output varies rapidly. Its visible output also fluctuates from night to night. Most of the light comes from a core, and that light has a continuous spectrum, without emission or absorption lines, and it is strongly polarized, suggesting that it originates from a synchrotron process. If the core light is blocked by an obscuring disk so that the fainter light from the outer parts of the object can be studied, it shows absorption lines and is in every respect what one expects from a giant galaxy; the redshift of the absorption lines is 0.07. The light from the core is similar to that from more well-known QSOs except that it lacks the emission lines most of them have. By the end of 1974 several such objects were known, at least one of which appeared to be a member of a cluster of galaxies having a redshift of 0.04. The cores of these objects have the spectral characteristics and variability one associates with quasars, but they appear to be the cores of galaxies. Whether they are the same as the QSOs with much larger redshifts is not known.

Although not all astrophysicists would agree, most probably would accept the following summary of the present state of knowledge about quasi-stellar objects: They are at cosmological distances, and their redshifts give accurate indications of their distances from us; they are related to Seyfert and N galaxies and probably to BL Lacertae objects and may be the nuclei of galaxies; whether they are the nuclei of special types of galaxies or whether the nuclei of many galaxies appear as QSOs during part of their lives is not known; the source of the energy being poured out in such prodigious amounts is not known.

QUESTIONS

1 Use the modern value of Hubble's constant and compute the distance to 3C 273. *Ans* 900 Mpc

2 Compute the velocity of recession of an object for which $z = 3.0$. *Ans* $v = \frac{15}{17} c$

3 If $z = 0.4$, what is the difference between the velocity of recession calculated using the approximate formula for the doppler effect and the velocity computed by the correct formula? *Ans* $0.08 c$

4 Explain the argument which leads to the conclusion that QSOs are small.

5 Why was an attempt made to find QSOs with blue shifts?

6 What is the significance of the studies of QSOs which are found near clusters of galaxies?

18
pulsars

Further evidence that the heavens contain objects previously undreamed of by either philosophers or astronomers came with the discovery in 1967 of the radio sources which have come to be known as *pulsars*. The name pulsar refers to the characteristic by which they were discovered—the emission of short radio pulses with precise and monotonous timing.

DISCOVERY

The fact that pulsars were not discovered by radio astronomers before 1967 is a result of the way radio telescopes are normally operated. What a radio astronomer looks for is essentially noise, and as the instrument looks at one place or another in the sky it sees more or less noise. If it is seeing more noise, then it is looking at a radio source. Because the sources send signals which have random variations in amplitude, however, and those random variations are of little if any interest, the electronic equipment attached to the telescope to amplify the signals usually is built in such a way that it averages signals over some short time, perhaps of the order of a second, and records a smoothed out signal so that all the minor and very rapid variations in intensity do not clutter up the record. The technical way to express that is to say that the circuits are built with long time constants. In a way the circuits act like the shock absorbers on an automobile, which let the passengers in the automobile rise and fall with hills and hollows, but protect them from all the small bumps and irregularities in the roadway. When riding in an auto one feels rises and drops that occur over distances of many feet but is totally unaware of small variations in height of the road that occur in 2 or 3 in. Similarly, the circuits with long time constants in radio telescope equipment "protect" astronomers from small variations in the intensity of the signals they receive and let them know only about the larger rises and drops produced by the sum of the small changes. Therefore short and sharp pulses of radio radiation will not show up on the record of the output of a radio telescope.

In 1967 one of the radio instruments at Cambridge, England, was modified so that it could see and record short variations.

The intention was to look for evidence of scintillation of the
radio waves coming from outside the solar system produced by
the solar wind. Scintillation of stars is produced by the passage
of cells of air of varying density across the column of light coming from the star to the observer's eye, and the streaming gas
from the sun which blows past the earth produces similar
twinkling of the radio sources which radio astronomers observe. When the Cambridge group looked at radio sources at
times at which they expected to find scintillation, they did find
it, but when they looked at the sky during a time that the earth
should have been blocking most of the solar wind and hence
the scintillation effect should not have been marked, they
found a source which emitted pulses of short duration. This
was the first pulsar found; news of its discovery was published
in February 1968.

The first pulsar sent out a radio-wave pulse lasting $\frac{1}{30}$ s every
1.33730109 s. During the $\frac{1}{30}$ s it was turned on, the source was
very bright (in the radio region, of course), but most of the time
it was sending no waves at all. The $\frac{1}{30}$ s is called the *pulse width*,
and the time from the start of one pulse to the start of the next is
called the *period*. One of the most remarkable features of pulsars is the constancy of their periods; most of them would make
better clocks than most of those used on earth.

During the spring of 1968 four pulsars were found, and by
December 1974, 141 were known. Their periods ranged from
0.033 to approximately 4 s.

CHARACTERISTICS

In the year after the announcement of the discovery of pulsars,
a great amount of information was collected about them. Pulse
shapes were studied, pulses were timed very precisely, positions of pulsars were estimated as well as possible, and two
pulsars apparently associated with supernova remnants were
studied with special care.

Pulse Structure

Each pulse has a width of many milliseconds, but often within
that pulse there are spikes which last no more than 100 μs. If
one argues that a spike must have a duration at least as great as
the time light requires to travel across the source of the radiation, the length of some of the spikes implies that the sources
have diameters of a few tens of meters. One total pulse has a
width of only 20 ms, and the source of that pulse must come
from an area no more than 6,000 km across.

One finds that the amplitude and the duration of the pulses

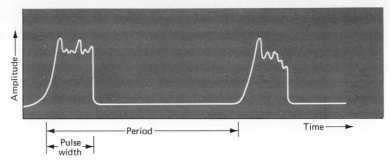

FIGURE 18.1 Pulses emitted by a pulsar, showing the pulse width, period, and the spikes within a pulse.

from a single pulsar vary, but the starting time of the pulses does not vary. A pulsar is like a band which is playing one note on the first beat of each measure. The note may be sustained longer one time than another, it may be played loudly or softly, and there may even be variations of other kinds during the time the note is being played; but the one thing which does not change is the beginning of the note at the first beat.

Locations

A question which must be answered before one can begin to guess *what* pulsars are is *where* are they. The ones we have found seem to be concentrated near the plane of the galaxy. They are definitely objects within our galaxy; the typical distance appears to be of the order of 700 pc. Two of them have been identified with visible objects, and we have additional information about these. The pulsar which has been studied the most is a star within the Crab nebula; it is observed to pulse in the visible region and the x-ray region as well as at radio wavelengths. The other pulsar related to a visible object is in the remains of a supernova in the constellation Vela. That pulsar pulses in the radio and x-ray regions, but it is not seen by visible light.

Crab Pulsar

The Crab nebula is the remnant of a supernova whose explosion was observed by Chinese astronomers in A.D. 1054. If we assume, as seems reasonable, that the pulsar is one of the results of that explosion, we know that it is only about 900 years old—young as astronomical objects go. The Crab pulsar has the shortest period of any pulsar found—about $\frac{1}{30}$ s—and in 1968 it was discovered to be slowing so that its period is increasing 36.48 billionths of a second per day. Several other pulsars are known to be slowing, but most of them are increasing their period at rates only of the order of 1 billionth second per day.

Photographs of the Crab nebula show a blue star embedded in the tendrils of hydrogen which give the structure its name, and that star had been observed to be unusual before its pulsating nature was observed. Photographs must be made with long exposures, however, and therefore variations in the brightness of the object within times of 1 s or longer are not detected. If, however, the Crab nebula is photographed through a shutter which blocks light for about $\frac{1}{30}$ s exactly at the times that the pulsar radio pulses arrive, then that star is not seen on the photograph. It goes completely dark during the time the pulsar is turned off, so that the pulsing is as complete in the visible region as in the radio region.

Since the only two pulsars which can be identified with other objects appear to be parts of the remains of supernovae explosions, it is tempting to believe that all pulsars are probably the remains of supernovae; indeed, that is now the general consensus. Their period increases slowly after they are formed, and the youngest we see (the Crab) has the shortest period. By the time its period has increased to 2 or 3 s, the gas remaining from the explosion will have dispersed until it cannot be seen, and observers at that time will have no clue to tell them that the pulsar was once part of a supernova.

The Crab pulsar has been observed to have a sinusoidal variation in its period, so that it is not so constant as it appeared at first. The time during which the variation goes from maximum through minimum and back to maximum is about $1\frac{1}{4}$ years; the variation in period is 5 or 6 ms. It has been suggested that a planet is orbiting the star, and as the planet and the star both move about the center of mass, the movement of the star toward and away from us causes the variation in period of its pulses.

Abrupt Changes in Period

In late 1968 astrophysicists thought that they understood something about the nature of pulsars, and why they slowed. The pulsar was pictured as a rapidly rotating star—a giant "top" in the sky—that slowed as it lost energy, just as a toy top slows as friction reduces its rotational energy. Then sometime between February 24 and March 3, 1969, the Vela pulsar suddenly speeded up. Its period decreased by 200 billionths of a second!

Astronomers who had been comfortably picturing a great spinning top, emitting pulses every time it turned, were terribly shocked when it speeded up, but there was no doubt about the fact. The Crab pulsar was also observed to speed up, and theorists soon offered a possible explanation for the increase. Then the Crab pulsar suddenly slowed! It has now been observed to abruptly slow down at least 5 times. Unfortunately, the explanation which had been devised to explain the speedups did not immediately explain changes in the opposite direction, and consequently the models had to be reconsidered. The current explanation of these abrupt changes, called *glitches*, will be discussed later.

NATURE OF PULSARS

Within a few months of the discovery of pulsars, rather general agreement was reached about their nature; it was arrived at by a process of elimination.

The most obvious and important clue about the nature of pulsars is the precise timing of the pulses: Something turns on the signal time after time after time with great regularity. And so we ask ourselves what kinds of physical mechanisms can produce that kind of regularity.

Oscillation Hypothesis

Perhaps the first thing we think of is oscillations. Pendulums oscillate with accurately repeated periods; the balance wheels of watches also oscillate very regularly; the tines of a tuning fork or the strings of a violin oscillate with constant frequency. And of course, some stars, such as the Cepheids, oscillate with constant frequencies.

Any elastic body has a natural period of oscillation which is of the order of

$$T \approx \frac{2\pi}{\sqrt{G\rho}}$$

where G is the universal gravitational constant and ρ is the average density of the body. For the earth, the period is approximately $1\frac{1}{2}$ h, and after a severe earthquake shock, the earth "rings" with oscillations of about that period. For the sun, the period is about 3 h, and variable stars range from approximately 3 h to 3 days. All these periods are much longer than the periods of pulsars, an observation which suggests that the density of pulsars must be higher than the density of the earth or of the sun.

We can take a typical value for the period of a pulsar and compute the density that would produce that period. The periods of the pulsars range from $\frac{1}{30}$ to over 3 s. To make the computation easy, let us take a period of 1 s as the basis for the computation. If we set $T = 1$ s and solve for the density, we find

that the density of the pulsar is of the order of 10^9 g/cm^3. That is
10^9 the density of water. The density of the earth is about
5.5 g/cm^3. The most dense thing we have discussed is a white
dwarf, and its density is only about 10^6 g/cm^3. That estimate
may be slightly low, but it cannot be pushed to 10^9. A pulsar
cannot be an oscillating white dwarf.

Under certain conditions it may be possible for the electrons
which in a white dwarf would provide the pressure to prevent
further contraction to be forced onto the protons in the nuclei
of the atoms in the star, producing a great number of neutrons.
A star in which that has happened is called a *neutron star,* and
neutron stars can have densities in the range 10^{11} to 10^{15} g/cm^3.
If that density is put into the equation which gives the period of
oscillation, the period is of the order of $\frac{1}{1,000}$ s or less. A neutron
star in oscillation cannot explain the observed periods of pulsars.

The natural question to ask is why there seems to be a jump
in densities between white dwarfs and neutron stars. Cannot
stars exist with densities anywhere between 10^6 and 10^{11} g/cm^3?
The answer (derived from theory, not observation or experiment!) is "no." There seems to be no probability that stable
matter can exist at intermediate densities, and the reason is that
in both the white dwarf and the neutron star range of densities,
the force which opposes gravitation and prevents further
collapse of the star is a force associated with the quantum nature of electrons in the one case and neutrons in the other. It
seems that matter cannot have densities between the places
where these two kinds of forces produce stability against gravitation.

The upshot of all this argument is that oscillation seems to
be eliminated as a possible explanation of the pulses.

Rotation Hypothesis

The other kind of motion which can produce great regularity is
rotation. If a fluid or elastic body rotates on an axis, there exists
a limit on its period; if the period becomes too short, the centripetal force is inadequate to hold the body together, and parts
of it fly off. The minimum period of a rotating body is given by
the same expression as the oscillation, $T \approx 2\pi/\sqrt{G\rho}$. We already know that if the densities of neutron stars are used, however, the period is much shorter than any known pulsar. Since
that is a minimum period, however, and a spinning neutron
star could have any longer period, the small value of the computed period is not a problem.

Because no one has proposed any other explanation for pulsars which can explain them adequately, they are believed to be

rotating neutron stars. The gradual slowing which several ex-hibit strengthens the belief that they are rotating objects rather than oscillating systems, for a rotating object should slow down as its energy is drained away, but an oscillating system which decreased in amplitude would also decrease in period.

One other rotating system should be mentioned before we agree to accept pulsars as neutron stars. Could double stars revolve about one another at the rates required for pulsars? Could, for example, two white dwarfs almost touching one another revolve once each second or even 30 times each second? Even if such rotations were possible, they could not explain pulsars, for white dwarfs or other dense bodies spinning about one another at those rates would radiate away so much energy in the form of gravitational radiation that they would slow down much more rapidly than pulsars actually do. Very close double stars can thus be eliminated from consideration, and we are back to spinning neutron stars.

NEUTRON STARS

But exactly what is a neutron star and how is it formed? We have discussed white dwarfs, which exist because of the peculiar behavior of electrons which sets a limit on the volume into which electrons can be squeezed. In a white dwarf many electrons are free from the nuclei of atoms, and the nuclei lead one life while the electrons exist as a gas within and among the nuclei. But what happens if some force stronger than the repulsion of the electrons forces them to occupy a smaller volume? The answer is that some of the electrons enter the nuclei and combine with protons to form neutrons, and that process continues until a substantial proportion of the protons within the star have been converted into neutrons. In the process the star shrinks and becomes more dense. A white dwarf is somewhat like one giant atom, but a neutron star is somewhat like a giant atomic nucleus. The density of the material in the neutron star is comparable to the density of atomic nuclei.

Formation

It is thought likely that the neutron stars which become pulsars—and perhaps all the neutron stars which exist—are formed in the final collapse of supernovae. As we explained in the chapter on stellar evolution, a massive star may finally reach a condition such that the temperature of the core, which by that time is heavily loaded with iron, is so high that photodissociation of nuclei begins, and within a time measured in seconds so much energy is soaked up by the destruction of

iron nuclei that the ability of the core to support the weight of
the upper layers of the star is lost. The core collapses to near
nuclear densities, and then the upper layers of the star crash
down upon it. In the sudden heating of the falling layers, they
are blown away, and the core is further compacted to neutron
star densities. After the outer parts have been blown entirely
away from the core, it is left naked, a neutron star.

Description

The radii of neutron stars are in the range 10 to 20 km, and their
masses probably are from about $\frac{1}{3}$ to 2 or perhaps 3 solar masses.
Typical densities might be of the order of 10^{14} g/cm^3. That density is almost meaningless, for we have nothing in common experience with which to compare it, but perhaps this kind of
comparison will be helpful. If 1 in^3 of the material from the interior of a neutron star were brought to the earth, it would
weigh about 10 billion tons! If all the mass of the sun, which
now fills a volume having a diameter of 800,000 mi, were compacted into a neutron star, it would have a diameter of approximately 6 mi.

Since the discovery of pulsars, theoreticians have done a
vast amount of work trying to understand what the material of a
neutron star would be like. If one had a sphere with some protons, many neutrons, some atomic nuclei, all compacted to the
densities at which neutron degeneracy prevents further collapse, what would it be like? Would it be solid? Crystalline?
Liquid? If it is crystalline, would it be easily fractured? Answering such questions is not easy, for they require extrapolation far
from the range of densities and pressures with which we have
direct experience, and it is not surprising that different theoretical studies have produced slightly different descriptions of the
"stuff" of neutron stars. It seems likely, however, that the neutron star has a crystalline layer, perhaps only a few hundred
feet thick, on the outside, and that it is liquid inside.

That outer crystalline layer may explain the glitches in
which the star appears to speed up. The argument goes like
this. Because of the rapid rotation of the star when it is first
formed, it is somewhat oblate, and the crust forms in that
shape. As time passes, however, the star slows its rotation rate,
and the effect of gravity is to make it more nearly spherical. If it
were liquid or had a crust with no strength, the adjustment
toward sphericity would occur gradually, but because the crust
is stiff, it does not respond to the forces tending to produce a
different shape until they have become strong enough to cause
a fracture in the crust. If the crust then rearranges itself so that
the diameter of the star is slightly decreased, the rotation rate

must increase in order that the angular momentum remain constant. The radius of the neutron star would have to decrease by only about 1 mm to account for the observed increases in speed.

This explanation for the glitches explains why stars sometimes speed up, but it does not explain the abrupt slowing changes. The best explanation for those is that they occur when matter falls onto the star, causing a slowing. Infall of matter would also explain the x-ray emission which is observed from some pulsars.

RADIATION MECHANISM

The spinning neutron star model explains the accurate timing of pulsars, but we still must explain the production of the pulse of electromagnetic radiation. To do that, we must consider another aspect of the neutron star—its magnetic field.

Magnetic Field

The neutron stars which we see as pulsars are believed to have magnetic fields of the order of 10^{12} G. Without defining the gauss, perhaps we can give the reader some feel for the significance of that number by saying that the magnetic field of the earth, the field which causes magnetic compasses to point toward the north, is about $\frac{1}{2}$ G. A very large magnet may produce a field of the order of 10^4 G, and the largest fields produced within a laboratory for any length of time are of the order of 10^5 G. The field which the pulsar is presumed to have is then at least 1 million times as strong as the strongest fields which can be produced in a laboratory.

The question of how such large magnetic fields could arise happens to be rather easy to answer. Magnetic fields have the curious property that they can be "trapped" within an electrical conductor. If a magnetic field is produced within a ring of a conductor and the ring is then squeezed to a smaller area, the field within becomes stronger, for currents are produced in the ring which strengthen the field inside. A magnetic field can be pictured in terms of lines, and the strength of the field is indicated by the number of lines passing through a square meter;

FIGURE 18.2 Magnetic field lines passing through a conducting ring. As the ring is squeezed to a smaller size, the lines are trapped within it, and the number of lines passing through the ring remains approximately constant (if the decrease in size occurs rapidly). The field strength is proportional to the number of lines per cross-sectional area, and therefore the field strength within the plane of the ring increases as the area decreases.

the stronger the field, the more dense the lines. The "trapping" can be visualized by thinking of the lines being squeezed into a smaller area, so that the number of lines is not changed but the density (lines per square meter) is increased. Extremely large magnetic fields have been produced for very short times by producing a field within an aluminum tube and then imploding the tube by detonating shaped charges of explosive to crush the tube inward; the magnetic field lines are trapped and their density is increased greatly. The same sort of thing is believed to happen in a collapsing star. Many stars have magnetic fields. The sun's field is modest, perhaps a few hundred gauss beneath the surface, but some stars have fields which are in the range of a few thousand gauss. In the abrupt collapse of the star, the field lines are trapped and squeezed into the core, which becomes the neutron star. The magnetic field strength in such a collapse would increase inversely proportional to the cross-sectional area of the star, so that if the diameter went from 800,000 to 8 mi, it would have decreased by 10^5 and the cross-sectional area by 10^{10}. The magnetic field would also have increased by about 10^{10}, so that even a very modest field before collapse would produce a huge field afterward.

Production of Radiation

The model of a pulsar pictures a neutron star possessing a large magnetic field, but the field is not along the axis of rotation. Rather, it makes some large angle with the axis of rotation so that as the star spins, the magnetic field lines extend out from the surface of the star and sweep around in space. The behavior of such a rapidly moving and strong magnetic field is not understood in any detail, and the exact mechanism for producing pulses of radiation is uncertain. One explanation suggests that near the star the rotating magnetic field produces electromagnetic waves which travel out, and in the gas some distance above the surface the waves pick up electrons and carry them along very much like surf riders are carried on the crest of a wave. The accelerated electrons then radiate the energy we see as their speeds approach the speed of light.

Another explanation, and probably the one preferred by most theoreticians trying to understand pulsars, compares the sweeping magnetic field to a rotating lighthouse beacon light. It is assumed that electrons are released from the surface of the star and that they move outward along the lines of the magnetic field. (Electrons would find moving *across* the lines of the field very difficult, but they could easily spiral about the lines and move outward.) Eventually, however, they reach a point at which their speed must exceed the speed of light if they move further out and still go around the star in the same period.

Since the electrons cannot move as fast as light, they must break away from the magnetic field. (The field lines also bend at what is called the *speed of light radius*.) The electrons moving out along the field lines radiate energy because of their accelerations, but because their speeds are very high, the radiation is directed into narrow cones ahead of them. Thus, if we are in the plane of their motion, we see a burst of radiation each time the electrons are coming toward us, but during the remainder of the time we see nothing. The analogy of a lighthouse with a revolving light is appropriate, and we will see only those pulsars whose light beams happen to strike us. The spike structure of pulses may be produced by bunches of electrons riding out together.

The Crab nebula glows from synchrotron radiation and by excitation of hydrogen, and some source of energy must keep it shining. The amount of energy it emits has been estimated, but its source was a mystery until the discovery of the pulsar. Now estimates of the energy the pulsar is losing, made on the basis of its rate of slowing, give almost exactly the same figure as the energy being radiated by the Nebula. It seems likely, therefore, that the energy for the entire Crab nebula comes from the pulsar, and that energy is carried to other parts of the nebula by radiation and by charged particles moving at high speeds. The source of the energy is the rotational energy of the neutron star, and as energy is radiated away, the rotation rate decreases slowly and steadily.

Cosmic Rays

In the early years of the twentieth century particles from outer space now known as cosmic rays were discovered striking the earth. Most of the particles are protons, but some are nuclei of elements heavier than hydrogen. Their energies extend over a wide range, but the most energetic have more energy than can be given to similar particles in any of the large accelerators man has constructed. The origin of the particles and the source of their energies has been a mystery since their discovery, but pulsars offer a possible explanation. The centrifugal pump mechanism of pulsars which carries electrons to speeds making possible intense synchrotron radiation and which sends energetic particles throughout the Crab nebula may also shoot protons and heavier particles across the galaxy to strike the earth. Estimates of the number of cosmic-ray particles which might reasonably be produced in this way by the pulsars known to exist agree as well as can be expected from such crude computations with the number which are known to strike the earth.

Because no pulsars are seen with periods greater than about 4 s, it appears that some aging effect must make them "turn off" after a time, or when the period reaches that value. What we know of them is consistent with the view that they begin life, after the collapse of a supernova, spinning very rapidly, and that with time they slow down. As we have said, the Crab pulsar is only 900 years old, and its period, now about $\frac{1}{30}$ s is visibly slowing. Other pulsars, about which we see no supernova remnants, have longer periods, and we presume that they are old enough that the gas which at one time surrounded them has diffused away into space. We do not know why old pulsars cease to pulse. Perhaps the magnetic field decays; perhaps they use up their supply of surface material which can provide the electrons required to make pulses.

X-Rays

Scattered about over the sky are sources of intense x-ray emission, and some of those x-ray sources are believed to be neutron stars. They are no longer acting as pulsars—if indeed they ever did—but the x-rays are being produced by an entirely different process. The sources are members of binary systems, and the best explanation of the x-ray emission is that matter is being transferred from a larger and more nearly normal star to the neutron star, and as it falls onto the neutron surface, much of its energy is radiated away in the form of x-radiation.

A BINARY PULSAR

In July 1974, a pulsar designated PSR 1913 + 16 was found by a group from the University of Massachusetts working with the Arecibo telescope. This pulsar is noteworthy for two reasons: it has the second shortest period found thus far (59 ms), and it is a member of a binary system. When the discoverers tried to determine the period to high precision, they found that it varies slightly in a cyclic manner, going through a cycle in $7\frac{3}{4}$ h. The pulsar moves on an orbit about another star, and as it moves toward us the period of its pulses is decreased, but as it moves away, the period is increased. The time for one complete orbit is the time of the cyclic variation, $7\frac{3}{4}$ h.

Analysis of the signals from this pulsar indicates that the sum of the masses of the two stars is somewhat more than 2 solar masses, that the separation of the two stars is about one solar radius, and that we are situated almost precisely in the plane of the orbit. It is almost certain that both stars in the system are neutron stars, but no pulses have been detected from the second star. Perhaps it is not a pulsar, or perhaps we are not in the path of its beam. The eccentricity of the orbit of the pulsar about its companion is 0.6, and the pulsar moves at approximately 300 km/s. The point of closest approach of the pulsar to its companion (periastron) precesses about 4° per year; this precession will be discussed again in Unit 20.

The processes that might produce a binary pair of neutron stars separated by only the radius of our sun are unknown. One can imagine two stars, separated by a much greater distance, which for some reason come closer together as they become neutron stars, or one can imagine one large star which fissions into two fragments, each of which becomes a neutron star. Probably the second process is more likely, but the details of such a splitting of a star have not been worked out.

Pulsars seem to have no cosmological significance as quasars have; they tell us nothing about the universe beyond our own galaxy. But they have demonstrated again that the universe can offer us surprises, and they have posed a challenge to those physicists interested in the structure of matter which may keep them busy many more years. Furthermore, the binary pulsar offers us a unique opportunity to test our understanding of gravitation. Because of the intense gravitational fields in which the stars move, effects which are small in the solar system are large there, and we may be able to distinguish between competing theories of gravitation by studying that particular pulsar.

QUESTIONS

1 How much will the period of the Crab pulsar increase in 1 year?   *Ans* 1.3×10^{-5} s

2 If the present rate of increase in period of the Crab pulsar continues, how long will be required for the period to increase from 0.033 to 2 s?   *Ans*   150,000 years

3 Suppose that a star had a uniform magnetic field of 100 G throughout its interior, and that all the magnetic field lines were trapped as the diameter of the star decreased from 800,000 to 8,000 mi. What would be the field strength? If the diameter decreased to 8 mi, what would be the field strength?

Ans 10^6 G; 10^{12} G

4 If a straight line were extended outward from a point on the equator of the Crab pulsar so that it swept out a circle in $\frac{1}{30}$ s, how far from the center of the circle would the line be moving at the speed of light?   *Ans*   1,600 km

5 No pulsars have been discovered at the positions of two known supernovae, Tycho's and Kepler's. Assuming that pulsars were actually formed in those two explosions, what might be the reason that we do not see them?

19
primordial blackbody radiation

The July 1965 issue of the *Astrophysical Journal* contained two papers which were closely related. One, by Dicke, Peebles, Roll, and Wilkinson of Princeton University, predicted the existence of blackbody radiation left over from the fireball stage of the universe. The other, by Penzias and Wilson of Bell Telephone Laboratories, reported the discovery of that radiation. The radiation, variously called cosmic blackbody radiation, primordial blackbody radiation, residual fireball radiation, microwave background radiation, etc., probably is the most significant discovery of recent times for cosmological theory.

PREDICTION AND DISCOVERY

In the late 1940s George Gamow and colleagues pointed out that if the universe began as a very hot ball of gas, as they thought likely, the blackbody radiation emitted at that time should still be present. If all space was filled with such radiation at one time, when the temperature was very high, the effect of the expansion of the universe would be to cool the radiation, so that now it should look like the radiation emitted by a blackbody at a very low temperature. Gamow had predicted a temperature now of about 50 K. At the time Gamow made the prediction, equipment capable of detecting such radiation was not available, and nothing came of the suggestion that the radiation might still be bouncing around space.

By 1965 Robert Dicke had arrived at a prediction of such radiation by a somewhat different route, but at the time his prediction was published, none of the authors knew of the earlier suggestion of Gamow, even though Gamow's work had been explained in popular books as well as in technical journals. Dicke's reasoning was based on the assumption that the universe undergoes cyclic expansions and contractions; that assumption has the attractive feature of eliminating the "beginning" by substituting an unending series of cycles. But in each expansion cycle, such as the one we are living in, hydrogen is converted into heavier elements, and unless the universe is to become overloaded with heavy elements, the temperature must rise high enough at each contraction—on the order of 10^{10} K—for the heavy elements to be broken down again to hydrogen. If the universe reached such a temperature at the time it was contracted to minimum size before the beginning of the present expansion (perhaps about 13 billion years ago), it would have been a ball of electrons, protons, neutrons, and radiation—the radiation emitted by a blackbody at 10^{10} K. A blackbody at that temperature produces a huge amount of radiation, and simple calculation from the Wien displacement law shows that the peak of the radiation curve is at approximately 3×10^{-4} nm, far down in the x-ray region. Since that time, however, the universe has expanded, and only a small amount of the radiation should have been absorbed by matter; most of it should still be present in space. The effect of expansion is to cool the radiation—to lengthen all the wavelengths so that now the peak is in the vicinity of 2 mm. One can describe the radiation now by telling the temperature of a blackbody which would radiate the same distribution of wavelengths, with the same peak of the curve. If we knew the temperature of the radiation at the beginning of the expansion, and exactly how much expansion has taken place, we could calculate the temperature of the radiation now. We are not able to make that computation with precision, but estimates of the temperature to be expected agree reasonably well with the measured temperature of the radiation which has been found; that temperature is 2.7 K.

EFFECT OF EXPANSION

One may visualize the effect of the expansion of the universe in this way. Imagine a box made of perfectly reflecting mirrors, and suppose that blackbody radiation from a hot source is admitted to the box, but then the box is closed so that there is no window. The radiation will be trapped, and it will bounce back and forth between the walls indefinitely. Now let the walls of the box be slid outward so that the volume of the box increases, still without opening any windows to the outside. As radiation strikes the moving walls, it undergoes doppler shifts to the red, and the wavelengths of all the radiation increase. But the wavelengths increase in such a way that the distribution among them still corresponds to the radiation curve for a blackbody; the effect of the moving walls as the box becomes larger is to change the radiation from that appropriate for one temperature to that appropriate for a blackbody at a lower temperature.

Discovery

Penzias and Wilson did not set out to look for radiation of cosmological significance; they were testing an antenna to be used with Telstar, the communications satellite. But the antenna picked up radio noise which had not been anticipated, and which did not seem to come from any particular part of the sky. At the wavelength being detected by their equipment, 7.4 cm, the intensity of the extraneous radiation was what would have been produced by a blackbody at a temperature of 3.5 ± 1 K.

Verification

By late fall of 1965, Roll and Wilkinson at Princeton had built and used a detector looking for radiation at a wavelength of 3.2 cm. They found that radiation at an intensity corresponding to a blackbody at 3.0 ± 0.5 K. Soon Cambridge astronomers reported that at 21 cm they had found a background radiation corresponding to a temperature of 2.8 ± 0.6 K. The best figure now, based on many measurements, is 2.7 K.

Figure 19.1 shows the curve of intensity versus wavelength for blackbody radiation at 2.7 K. This curve differs in one important respect from the blackbody curves discussed in Unit 5, but it is being used here in the form most often used in discussions of the microwave background. The difference is that the quantity plotted along the vertical axis here is I_ν rather than I_λ, which was used in Unit 5. I_λ is the radiation power per unit wavelength range, perhaps per nanometer; I_ν is the radiation power per frequency range—actually per Hertz. The measurement per Hertz is the natural one to make with microwave

FIGURE 19.1 The radiation curve for a blackbody at 2.7 K. The region within which measurements can be made easily from the ground lies between the light lines.

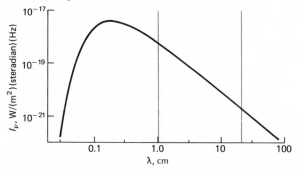

equipment, whereas the measurement per wavelength range is a natural one to make with optical spectrometers.

OBSERVATIONS

Dotted lines on the figure indicate the "window" within which observations can be made from the ground. Most measurements have been made from the ground, although some have been made from balloons flying above much of the atmosphere, and a few have been made from rockets. At wavelengths greater than about 20 cm, the galactic background noise obscures the signal sought, and therefore measurements can be made only with the greatest difficulty beyond that point even from above the atmosphere. Below about 1 cm, atmospheric absorption sets the limits on the observation.

Absolute Measurements

Because the measurements required are absolute measurements, they are difficult to make, and atmospheric absorption or galactic emission are particular problems. Radio telescopes normally look for sources which stand out from the background noise because they are brighter than the surrounding area, and so a continuous and steady background of noise can be rejected so long as it does not overwhelm the signal being sought. Measurements of the microwave background, on the other hand, must measure the actual intensity of the radiation being received, not just variations, and that is much harder to do. The most common measuring instrument uses a rather small horn, perhaps only a few centimeters across, to look at the sky. The instrument contains a reservoir of liquid helium, and the element which actually detects radiation is caused to look alternately at the sky through the horn, and then at the liquid helium, so that the signal which it produces is a comparison of the radiation being received from the sky with the radiation being emitted by the boiling helium, which is at a temperature of about 4.2 K. Such equipment is not easily placed into a satellite or sent up in rockets. Consequently no good measurements are available in the wavelength range below 0.1 cm.

IS IT BLACKBODY?

The discussion thus far has assumed that the radiation being detected is blackbody in nature, and its identification as residual radiation from the fireball stage of the universe depends upon its having that nature. But is it really blackbody radiation? What is implied by the statement that it is blackbody in character is that if the intensity of the radiation is plotted as a function of wavelength, the curve follows the curve shown in Fig. 19.1. The most critical part of the curve so far as a test of the blackbody nature is concerned is the peak at about 2 mm and the drop at shorter wavelengths. But that is precisely the region where measurements are difficult to make, and where no good data are yet available.

Direct Measurements

Numerous groups around the world have made measurements from the ground over the range from about 75 to 3 cm, and all are in good agreement with a temperature of 2.7 K. The few measurements which have been made from balloons and rockets are also in agreement with that figure. But in the critical part of the spectrum, direct measurements have not been made, and we have only indirect observations which serve to set an upper limit on the curve.

Upper Limit from Interstellar Molecules

Nature has provided us with thermometers lying between the stars in the form of cyanogen (CN) molecules. In the light coming from at least 10 different stars in separated parts of the galaxy we see absorption lines produced by the CN lying between us and the stars. A molecule such as CN can absorb and emit radiation by three different kinds of changes within itself.

1 Its electrons can be lifted to higher levels or can drop down to lower levels. They are lifted up by the absorption of energy in some form—radiation or collision—and when they drop down they radiate away the energy. That radiation is likely to fall in the visible or ultraviolet part of the spectrum, and the absorption of starlight which reveals the presence of the molecules involves the lifting of an electron to a higher level, thereby greatly increasing the energy of the molecule. The subsequent reemission of the energy does not interest us, for it is very unlikely that photon comes our way.
2 The molecule can also change its energy by changing its condition of vibration. The carbon and nitrogen atoms act very much as if they were connected by a spring, and they vibrate along the line joining them. The radiation involved in vibrational changes is likely to fall into the infrared part of the spectrum.
3 The molecule can rotate. If we ignore the vibration, the molecule looks much like a dumbbell, with the carbon atom at one end and the nitrogen atom on the other. Any time its ro-

tation rate increases, it has absorbed energy from some source, and for the rotation rate to decrease, the molecule must radiate away energy. That radiation lies in the microwave segment of the spectrum, for the energy changes are very small for changes in the rotation rate.

If a molecule is left to itself long enough, it falls into a ground state, in which it has the minimum of energy in electronic, vibrational, and rotational forms. (That does not mean that it has no vibration and that it is not rotating; it may not be able to lose all the energy in those forms even in its lowest possible state. This is a restriction which nature imposes and quantum mechanics explains.)

The absorption lines produced by the CN molecules in starlight indicate that most of the molecules are in their ground state so far as rotation is concerned, but some are in the first rotational excited state and a smaller number are in the second rotational excited state. Because the molecules are dispersed so thinly, it is very unlikely that the number observed to be in excited states have acquired that extra energy by collisions, and so we must suppose that they have absorbed radiation. And because an excited molecule will radiate away that energy and drop back down to ground level, the source of excitation must be acting on the molecules constantly so that some are always in the excited levels. Molecules in the ground state are lifted into the first excited state by absorption of a photon of wavelength 0.26 cm, and they are lifted from the ground state into the second excited level by absorption of a photon at 0.13 cm.

From the relative intensities of three absorption lines which differ only in the ground state of the molecule, one can compute the fraction of the molecules in ground, first excited, and second excited states. Then one can compute the intensity of radiation at 0.26 and at 0.13 cm which must be bathing the molecules to produce those numbers in the higher-energy states. The computation for the 0.26-cm transition indicates that the intensity of radiation in space at that wavelength must be little if any more than that predicted by the blackbody curve for 2.7 K; the computation for 0.13 cm indicates that at that wavelength the intensity is not greater than would be predicted by a curve for 4 K. This computation therefore sets an upper limit on the curve at 0.13 cm, where no direct measurements have been reported.

A similar measurement has been made of the excited states of HCN which exists among the stars, and it also is consistent with a blackbody temperature of 2.7 K. Therefore, although we do not have measurements on the short side of the 2-mm peak, where we would like to have them, we have upper limits in that range which at least tell us that the curve does not go up sharply. It is an interesting fact that the excitation of CN lying between us and some stars was discovered in the 1930s, but at that time the cause of the excitation was not suspected.

Other Explanations

Before we can be confident that the radiation being found is actually residual from the early hot stage of the universe, we must consider other possible sources and determine whether any of them could produce the same effect. Several astrophysicists have considered the problem, and one small group of cosmologists have had particular incentive to carry out such investigations. The discovery of fireball radiation was a terrible blow to the advocates of a steady-state model of the universe, and they immediately began to try to find alternative explanations for the radiation. The steady-state theory does not permit the existence of a fireball at some time in the past, and if the steady-state model is to be credible at all, some other explanation for the 2.7-K radiation must be found.

Three possible explanations for the radiation other than the fireball stage of the universe have been carefully explored.

1 If the dust of interstellar space is heated by starlight, it should radiate away that energy, perhaps as blackbody radiation.
2 Multiple sources, at differing temperatures, may combine to produce what appears to be an isotropic sea of radiation at 2.7 K.
3 Electrons, moving freely through an ionized intergalactic gas, come close to protons and are accelerated but not captured; the accelerated electrons radiate away some of their energy, and that is the energy we are seeing. Such an interaction, in which an electron comes near a proton without being captured to produce a neutral hydrogen atom, is called a *free-free transition*.

Detailed computations of all of these have been unable to explain the radiation actually observed, and thus far the only satisfactory explanation is the original one—that the radiation is a fossil remaining from the time when the universe was very hot and dense.

ISOTROPY

The 2.7-K radiation is amazingly isotropic, and that fact may be as important as its very existence. The simplest way to observe the isotropy is to point a detector at the sky and watch its output for 24 h, while the rotation of the earth carries it around a circle. The variations in intensity of the radiation being seen at a particular wavelength are expressed in terms of the temperature change required to produce that variation in intensity, and the variations seen when different parts of the sky are observed are less than the equivalent of 10^{-3} K. What slight variations are seen probably are produced by the motion of the earth through the sea of radiation rather than by anisotropies in the radiation itself. In fact, an attempt has been made to detect the earth's movement through the radiation, and it may actually have been seen. Three different motions are involved: our motion about the center of the galaxy as a result of the galactic rotation, our motion toward Andromeda because the Andromeda galaxy and the Milky Way probably revolve about one another, and the motion of the local group of galaxies toward the giant galactic cluster in Virgo. The last motion is probably the most important, and if the motion of the earth because of these movements can be seen, it will show up as a slight increase in the effective temperature of the radiation coming from in front of us and a corresponding decrease in the effective temperature as we look behind us. The slight effect which has been seen suggests that the sun is moving through the microwave radiation sea at about 350 km/s.

Cosmological Implications

From a cosmological standpoint, the isotropy of the radiation has profound implications, for it tells us something about the nature of the universe in its early stages. The radiation reaching us now has traveled for a substantial part of the life of the universe without interacting in any significant way with matter. Far back in the distant past, however, it did interact; it was scattered by matter. The isotropy we see now indicates that the universe itself—the matter that last scattered the radiation—was isotropic at the time of that interaction. If we suppose that the same observations would be made anywhere in the universe, then the early universe was not only isotropic but also highly homogeneous.

In developing cosmological models, astronomers usually assume that the universe is isotropic and homogeneous because without such simplifying assumptions the models cannot be worked out, but it has always been supposed that the assumption was a so-so approximation to reality. Now the isotropy of the cosmic blackbody radiation indicates that what has been considered an assumption is in fact a very good description.

Indeed, the isotropy and homogeneity implied by observations of the blackbody radiation are so great that they pose another problem. Let us turn our minds back to the beginning of the universe and consider how the radiation acquired its present form. (We will leave until later a consideration of the first few hours and even the first few years of the universe.) Several thousand years after the expansion had begun, the temperature of the matter in space was dropping toward 4000 K. At that time most of the matter was ionized, so that it consisted primarily of electrons, protons, and helium nuclei. The radiation which filled all space was coupled to the matter in the sense that because electrons scatter radiation very efficiently, variations in the temperature of the matter would have produced variations in the temperature of the radiation, and variations in the density of one would have produced corresponding variations in the density of the other. As the temperature dropped to about 3000 K, the hydrogen and helium nuclei combined with electrons to form neutral atoms, and because the neutral atoms scatter radiation far less efficiently than electrons, the matter and radiation are said to have *decoupled* at that time. From that time on, matter and radiation cooled at different rates, and either one could develop density variations without affecting the other.

The isotropy of the radiation we see indicates that at the time of decoupling, the matter in the universe was a thin, smooth gas; it was spread very uniformly; almost devoid of clumps or large density variations. Any such variations would show up now as anisotropy of the radiation. For galaxies to form, however, the matter had to form into clumps. The processes that produced clumping so soon from the uniform gas which existed at the time of decoupling are not known, and understanding why that clumping occurred is an important concern for the cosmologist attempting to describe the formation of galaxies.

SIGNIFICANCE OF THE RADIATION

All the information we have about the microwave background radiation now indicates that it is blackbody in character, but the crucial measurements are yet to be made. And thus far no explanation has been found for such blackbody radiation other than that it is residual from the early hot stage of the universe. If later work firmly establishes both these conclusions, which now remain slightly tentative, then this radiation will be the most important discovery for cosmology since the discovery of the Hubble expansion.

The radiation tells us that the universe once had a very hot phase; it thus eliminates rather conclusively the steady-state model, as well as some other models which picture the universe evolving from a cold gas. The radiation also tells us something about the conditions existing long before galaxies or quasars formed. It literally tells us of conditions in the universe before "the morning stars sang together."

Perhaps most important of all, this radiation tells us that on the large scale, the universe as seen from the earth is highly isotropic. Unless we wish to suppose that we are in a unique position in the universe, in violation of the cosmological principle, that implies that the universe is also homogeneous on the large scale. Those are rather grand and sweeping conclusions to be drawn from the fact that a radio receiver pointed at the sky sees noise that is about the same in all directions.

QUESTIONS

1 Why must the measurements of the cosmic background radiation be absolute measurements?

2 Where are the CN molecules located which are used to check the blackbody curve of the background radiation?

3 How can observations in the visible region of the spectrum of the CN molecules give information about the intensity of radiation in space at wavelengths in the millimeter range?

4 Why should motion through the sea of radiation produce an effect which is interpreted as an increase in temperature when one looks forward?

5 Why does the observed isotropy of the radiation plus the assumption that we are not in a unique position in the universe imply that the universe is homogeneous?

20
geometry of space and space-time

To understand the concepts underlying cosmological models and to understand the idea of black holes, one must have some familiarity with the notions of general relativity, and this chapter is intended to provide that familiarity.

Einstein developed two theories of relativity: the special theory, published in 1905, and the general theory, published in 1915. The special theory applies to bodies moving at constant speeds, observed at close distances, and it neglects gravitational effects. It is rather simple mathematically. The concepts seem strange, and some of its predictions seem contrary to common sense (because our common sense is based on experiences with objects moving at speeds that are very low compared to the speed of light), but many of its results can be worked out with high school algebra. General relativity, on the other hand, is complicated mathematically. It provides an alternative to Newton's theory of gravitation, and so it can be used as a foundation for models of the entire universe, but the mathematical details are far above anything which can be presented here.

General relativity explains the effects which we have heretofore called gravitation in terms of a curved space-time rather than in terms of the attraction of one body for another. In order to explain the ideas of relativity, we must explain curvature, space-time, curved space-time, geodesic paths, etc.

CURVATURE

The idea of curvature will be explained for curves and surfaces, where it can be visualized, to assist the reader in forming some notion of the meaning in three- and four-dimensional spaces where visualization is difficult.

Curve on a Plane

A curve drawn on a plane surface has a curvature associated with each point, and the curvature is defined to be the reciprocal of the radius of what is known as the *osculating circle* at

353

that point. Figure 20.1 shows a curve and a point marked *P* at which the curvature is to be determined. Figure 20.2 shows the same curve and point, with two other points, labeled *A* and *B*, one on either side of *P*. Through any three points not in a straight line a circle can be drawn, and Fig. 20.3 shows the curve with its points and the circle drawn through those points. Now if points *A* and *B* are allowed to slide along the curve toward *P*, and if at each of their positions a circle is drawn through the three points, a series of circles results, as shown in Fig. 20.4. The limiting circle which this series of circles approaches as *A* and *B* come into coincidence with *P* is called the osculating circle; it and the curve have three points in common even though those points are one on top of the other. Now the radius of the osculating circle is measured as in Fig. 20.5, and its reciprocal is the curvature of the curve at point *P*. It should be noted that the curvature of a circle is just the reciprocal of its radius and is the same at all points on the circle, and the curvature of a straight line is zero because the radius of the osculating circle is infinite.

Gaussian Curvature

The extension of this concept of curvature to a surface leads to what is called the gaussian curvature, defined in this way: Through the chosen point *P* on the surface a plane is passed so that it cuts the surface; the intersection of the plane and the surface generates a curve which at every point has a curvature as defined above. If the plane is rotated about the perpendicular to the surface at *P*, it cuts many curves, and unless the surface is a sphere, not all those curves have the same curvature. If the curves of maximum and minimum curvature are chosen, the product of their curvatures is called the *gaussian curvature* of the surface at the point *P*.

FIGURE 20.1 Plane curve and a point *P* at which the curvature is to be determined.

FIGURE 20.2 The same curve with points *A* and *B* labeled on either side of *P*. Points *A* and *B* are chosen arbitrarily.

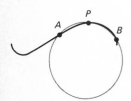

FIGURE 20.3 The only possible circle drawn through points *A*, *P*, and *B*.

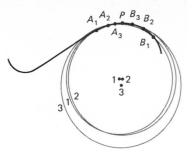

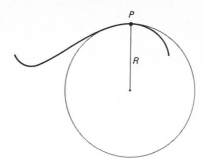

FIGURE 20.4 Successive circles defined by the three points as A and B approach P.

FIGURE 20.5 The osculating circle at point P. The curvature of the curve at point P is defined to be the reciprocal of the radius of the osculating circle, or $k = 1/R$.

Example 1 A surface is formed by rolling a sheet of paper to form a cylinder. The curve of maximum curvature is formed by the intersection with a plane perpendicular to the axis of the cylinder (Fig. 20.7), and the curve of minimum curvature is a straight line (curvature = 0) formed by the plane passing through the axis. The product of the two curvatures is zero, as is also the curvature of a plane.

Example 2 For a sphere, all intersection curves are the same, and each has the curvature $1/r$ where r is the radius of the sphere. Consequently the gaussian curvature at any point on a sphere is $1/r^2$. (The planes which cut the surface must be perpendicular to the surface, and hence all the curves on a sphere are great circles.)

FIGURE 20.6 A plane intersecting a surface produces a curve which at every point has a curvature.

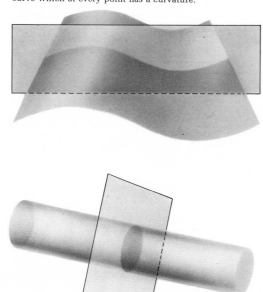

FIGURE 20.7 When a plane intersects a cylinder such as might be formed by rolling a sheet of paper, the curve of maximum curvature (smallest radius for the osculating circle) is produced when the plane is perpendicular to the axis of the cylinder.

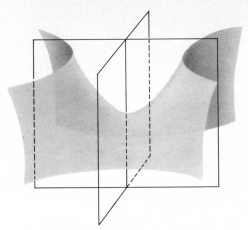

FIGURE 20.8 On a saddle-shaped surface the curves
of extreme curvature made by intersection with planes
bend in opposite directions, so that the product of
their curvatures is a negative number.

Example 3 For a saddle-shaped surface, the maximum and minimum curvatures will be for intersection curves as shown in Fig. 20.8. Since one of the circles is above the surface and the other is below, one is said to have positive curvature and the other negative, and the resulting gaussian curvature is a negative number.

Intrinsic Property

A most important aspect of the curvature of a surface is that it is intrinsic to the surface; i.e., it can be determined from measurements made entirely on the surface. It is easy for us, looking from a vantage point above and outside a surface, to see something of its curvature—for example, to see that a hilltop has positive curvature, a dish-shaped depression in the earth has positive curvature, a long valley between ridges of mountains has zero curvature, and a pass through mountains which is saddle-shaped has negative curvature. But it is equally possible for "flatlanders," living only on the surface and unable to view curves in three-dimensional space, to determine curvature. Two methods will be suggested by which the determination might be made.

Interior angles of triangle The first method is the measuring of the interior angles of a triangle. Everyone knows from high school geometry that the sum of the interior angles of a triangle is 180°, but what everyone may not realize is that that is true only for a triangle drawn on a plane, or, more precisely, on a surface of zero curvature. On a plane sheet of paper draw a triangle; now roll the sheet into a cylinder and note that the

FIGURE 20.9 A triangle drawn on a sheet of paper retains the same angles when the sheet is rolled into a cylinder.

angles of the triangle have not changed. If we want to draw a triangle on a cylinder which is already rolled, we may be unsure what is to replace the "straight line" which is so clear on the plane, but the proper replacement is the geodesic, which in this case is the shortest distance between two points. If three points are placed on the surface and threads are stretched between them, the threads will define the geodesic lines, and the triangle made by the threads will have interior angles adding to 180°.

If the same sort of triangle is drawn on a sphere, as, for example, a globe of the world, the results are quite different. It is impossible to draw the triangle on a sheet of paper and then to transfer it to the globe because the paper will buckle and hump; it cannot lie flat on the surface. The triangle must be drawn directly on the sphere, using the string technique to find geodesic lines. The interior angles of any triangle so drawn will add to more than 180°, a fact which can be visualized easily by looking at some particular triangles. Imagine a triangle drawn, as in Fig. 20.10, by longitude lines from the north pole to the equator and a section of the equator. The longitude lines are perpendicular to the equator, and so the triangle has two angles of 90° each; obviously the three angles add to more than 180°. That is true for any triangle drawn on a sphere, and the excess above 180° is proportional to the area of the triangle.

A triangle drawn on a saddle surface, as in Fig. 20.11, has interior angles less than 180°, and again the difference is proportional to the area of the triangle. Both cases can be described by the same formula. If S is the sum of the interior angles of the triangle, measured in degrees, A is the area of the triangle. and C is the curvature of the surface, then

$$(S - 180°) \frac{\pi}{180°} = CA$$

If the angles are measured in radians, this becomes

$$S - \pi = CA$$

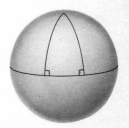

FIGURE 20.10 On a sphere, the interior angles of a triangle add to more than 180°. The triangle shown has two 90° angles.

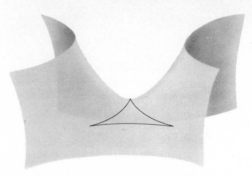

FIGURE 20.11 A triangle drawn on a saddle-shaped
surface has interior angles which add to less than 180°.

For a surface of positive curvature S is larger than 180°; for a surface of negative curvature it is less.

The conclusion about the sum of the interior angles of triangles can be summarized thus:

For a surface whose curvature is	The sum of the interior angles of a triangle is
Zero	Equal to 180°
Positive	Greater than 180°
Negative	Less than 180°

Circumference of circle The second method for determining something about the curvature of a surface from measurements made on the surface is to measure the ratio of the circumference of a circle to its radius. A circle is defined as the locus of points equidistant from a center, and the distance from the center to the locus is the radius. Thus one may draw circles by tying a pencil to a pin by a length of thread, sticking the pin into the surface, and moving the pencil everywhere possible with the string fully extended *on the surface.*

On a plane, the circle has its usual properties which are familiar from plane geometry, and the circumference is 2π times the radius. The ratio of circumference to radius is therefore 2π. If the same procedure is used to draw a circle on a cylindrical surface, having zero curvature, the same result is obtained, a fact which can be seen by realizing that the circle could first be drawn on a sheet of paper and the sheet then matched to the cylinder.

If a circle is drawn on a sphere, the ratio of circumference to radius is less than 2π. This can be visualized easily by thinking

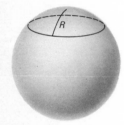

FIGURE 20.12 On a sphere, the circumference of a circle is less than $2\pi R$.

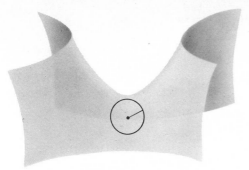

FIGURE 20.13 On a saddle-shaped surface, the circumference of a circle is greater than $2\pi R$.

of a globe of the earth and placing the center of the circle at the north pole. If the radius is long enough that the circle is the equator, the circumference is 4 times the radius, so that the circumference/radius ratio is 4. If the radius is longer, the circumference becomes smaller, so that the ratio decreases, approaching zero as the circle comes closer to the south pole. Even for small circles, near the north pole, however, the circumference is always less for a given radius than if it were on a plane.

The circle on a saddle surface is harder to visualize, but it has a circumference greater than 2π times the radius.

The effect of the curvature of the surface on a circle drawn on it can be summarized in this way:

For a surface whose curvature is	The ratio of circumference to radius of a circle is
Zero	Equal to 2π
Positive	Less than 2π
Negative	Greater than 2π

In at least these two ways, the curvature of surfaces in a region can be determined without any recourse to the appearance of the surface as it appears from a position above or outside it; because such a determination can be made, the curvature is said to be an *intrinsic* property of the surface.

This notion of curvature can be extended to a three-dimensional space, but visualization becomes difficult if not impossible. This idea will be discussed at more length in the last section of this chapter when geometry is considered as an experimental science.

SPACE-TIME

Before talking about the curvature of space-time, we should make clear the meaning of space-time itself.

In everyday experience, we think of space and time as being entirely different sorts of things, and we distinguish sharply between separations in space and separations in time. Two corners of the room are separated by a distance which is purely spatial; no time is involved in the idea of their separation or in the measurement with a meter stick of the distance from one corner to the other corner. On the other hand, if I snap my fingers twice as I sit still—snap, snap—the two actions are separated by a time interval, and distance or space does not enter the description. Even if I go to one corner of the room and snap my fingers and then go to the other corner and snap them again, we can talk about the time interval between the two snaps and the distance between the points at which the snapping took place in a completely unambiguous way. It is as if

Space is Space and Time is Time,
And never the twain shall meet.

But the special theory of relativity, published by Einstein in 1905, forces us to realize that this neat separation is an illusion—an illusion which we are permitted in normal life because we move at such low speeds. If we habitually moved at speeds which are a significant fraction of the speed of light, we would know intuitively that space and time are not uniquely separated because we would be familiar with disagreements involved in measurements of the two.

Instead of snapping the fingers, think of flashing two photographic flashbulbs. Let them be separated by some distance, and let one flash after the other. An observer on the earth can measure the distance between the bulbs, and he can measure the time between their flashing, and he finds two numbers which he can call the time and the distance; for simplicity of discussion, let us designate these Δt and Δx. If a man in a spaceship passing the earth at high speed observes the two flashes of light, he can also measure the distance between the sources and the time between the flashes, but he will not find the same Δt and Δx. Another observer moving relative to both the first two will find still another pair of measurements for the spatial and temporal separations of the flashes.

What we must realize is that attention must be focused on *events*, and that events are separated in time and space. But since not all observers can agree upon the separation in space or time, we must recognize that events take place in a space-time which is the same for all observers but which is divided into space and time differently by different observers depending upon their motion.

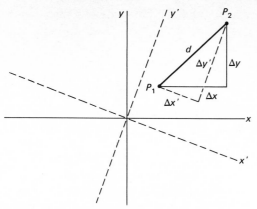

FIGURE 20.14 The distance d between points P_1 and P_2 is the same whether it is computed from Δx and Δy or from $\Delta x'$ and $\Delta y'$. By the Pythagorean theorem, $d^2 = (\Delta x)^2 + (\Delta y)^2 = (\Delta x')^2 + (\Delta y')^2$

Interval

Each of the two or three men who saw the two flashbulbs flash measured a different separation on the ground and a different time separation, but each can combine his measurements to produce a new number that they can agree upon. If we define the interval, designated Δs, by the equation

$$(\Delta s)^2 = (c\Delta t)^2 - (\Delta x)^2$$

then all observers of the same two events will compute the same interval between them. The interval thus plays the same role in space-time that the distance does in ordinary geometry. If two points are described by two different sets of coordinates, x and y or x' and y', the difference in the x coordinates and the y coordinates of the points will depend upon which coordinate system is used, but the distance between the points will be the same for both systems.

In ordinary space a point is described by three numbers, and in the simplest type coordinate system commonly used in mathematics, those numbers are called the x, y, and z coordinates. Events are specified by the addition of another number, the time t. From a formal standpoint, therefore, one can speak of a four-dimensional space in which events occur, and each event is uniquely specified by giving the position and the time of its occurrence, or by giving its x, y, z, and t coordinates.

Curved Space-Time

Visualize two black ants on the equator of a globe separated by a distance of 2 in. Let the ants begin crawling straight north, and imagine their measuring the distance between them when they begin and after they have crawled half way up the globe. What they discover, of course, is that they are coming closer together, even though each is crawling on a straight line. If we suppose that the ants have some scientific knowledge, we can guess the explanation they will invent for their decreasing separation: Some force must be acting on us to pull us together, perhaps some form of gravitation. If the ants are scientifically trained, however, they will suspect that the force acts on them only because they are black ants, and so they will ask two red ants to make the same trip, starting on the equator from points 2 in apart and crawling straight north. The red ants, of course, find the same result, that is they come closer together at an increasing rate. All that proves is that the mysterious force does not depend upon the color of the ant, but it may still be specific for ants, and so the scientist ants invite the participation of two ladybugs. When the ladybugs begin at two points 2 in apart on the equator and crawl straight north, they come together just like the ants did, and finally the ants feel justified in stating a universal law: All insects crawling north from the equator are attracted to one another by a force which increases as the distance between them decreases.

Looking at the globe from outside, we feel justified in commenting: No force is pulling those crazy insects together; they are coming together because they are on a curved surface, and no other explanation is needed. In fact, we might point out to the insects, if we could speak with them, that the very fact that the same amount of "pulling together" is observed for all sorts of insects should make them suspect that the effect is a geometrical one rather than the result of a force. Most forces, such as the electrical force, the magnetic force, the nuclear force, etc., do not affect all objects in the same way; hence any presumed force which does affect everything in precisely the same way is probably not a force at all, and a geometrical explanation might be inferred.

Uniform response to gravitation This is the sort of reasoning Einstein used in proposing that gravitation be explained in terms of curvature of space-time. The most striking property of gravitation is that it affects all matter in precisely the same way. The experimental test of that property is the fact that all falling objects have the same acceleration. That this is true to a very high degree of precision was first demonstrated by Eötvös in

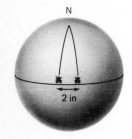

FIGURE 20.15 As black ants crawl north on a globe from starting positions 2 in apart, they come closer together.

1890 and then by Dicke in 1964. Dicke's experiment proves that two very dissimilar materials falling freely have the same acceleration to within 1 part in 10^{11}! Gravitation is thus unlike any other force known in physics in that it produces precisely the same effect on all matter.

This point can perhaps be better appreciated if it is stated in another way. Think of an artificial satellite in orbit 1,000 mi above the surface of the earth. If no rockets are firing, the satellite continues indefinitely in the same orbit, round and round the earth at a constant speed. If an astronaut inside holds a pencil up in the air and releases it, the pencil remains in the same position, also going around the earth in the same path, at the same speed. If a 1-lb container of waste materials is gently released to the outside, it follows the same orbit as the multiton space ship, going around the earth at the same speed. The important point is that the path followed is completely independent of the size, the mass, and the composition of the object; it depends solely upon the distance from the center of the earth and its initial direction and speed. Objects in orbit act as if they were following tracks in space like the longitude lines on the globe which the insects followed; the initial position and speed determine which line they are on, but from that time they follow a geometrical path, and the same path is there for all objects.

The exact analog of the insects crawling on the globe is two objects separated by a short distance which are dropped toward the earth. As they fall, they come closer together, and in conventional terms we explain that fact by saying that they are being attracted toward the center of the earth by gravity. Einstein suggested that we say that they are moving in a curved space-time, and that they come closer together because of its curvature. The earth obviously has some effect, but its effect is to determine the curvature of the space-time in its vicinity.

Tests of space-time curvature   From this point of view, a falling object or a satellite "tests" space-time and reveals its properties. Because any satellite in the vicinity of the earth is somewhat affected by the atmosphere of the earth and the atmosphere of the sun which extends to this distance, a satellite with a "conscience" has been proposed to accurately test the properties of space-time. The satellite would be equipped with small jets which could change its direction and speed slightly, and on the inside, it would have another body floating free of the outer satellite casing. The inner "conscience" satellite would follow its natural path in space-time, and if atmospheric drag caused the shell to slow or to deviate from the natural path, sensors inside would detect the fact that the conscience was no longer in the center of the shell, and jets would be fired to move the shell so that the conscience was again centered. The inner body would have no contact with the outside world, being completely enclosed. Its motion would be determined entirely by the nature of the space-time it felt, and it would move in a geodesic path in that space-time. (Geodesic path will-be explained more fully in a later section.) The geometrical nature of the phenomenon is emphasized by the fact that the conscience satellite could have a mass of 1 g or 1,000 kg, that it could be made of gold or aluminum or wood or glass, that it could be spherical, cubical, or any other shape, and that none of these changes in its characteristics would have the slightest effect on the path it would follow. Apparently its path is determined not by any of its properties but by the nature of the space in which it finds itself.

GENERAL RELATIVITY

The basic ideas of general relativity can be expressed in terms of two prescriptions: a prescription for determining the path of a particle moving under the influence of no forces or of only gravitational forces in a curved space-time, and a prescription for computing the curvature of space-time.

Field Equations

The properties of space-time in a region are determined by the values of 10 functions or numbers called the *metric coefficients*. Einstein suggested a connection between those coefficients and the density of mass, energy, and momentum in the region, and this relation comprises the prescription for finding the curvature of space-time. The relation consists of 10 equations, and only in a few special cases have the equations been solved exactly. For most problems, approximations must be used, and there is no hope of having complete and exact solutions of the sort one is accustomed to in algebra. One of the special cases for which an exact solution is known is that of the space around a spherically symmetric star whose rotation is negligible; the solution applies reasonably well to the sun. This is known as the *Schwarzschild solution* after the mathematician and astrophysicist who first proposed the solution in 1916. The complexity of the mathematics involved in this part of general relativity is so great that no more will be said about it.

Before giving the prescription for determining the path of a particle in curved space-time, we must define two terms. The *world line* of a particle is the path it follows in space-time. If the object remains at rest, its world line is a straight line, with the x, y, and z coordinates remaining constant and only t changing. If the object is moving, then all the coordinates may change, and in space-time its path is some sort of curve. No matter what the path may be, the world line is the line made up of the succession of points in space-time occupied by the object.

Geodesic

The other term which must be defined is *geodesic*. In two or three dimensions, one often defines a geodesic as the shortest distance between two points. In four dimensions of space-time, the same idea is true but distance has a different meaning, and the geodesic must be defined in terms of the *interval*, the quantity which takes the place in space-time of distance. Suppose that two points on a surface are chosen, and the problem is posed of finding the geodesic which passes through the points. Many paths can be drawn through the points, and the distance along each path can be measured. That path for which the distance is the least is the geodesic. Mathematical techniques are available for picking out from all the possible paths the one which gives the least distance, and the method is to find the path which has the property that the distance measured along any other path which differs from the chosen one by only an infinitesimal amount is the same as along the geodesic. Because the distance does not change if the path is changed an infinitesimal amount, it is said to have a *stationary value along the geodesic*. In the problem of the distance between two points, the stationary value of the distance yields the minimum distance, but in other problems the stationary value can be a maximum or neither the minimum nor maximum.

Another example of the determination of an extreme value along a path may make the idea somewhat more clear. One way to describe the behavior of light as it goes from one medium to another is this: A light ray takes the path from point A to point B which makes the time of travel the minimum. Figure 20.16 shows point A chosen in air and point B chosen inside a container of water. If a small light bulb is placed at A, light reaching B will have traveled by only one path. To find that path, one can draw all possible paths which the light might

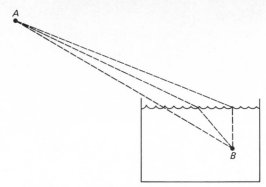

FIGURE 20.16 Light leaves point A and reaches point B. Three possible paths are shown; the light chooses the path which makes the time of travel from A to B a minimum.

take as the figure suggests, and then compute the time of travel from A to B taking into account the fact that in water light travels only about three-fourths as fast as it does in air. That path which makes the time the least (along which the time has a stationary value) is the path actually followed by the light. (The same path would be followed if the light bulb were placed at B and the light were observed reaching A.)

In the most general case, two points in space-time are separated by distances in x, in y, in z, and in t. The square of the interval between the two points is

$$(\Delta s)^2 = (c\,\Delta t)^2 - (\Delta x)^2 - (\Delta y)^2 - (\Delta z)^2$$

A geodesic in space-time is a curve along which the interval has a stationary value.

Motion of a Particle

Now the prescription for the motion of a particle can be given: Particles moving under the influence of no forces or of only gravitation have world lines which are geodesics; if a particle has a world line which is not a geodesic, it must be influenced by some force which is not expressed in the metric coefficients, i.e., which is not gravitation. Particles acted on by electrical forces, magnetic forces, or pulls (as by a rope), follow world lines which are not geodesics, but freely falling objects near the earth, satellites around the earth, planets, stars, and light rays in empty space follow geodesic paths in space-time. Figure 20.17 shows a schematic view of the world line of the earth during the course of 6 months.

Light travels on a special kind of geodesic—one on which

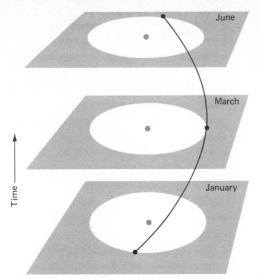

FIGURE 20.17 The world line of the earth for 6 months represented with one of the spatial dimensions suppressed and replaced by time. If the earth's orbital plane is thought of as moving "upward" in time, then the earth takes a corkscrew path in space-time.

the interval is zero. Because the interval is zero, the path is called a *null geodesic*.

Tests of General Relativity

General relativity is perhaps the most elegant physical theory ever devised, and because of the indifference of gravitational effects to the nature of the object upon which they act it seems to be almost logically necessary. The physicist, however, cannot feel happy with a theory unless it can be tested against experiment, and consequently much attention has been given to the few cases known where the general theory of relativity predicts a different result of an experiment than is predicted by newtonian gravitation theory. Four tests are known, three of which were suggested by Einstein and one which has been devised within the past decade.

Gravitational redshift The first of the tests is the gravitational redshift. The theory predicts that if light leaves a region of high gravitational field and is observed in a region of lower field, spectral lines will be shifted toward the red by an amount proportional to the difference in the fields. Thus light originating on the sun and observed from the earth should have a slight shift toward the red. This effect has been observed on the sun, but it is very small and hard to detect because of the effects of thermal motions and gas turbulence which produce doppler shifts greater than that produced by the gravitational effect. The effect has been observed in the light coming from white dwarfs, where the effect is much larger than from the sun. The most precise measurement of this effect was carried out on the surface of the earth, however, when the gravitational redshift was measured as light moved 75 ft up or down a tower. The measurements were made by Pound and Rebka (reported in 1960) and by Pound and Snider (reported in 1965). Even though the gravitational redshift has been fully confirmed, it gives us little reason to feel confident about general relativity, because the redshift does not depend upon the details of the general theory. Actually, the redshift can be predicted without any reference to general relativity,* and so although a failure to find the redshift would have caused problems for the general theory, a success in finding the redshift offers little support for the mathematical superstructure of general relativity.

Deflection of light   The second test of general relativity is the deflection of light passing the sun. If we compute the path (null geodesic) of a light ray from a distant star which passes very near the rim of the sun, we find that the ray is bent enough that the star should appear to be displaced 1.75 seconds of arc from the position it would occupy if the sun were not near the path. If we assume that light consists of particles having mass (the magnitude of the mass is unimportant because it cancels from the equations), we can compute the deflection of the light particles because of the gravitational attraction of the sun using newtonian theory of universal gravitation, and the result is that the deflection of light should be just half what general relativity predicts. The observation of the effect can be made with visible light only during a total eclipse of the sun so that stars near the solar disk can be seen, and the first confirmation of this prediction came during a solar eclipse in 1919. Photographs of the sky are made during the eclipse; several months later, when the sun is not in the sky at the time the same stars are above the horizon, photographs are made of the same stars again. Comparison of the two photographs shows displacement of the images of those stars near the sun, and the displacement is what has been predicted by the general theory, at least to within about 10 percent. This measurement was repeated during the eclipse of June 1973, and again the results were within 10 percent of the predicted values.

---

*Beryl E. Clotfelter, "Reference Systems and Inertia," pp. 95–98, Iowa State University Press, Ames, Iowa, 1970.

Since 1969 observations of the bending of "light" coming

near the sun have been made using a quasar as a source and radio waves instead of visible light. On October 8 of each year the sun passes between the quasar 3C 279 and the earth. Another quasar, 3C 273, is near enough to 3C 279 that their apparent separation can be monitored by a pair of radio telescopes operated as an interferometer, and as the sun comes near the line of sight of one so that it seems to shift position in the sky, that apparent motion can be seen. The results of this experiment are in close agreement with the predictions of general relativity.

Another measurement of this type was made in the spring of 1974, as the sun passed through a group of three quasars and very near one of them. The relative positions of the three radio sources were observed for several days using a radio telescope array which had one antenna 35 km from the other elements. The long base line of the interferometer permitted measurements of the deflection which were believed accurate to within 1 percent. The deflection observed agreed with that predicted by the general theory to within about 3 percent.

Observations of stars during total eclipse involve problems which cannot be easily overcome (atmospheric turbulence because of the sudden temperature change during eclipse, atmospheric distortion, clouds, etc.), but none of these is a serious impediment to the observation of quasars, and the radio observations almost certainly can be made more accurate than visible light observations.

Precession of the perihelion of Mercury The third of the classical tests of general relativity is the anomaly in the precession of the perihelion of Mercury. (The perihelion is the point of closest approach to the sun.) Mercury does not precisely retrace its path on successive trips around the sun, and a line joining its perihelion to the sun rotates in space. The same is true of the other planets, but the effect is much more noticeable with Mercury than any other planet because of its nearness to the sun and the relatively high eccentricity of its orbit. This rotation is called the precession of the perihelion, and it amounts to 532 seconds of arc per century. Most of the effect is caused by the attraction of other planets, and their contribution can be computed; it is 489 seconds of arc. The difference, 43 seconds of arc per century, was recognized before the introduction of general relativity, but no explanation was known for it. The general theory predicts precisely this much extra precession. Since this does require all the machinery of the general theory, the agreement is impressive.

A similar precession has been observed in the binary pulsar

described in Unit 18, but because the gravitational field in which the pulsar moves is so much stronger than the field in which Mercury moves, the pulsar precession rate is 4 *degrees per year* instead of 43 *seconds per century*. The relativistic effect appears to be more than 30,000 times greater in the pulsar system than in the sun-Mercury system. For weak gravitational fields, newtonian gravitation theory is adequate, but the stronger the field, the more inadequate is newtonian theory and the more important are relativistic effects. The fields produced in the binary pulsar system are so intense that many relativistic effects in addition to the precession of the periastron are expected, and that system may provide the most sensitive tests yet found of general relativity.

Time delay   The fourth test of the general theory is a delay in radio signals which pass near the sun. In the spring of 1970, as Mariners 6 and 7 went behind the sun, signals were sent from the Jet Propulsion Laboratory's Goldstone antenna telling the satellites to send a return signal. The total time for the round trip of the radio signals was about 43 minutes, but it was 204 μs more than it would have been if the sun had not been near the path. The general theory of relativity predicts that the signals should be delayed by 200 μs; the agreement is satisfactory considering the difficulty of making the measurement.

If any of these tests failed, the general theory of relativity would have to be revised or discarded, but the success of the tests does not prove that the general theory is correct. The theory is so complex and can be checked in so few ways that it is impossible to say that other theories may not be devised which make the same predictions in these four experiments or that other tests will not be devised which will prove the theory inadequate. Variations of general relativity have been proposed, and one, the Brans-Dicke theory (devised by R. H. Dicke and C. Brans) has been extensively developed and widely considered, but experimental evidence is now running heavily against it and in favor of the original Einstein form of the theory. At the present time (early 1975) there are no reasons to seriously question the accuracy and adequacy of general relativity.

GEOMETRY AS AN EXPERIMENTAL SCIENCE

Thus far the discussion has been entirely of the *geometry of space-time*, where curvature is postulated as an explanation of what is otherwise described as gravitation. This says nothing, however, about the geometry of ordinary three-dimensional space, and to that we now turn. Three-dimensional space can have zero curvature (be euclidean) or have positive or negative curvature even when four-dimensional space-time is curved.

Do we live in a euclidean world? Stated another way, if one draws a triangle by joining three widely separated points, do the interior angles add to 180°? Until about the middle of the nineteenth century, when the work of Bolyai, Lobachevsky, Riemann, and Gauss was published, everyone would have answered "yes" to that question and indeed thought the question itself insane. To Immanuel Kant nothing was more obvious than that euclidean geometry is the only geometry the mind can conceive; but Kant was wrong, and we know that noneuclidean geometries are possible and are self-consistent. Whether they describe the real world is a question which can be answered only by experiment.

Gauss's Measurement

The first man to try to answer the question experimentally was Karl Friedrich Gauss, who measured the interior angles of a triangle formed by three hilltops. No deviation from 180° was found, but that proves nothing, for such an experiment is like trying to determine the curvature of the earth by measuring the angles of a triangle on the ground with sides a few inches long. Perhaps if one could use a triangle with corners in distant galaxies the results would be different, but it is difficult to go to those galaxies to make the observations, and that is not a practical way of seeking an answer to the question.

One direct method and at least two indirect methods can be used to try to determine whether our physical space has zero, positive, or negative curvature. The direct method is an extension of the argument given in the first section of this chapter about the ratio of the circumference of a circle to its radius. In three-dimensional space the analogous measurement is the ratio of the area of a sphere to the square of the radius. In euclidean space, the area of a sphere is $4\pi r^2$; such a space has zero curvature and is said to be "flat," in analogy with the two-dimensional plane. In a space with positive curvature, the area of a sphere is less than $4\pi r^2$; and in a space with negative curvature, the area is more than $4\pi r^2$. One can make a summary table as before:

If the curvature of space is	The ratio of area of a sphere to the square of its radius is
Zero	Equal to 4π
Positive	Less than 4π
Negative	Greater than 4π

Now if one only had a way to attach a long string to the earth or the sun and to sweep out the surface of a sphere, and could then measure the area of that surface and compare it to the square of the length of the string, one would know something about the curvature of the space of our universe.

Counts of Galaxies

Marking out a sphere with a string many million light years long is hardly practical, but perhaps similar information about the ratio of the surface area of a sphere to the square of the radius can be obtained in a more realistic way. On the very largest scale, it appears that galaxies and clusters of galaxies are uniformly distributed in space. If we could mark out a thin spherical shell whose center is at the earth, therefore, we should expect to find that the number of galaxies within the shell is proportional to the volume of the shell. But if the shell is thin compared to the radius of the sphere, its volume is for all practical purposes proportional to the area of a sphere of that radius, and hence the number of galaxies within the shell is proportional to the area of the sphere. If we could count the galaxies within thin shells at differing distances, we could assume that the numbers of galaxies give an indication of the areas of the spheres. The only way to count galaxies within a shell is to assume that on the average galaxies have about the same intrinsic luminosity and to relate apparent luminosity to distance. Equations can be derived which take account of the time lag in the observations caused by the expansion and predict the variation of number, seen within a magnitude range, with magnitude. However, the differences between the three types of curvature make effects which are so slight that at the extreme range of the 200-in telescope no decision can be made. When radio galaxies are counted with radio telescopes, the counts can be extended considerably farther than is possible with the optical telescope, and such a project may have some slight hope of success.

Indirect Methods

The indirect methods involve measuring either of two other constants. The equations of general relativity combined with certain reasonable assumptions about the nature of the universe permit the prediction of relations among the curvature of

FIGURE 20.18 The volume of a thin spherical shell is approximately proportional to the area of the sphere, and if galaxies are uniformly distributed in space, the number within the shell is proportional to the area of the sphere.

space (assumed constant throughout the universe), the Hubble

constant, the average density of matter in space, and the deceleration of distant galaxies. If the average density of matter in
the universe is equal to or greater than about 5×10^{-30} g/cm^3,
space has a positive curvature and the universe is "closed."
(See page 405 for a detailed explanation of a "closed" versus
an "open" universe.) Visible matter seems to have a density of
approximately 10^{-31} g/cm^3, although there are reasons to suspect that that figure is too low. Whether enough matter is
present in the form of hydrogen between galaxies, black holes,
or other forms to close the universe is not yet known.

If the universe began with a big bang, and matter began
rushing out from an initial dense state, the mutual gravitational
attraction of the material should have had a slowing effect on
the expansion, and the expansion should be proceeding more
slowly now than it did originally. This slowing is described in
terms of a deceleration parameter designated q_0, and it also is
related to the curvature of space. If space has a positive curvature so that it is closed, the deceleration should be great enough
that eventually expansion will stop and the process will be
reversed, with everything falling back together. One method of
measuring q_0 is to look for deviations from linearity of the dis-
tance-magnitude relation, i.e., the indications that Hubble's
constant is not really a constant but has changed with time. Unfortunately, the measurements are not sufficiently precise yet
that we can compute the value of q_0 to within the limits needed
to determine the curvature of space. These indirect measurements will be considered again in the final unit.

If the curvature of the space of our universe is zero or negative, space is infinite and open. If the curvature is positive,
space is closed, and it has a finite volume. If the latter case is
the actual one, is there an edge to space, and is there a region
outside space? Could one conceivably travel far enough to
reach the edge or boundary? The answer is "no," for space is
boundless even though it may have finite volume. Just as the
traveler on the spherical earth who did not know of the shape
of the earth could travel forever without ever finding an edge,
always seeing land or sea stretching ahead of him just as before,
so a traveler through our space would always see uniform space
stretching ahead of him. Perhaps he might eventually return to
his starting point, but he would never find a boundary or a wall
to stop him or to offer a view of what is outside space.

QUESTIONS

1 Suppose that one took a series of points on the surface of an
egg, from the center (tip) of the small end, along the side, to the

center of the large end. If the gaussian curvature is computed at each point, are all the curvatures of the same sign? How do the magnitudes change as the point moves from the small end to the large end?

2 Suppose that you knew the distances between pairs of cities a great distance apart, as, for example, New York, Los Angeles, Cape Town, Buenos Aires, etc. If you knew the distance between only one pair, they could be plotted on a flat sheet of paper with the separation of the points proportional to the actual distance. Could you plot three on a flat sheet? Four? What is the maximum number that could be plotted on a plane?

3 On the surface of the earth, what must be the area of a triangle so that its interior angles add to 190°?

$$Ans \quad \approx 300,000 \text{ mi}^2$$

4 If two events occur, separated in time (as seen in one coordinate system) by 10^{-6} s and in distance by 100 m, what is the square of the interval between them? $\quad\quad Ans \quad 8 \times 10^4 \text{ m}^2$

5 Describe the world lines of (a) a chair sitting motionless, (b) a person walking at constant speed in a straight line, (c) an automobile accelerating on a straight road.

black holes

One can classify the objects of nature on the basis of the forces which maintain them, and black holes, if they exist, are produced by the failure of all other forces to balance gravitation.

FUNDAMENTAL FORCES OF NATURE

Physicists recognize four fundamental types of interactions among the elementary particles which make up the world: *gravitation, electromagnetic interaction,* the *strong nuclear interaction,* and the *weak nuclear interaction.* In many respects these interactions express themselves as forces between particles, and gravitation produces by far the weakest of the four forces. For example, the gravitational attraction between two protons is quite insignificant compared to their electrical repulsion if they are a centimeter apart, and it is equally insignificant compared to their strong nuclear attraction if they are 10^{-13} cm apart. The strong force holds nuclei together, and the weak nuclear interaction shows itself in certain kinds of nuclear decay. Weak though the weak nuclear interaction may be, it is still stronger than gravitation when the effects of the two interactions are compared for particles separated by short distances. The other three forces have certain limitations which gravitation does not share, however, and because of that gravitation can be far more important than the others under certain conditions. Both the strong and the weak nuclear forces are limited in range. If nuclear particles are less than about 10^{-13} cm apart, the nuclear forces act on them very strongly, but if they are considerably farther apart, those forces are totally unimportant. Both gravitational interaction and electromagnetic interaction also decrease with distance, but far more slowly, so that the forces are still important at very great distances. Because the electromagnetic force acts on electric charges, and charges come with both positive and negative signs, the electromagnetic force can be shielded. Under normal conditions we are unaware of the fact that all matter consists of electrically charged particles, acted on by very strong electromagnetic forces, because matter usually has so nearly the same number of particles of both signs that their effects cancel as viewed from the outside. Two positive charges placed a few centimeters apart repel one another because of the electromagnetic interaction, but if one charge is enclosed in a metal box, it will be impervious to the effect of the other charge because of a shielding effect. That effect is possible because the metal contains electrons which are free to move, so that they can place themselves near the outer positive charge and prevent its effect being felt inside the box. By contrast, gravitation arises from mass, which does not have a negative form, and consequently shielding does not occur. The idea of gravitational shields was used in an early science fiction story by Jules Verne, but they are not likely ever to become reality. Gravitation owes its unique role, as the dominant force in the universe, to the twin facts that it acts over long distances and that it cannot be shielded.

Planets

Consider a large cloud of gas and dust in space, which begins to contract because of mutual gravitational attraction. Each particle attracts every other particle, regardless of how far away it may be or how many other particles may lie between them. As the size decreases, all the mutual attractions increase, for the attraction obeys the inverse square law (force is proportional to $1/r^2$). Suppose that the cloud has a mass of 10^{24} kg—somewhat less than the mass of the earth. Eventually the contraction stops, and all the particles come to equilibrium. What force is now balancing the gravitational attraction? It is electromagnetic force. In a planet or a smaller body, the force which stops its contraction or prevents further gravitational contraction is the repulsion of electrons for one another—the effect which determines the nature of ordinary matter familiar from everyday experience.

Stars

But suppose that the cloud of material has a mass of 10^{30} kg—somewhat less than the mass of the sun. Then a different sort of force develops, for the interior becomes hot as the cloud contracts, and the thermal motion of the particles provides the force which balances gravitational attraction. That force depends on the electromagnetic interaction, which lets the particles feel one another at large distances. Electromagnetic forces in cold bodies, such as act within a planet or in the material which makes up a chair, usually produce crystalline materials, but the thermal energy of the interior of a star produces random motions of the particles of the matter at high speed and over long distances. We presume that most of the material of the universe is in bodies whose resistance to further compression by gravitation is produced by high temperatures.

White Dwarfs

The time comes, however, in the life of a star when the nuclear fires at the core burn low, and thermal energy no longer can provide adequate opposition to the gravitational forces acting inward. The star shrinks, and it passes the point at which electromagnetic forces might stop its collapse as normal matter, for a star has too much mass, and the gravitational forces are too great for it to exist with the electromagnetic forces maintaining the densities and structures of matter we are familiar with. If the mass of the star is below the Chandrasekhar limit, about 1.2 solar masses, its shrinking is stopped by a new effect, electron degeneracy, and it is a white dwarf. Once it reaches that state, it can remain there forever, for the forces which arise from electron degeneracy depend not at all on the temperature of the atomic nuclei within the star. As it radiates away the energy it can lose, nothing else about it changes, for the gravitational forces that would have caused it to shrink further have met their match.

Neutron Stars

If a star has a mass above the Chandrasekhar limit, it cannot become a white dwarf when it has exhausted its fuel, and it is likely to explode, as we have already discussed. One result of the explosion may be a neutron star. For our present purposes, another method of forming a neutron star will be more enlightening, although we do not know whether it ever occurs in nature or not. Suppose that a white dwarf has a mass very near the Chandrasekhar limit, and suppose further that matter continuously falls onto the star, so that its mass steadily increases. As the mass increases, the radius of the star decreases, and finally it is pushed over the brink. Once again gravitational forces take control, and the star shrinks. Electron degeneracy has been overwhelmed and is not able to balance the force of gravitation tending to cause the star to shrink. As it shrinks, electrons are forced to combine with protons to form neutrons, and eventually another effect becomes important—neutron degeneracy. When the forces associated with neutron degeneracy can counterbalance gravitation, the body is a neutron star, and again it is stable. If no more mass is added, it can exist forever as a neutron star.

378 Black Hole

But suppose that more mass is added. Eventually the neutron star reaches a maximum limit of mass, and the forces of neutron degeneracy can no longer sustain the crushing weight of its mass. It is not known what that limit is with the precision the Chandrasekhar limit is known, but it probably is between 2 and 3 solar masses. Whatever the limit may be, once it is passed no force exists to stop the shrinking caused by the gravitational force. As the star becomes smaller, gravitation becomes stronger (r becomes smaller and hence $1/r^2$ becomes larger), and the star continues in uncontrolled and unchecked gravitational collapse. It becomes a black hole.

Do not confuse a blackbody with a black hole. A blackbody absorbs all the radiation striking it, but it also emits radiation. It can be approximated quite well by an object covered with black paint or lampblack or by a cavity with a small hole in the side. A black hole absorbs all the radiation striking it, but it radiates nothing. Neither matter nor radiation can escape from it. A black hole cannot be approximated by anything from familiar experience. (In 1974 Stephen Hawking suggested that the statement that a black hole radiates nothing is incorrect. He proposed that photons can be formed immediately outside the boundary of the black hole and that they can carry away part of the energy of the hole. The smaller the mass of the black hole the more it radiates; a hole made by the disappearance of M solar masses radiates like a blackbody of temperature approximately $10^{-6}/M$ K. A black hole with the mass of the sun would act like a blackbody of temperature 10^{-6} K; a hole with the mass of 1 million suns would have an effective temperature of 10^{-12} K; a hole with the mass 10^{-6} solar mass would have an effective temperature of 1 K. The radiation from holes as massive as the sun would be negligible, but if holes with masses as low as 10^{12} kg formed early in the expansion of the universe they would have "evaporated" by now.)

Schwarzschild surface A black hole is a region in which space-time curvature has become so great that light cannot escape from the region. It is bounded by a surface called the *absolute event horizon,* or *Schwarzschild surface,* which has a radius proportional to the mass of the matter which has fallen into the hole; the radius is given by $r = 2Gm/c^2$, where G is the universal gravitational constant, c is the speed of light, and m is the mass which has fallen through the horizon. Nothing—matter, light, or anything else—comes out through the event horizon, but both matter and radiation can go inward through it. And as matter falls in, the size of the surface which marks the boundary of the hole increases.

If mass is added to the neutron star until the forces maintaining its size are overwhelmed, it begins shrinking and the density climbs higher and higher. The gravitational field at the surface becomes stronger and stronger, or, expressed in proper terminology, the space-time curvature becomes greater and greater. As the surface of the star shrinks through the absolute event horizon, light emitted from it ceases to reach the outside world; it is all bent back upon itself. So far as observers outside the event horizon are concerned, the star has disappeared. Within the horizon, the size of the star continues to decrease, and general relativity theory predicts that the density becomes infinite as the volume becomes zero; the matter has been crushed out of existence. This point will be discussed in more detail later.

Relative sizes   Perhaps we can gain some idea of the position of black holes on the scale of size by considering the sun, or stars of the same mass as the sun, in different states. At the present time the sun has a radius of about 7×10^5 km—about 400,000 mi. When it becomes a red giant, it may increase in radius by a factor of perhaps 250 or more. When it becomes a white dwarf, it will shrink in radius by a factor of about 100, so that it will have a size comparable to that of the earth. The radius of the white dwarf sun will be something like 7,000 km. A neutron star having the mass of the sun has a radius of about 10 km—6 miles. But if the mass of the sun formed a black hole, the radius of the absolute event horizon would be 3 km.

Bodies like the earth are not likely to ever suffer the crushing which would convert them into black holes, but it is interesting to note by way of comparison that if the earth were compressed until it formed a black hole, it would have a radius of 9 mm!

FORMATION OF A BLACK HOLE

Obviously if the mass of the earth were compressed into a sphere of radius 9 mm or if the mass of the sun were compressed into a sphere of radius 3 km, the densities would be enormous. It should not be inferred from this that high densities must always precede the formation of a black hole, for as the mass being compressed increases, the density which must be reached decreases. If a mass of 10^9 solar masses increases in density and decreases in volume until it forms a black hole, the density at the time everything disappears within the absolute event horizon is only 1 g/cm^3, the same as the density of water. It is possible that black holes of such mass exist at the center of galaxies or of star clusters, and they could have been formed by material which never reached unusual densities until it had fallen within the Schwarzschild surface.

Suppose that one could observe (from a safe distance) the formation of a black hole by the collapse of a body having a mass a few times that of the sun. What would be seen? The probability of our seeing such a collapse is slight, both because the star shortly before the onset of collapse is small enough to be faint, and because the collapse occurs on a time scale of seconds; one does not have months or even days to notice that something odd is occurring and set up an observational program.

First critical point For a star like the sun, photons leaving the surface headed directly up move on straight lines, but photons which leave at other angles are bent slightly, very much like the photons which come from distant stars and pass near the rim of the sun. For the sun, however, the effect is negligible. In the case of a neutron star the effect would be rather great for photons which left the surface approximately tangential to the surface. As a star went into uncontrolled collapse, the bending of the paths of photons leaving in any direction except straight up would increase, until the radius of the star reached the value 1.5 times the radius of the Schwarzschild surface (also called the *gravitational radius*), at which point those photons leaving tangential to the surface no longer can escape. Until the surface of the star reaches that point, all the photons finally escape, but when the radius of the star reaches the first critical point, $1.5 \times 2Gm/c^2$, some of the photons are trapped in a cloud, from which they rapidly leak away. Because the photons leaving the surface tangentially are trapped at that point but those slightly outside the point leak out, the radius of the star appears to remain fixed, but to dim very rapidly.

Gravitational radius Within less than a second the star shrinks to the gravitational radius beyond which no photons—even those leaving straight up—escape. Figure 21.1 is a representation of photon paths for stars at various stages of development. Photons emitted just before the star's surface passed the gravitational radius would leak away slowly, so that an observer watching from a distance would see the center of the star dim rapidly but not abruptly. The observer would see nothing of the star's behavior within the gravitational radius.

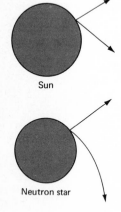

Sun

Neutron star

Star collapsed within the
Schwarzschild surface

FIGURE 21.1 Light rays leaving three kinds of stars; the stars are not drawn to scale. From the sun all rays leave on almost straight lines. From a neutron star rays moving radially take straight lines, but those more nearly tangential are bent. From a star within the Schwarzschild surface, all rays turn and return to the star. The light circle represents the Schwarzschild surface.

Inside View

If we were riding on the surface of a collapsing body, we would feel nothing unusual as we passed through the gravitational radius, but it would be a point of no return for us. Once we passed through that point, we would have no possibility of escaping to the outside. As the collapse proceeded, we would be acted upon by tidal forces of ever increasing strength; our feet would be drawn down with greater force than our heads, and we would be stretched. Eventually, as the collapse reached the *singularity*, the place at which gravitational forces become infinite, the tidal forces would become enough to deform even the elementary particles and squeeze them out of existence.

Infinities in Theory

When a theory predicts the existence of infinities—infinite density, infinite force, etc.—physicists tend to be suspicious. In the past, such problems in theories have always indicated that something important has been neglected or that the theory was being pushed beyond its realm of applicability. Many physicists now suspect that the infinite densities predicted by the general theory of relativity are not reached in nature, and the most likely reason is that quantum effects, which are ignored in the general theory, must come into play when space-time curvatures become very great. Thus far attempts to develop a combination of quantum theory and general relativity applicable to the region at the center of a black hole have not been entirely successful, but work continues.

SEARCH FOR BLACK HOLES

Although we may have doubts that the extreme conditions predicted at the very center of the black hole exist just as the general theory of relativity seems to predict, we have no reason to doubt that the theory's predictions of the possibility of black holes are accurate. We see many stars whose masses are far above the maximum for white dwarfs and for neutron stars, and although most of them may explode and throw away enough mass to bring the remainder below neutron star limitations, it seems very likely that some, even if they explode, have cores so massive that they cannot stop as neutron stars and undergo catastrophic gravitational collapse. If the formation of black holes is possible, we must consider how they might be detected and search the heavens for evidence that they exist.

Properties of a Black Hole

As matter collapses to form a black hole, and as other matter later may fall into the hole, it loses all its characteristics except three: angular momentum, electric charge, and mass. In other words, black holes made by the collapse of hydrogen and those made by the collapse of iron would be exactly the same if in both cases the angular momentum, electric charge, and mass were the same. Usually the mass which collapses is electrically neutral, and so the charge is unimportant, although if an electric field surrounds the collapsing matter (because of an excess of either positive or negative charge) that field still exists outside the black hole. The angular momentum is more complicated, and we shall defer a consideration of the effects it might produce for a short time. The most important characteristic of the black hole is determined by the mass of the matter which has fallen through the Schwarzschild surface. The mass determines both the radius of that surface (and hence the effective edge of the black hole) and also the gravitational field felt outside. The gravitational field outside the hole is the same as that produced by the mass which disappeared within the hole. If a planet is circling a star when the star suffers catastrophic collapse and disappears as a black hole, the planet still feels the same gravitational force (or sees the same curvature of space-time) and continues to follow the same orbit.

Observing a Single Hole

If a black hole, continuing the orbit around the galaxy of the star from which it formed, should drift across our line of sight to another star, its intense gravitational field would bend the light of the more distant star and perhaps even absorb it for a time. But the effects would be pronounced only for light passing near the Schwarzschild surface, and that has such a small size (radius = 30 km for a star 10 times as massive as the sun) that the probability of seeing such an effect is almost zero, and the probability of recognizing it if it were seen is almost certainly zero. We are not likely to observe black holes wandering singly among the stars.

Binary Systems

Our only hope of recognizing black holes, if they exist, seems to be to find binary systems in which one member has collapsed to a black hole. Because the gravitational effects outside are the same as if the mass were still present, Kepler's laws of motion of binary stars apply even if one of the members has collapsed, and so in ideal cases we should be able to determine the mass

of the dark member of a pair. And of course if one member of a
pair is a black hole, it will be dark.

Eclipsing binaries The first systems studied as possible
sites of black holes were eclipsing binaries. If a black hole were
bare in an eclipsing binary system, it could not be seen, for it
would be much too small to cause a decrease in light when it
came in front of the primary. But if the black hole were surrounded by a cloud of orbiting material, probably remnants of
the outer part of the star whose core formed the hole, then that
orbiting cloud could cause a dimming of the primary which
would be observable. If the secondary has a mass high enough
to make it a possible black hole candidate, and if its luminosity
is anomalously low for its mass, it could be a black hole.

Two systems were proposed as possibly meeting these criteria: Epsilon Aurigae and Beta Lyrae. Epsilon Aurigae is an
eclipsing binary with a period of just over 27 years; the primary
eclipse lasts about 700 days, but during much of that time about
40 percent of the primary's light is transmitted. The eclipse
suggests that a large cloud of dust, orbiting a massive but dim
star, eclipses the primary star. The dim secondary could be a
black hole, but other explanations are also possible for it.

Beta Lyrae has proved a difficult binary to analyze, but in
late 1971 an analysis was published which suggested that the
fainter star is approximately 20 times as massive as the sun, but
4 magnitudes fainter than a star of that mass should be. Observations in the ultraviolet made by the Orbiting Astronomical
Observatory added to the picture, indicating that the faint component is very hot and blue, and that it is surrounded by a
cloud of dust. That might be a possible appearance of a black
hole surrounded by dust if the dust is falling into the hole. Dust
near the Schwarzschild surface could be compressed and
heated very hot, so that it might look much like the surface of a
hot, small star.

Because the existence of black holes is still in doubt, and
because in any case a diagnosis of a black hole is an exotic and
unlikely way to explain the peculiarities of astronomical objects, astrophysicists are reluctant to accept that explanation
unless all other possible explanations can be eliminated. In the
case of the eclipsing binaries that is not possible, and astronomers trying to demonstrate the existence of black holes
beyond a serious doubt must look elsewhere.

X-ray sources Attention that has shifted from the eclipsing
binaries has gone to certain x-ray sources—particularly one
known as Cygnus X-1 (the name meaning that it is the first
x-ray source found in the constellation Cygnus). The x-rays
seem to come from a spectroscopic binary system, in which one
of the members cannot be seen. The visible star, an O-type star,

is designated in catalogs as HDE 226868. Observations of the visible star indicate that mass is flowing from it to the invisible companion. X-rays might be produced by mass falling onto a white dwarf, a neutron star, or a black hole, and the choice among the three depends primarily upon the apparent mass of the secondary. Calculation of the mass, in turn, depends primarily upon measurements of the distance to the system. The period of the binary rotation is known—5.6 days.

By early 1974 measurements indicated that the distance to the system is at least 8,000 light years and that the masses of the primary and secondary are 30 and 6 solar masses. If the dark body has a mass 6 times that of the sun, it most certainly is not a white dwarf, and it almost certainly is not a neutron star; by elimination it seems likely to be a black hole.

The heating of gas falling onto a white dwarf or neutron star and consequent emission of x-rays seems plausible, but what mechanism could produce x-rays from matter falling into a black hole? Clearly whatever it is must act on the matter before it passes the Schwarzschild surface, for once it passes that position, nothing leaving it can ever escape back to the normal universe.

Figure 21.2 shows the model which has been developed by Kip Thorne to explain x-ray production. Matter flowing from the bright and hot star toward the black hole has some angular momentum, and it goes into orbit about the condensed star, forming a rotating ring of material. At the inner edge, matter is gradually drawn from the ring into the hole. Because of the intense gravitational field near the inner edge of the ring, the material flowing into that region is compressed and heated until it becomes a source of x-radiation.

Other x-ray sources are known which may be candidates for black holes, but the first one to be examined sufficiently closely that its black hole character seems rather well established is Cygnus X-1.

GENERAL ARGUMENTS FOR EXISTENCE OF BLACK HOLES

All the discussion thus far of black holes has been based on the simplest possible solution of the equations of general relativity. That solution describes a black hole which has no angular momentum and in which the in-falling material at the time the

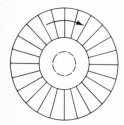

FIGURE 21.2 Two views of a ring of matter circling a black hole. Near the inner edge of the ring the material is compressed enough that it becomes very hot; hot spots in that thin ring may be sources of x-rays.

hole is formed has spherical symmetry. The restriction to spherical symmetry can be argued away rather easily because if a star which is collapsing is not spherically symmetric, the extra energy it may have which would normally take the form of oscillations will be radiated away in the form of gravitational radiation. The result will very quickly be the same as if the star had been symmetric.

If the star is rotating at the time of collapse, its final state is not the same as for the nonrotating star. The model we have been discussing does not apply in detail, although all the important results are the same or similar. The hole formed by a rotating star is described by two surfaces, rather than the single Schwarzschild surface which characterizes the spherical hole.

The cases of the rotating and/or nonspherical stars pose another problem, however, which deserves some consideration. The theory predicts that if a spherical, nonrotating body collapses past the point that thermal and strength-of-material forces can sustain it, it will find no other stopping points, and its collapse to a singularity in which matter is crushed out of existence is inevitable. Whether the same inevitability applies to a more realistic body, rotating and perhaps not perfectly spherical, is a more difficult question to answer.

Answers to such questions have been sought by general and abstract arguments rather than by detailed computations based on specific models. The men whose names are most closely associated with such arguments are Roger Penrose and Stephen Hawking, both English astrophysicists and cosmologists.

Penrose's Theorem

One result of this kind of general analysis is of such importance that it will be mentioned, although no attempt will be made to indicate the method of proof. The result is known as *Penrose's theorem*, and it can be stated as follows:

If three conditions are satisfied—(1) the future direction of time is always distinct and distinguishable from the past, (2) the density of mass plus energy in the universe is never negative, and (3) at some moment before the collapse began the universe had infinite volume—these conclusions follow:

1 For a spherical collapse, a singularity develops in space-time and matter is crushed out of existence.
2 For a nonspherical collapse, a singularity develops, but matter involved in the collapse may possibly escape crushing and may emerge in another universe. One possibility for its emergence is that it will reappear explosively at another point in space-time in our own universe; such explosive appearance of matter has been given the name *white hole*.

The first two conditions seem almost certain to be satisfied, but whether the universe has ever had infinite volume is not known. It is considered likely that the conclusions follow even if that third condition is not satisfied, but the theorem has not been proved without use of condition 3.

The black hole also enters into cosmological speculation in a different way. If the curvature of our three-dimensional space is positive, then it seems likely that the entire universe lies inside its Schwarzschild radius; in other words, we live inside a black hole!

Regions of space-time that close upon themselves, forming pockets cut off from the rest of the universe; matter falling into a black hole at one point in space and time and explosively reemerging at another position in space and/or time to form a white hole; matter going down a black hole in our universe and emerging in another universe; another universe??????. These are some of the aspects of black holes which have made them exciting topics for scientists, science writers, and nonscientists of all professions. And those astronomers and astrophysicists looking for them out there among the stars are bolstered by what appears to be a fundamental principle, derived from repeated discoveries of things which seemed possible but unlikely, and stated in this form: Whatever is not forbidden is required. So far as we know, black holes are not forbidden by nature or any theory, and so we have considerable confidence that they exist. All that remains is to find them.

QUESTIONS

1 Compute the Schwarzschild radius for masses of (a) 10^{24} kg, (b) 10^{30} kg, and (c) 10^{40} kg. Compute the average density of bodies of those masses as they reach the Schwarzschild radius.

> *Ans* (a) 1.5×10^{-3} m, 7×10^{31} kg/m^3, (b) 1.5×10^3 m, 7×10^{19} kg/m^3,(c) 1.5×10^{13} m, 0.7 kg/m^3

2 If a black hole with the mass of the sun came within 100 AU of the sun could it be seen? Would any effects be noticeable from the earth?

3 Can we be sure that the sun is not one member of a binary pair, the other member of which is a black hole 100 AU from the sun with the same mass as the sun?

4 In order to compute the mass of the dark member of the Cygnus X-1 pair, the mass of the visible star must be known. Its mass is estimated from its spectral type. Explain how the spectral type can be used to determine mass.

5 If a soft rubber ball fell into a black hole, what forces would act on it to deform it as it neared the Schwarzschild surface? What shape might it take as it approached the surface?

22
matter–antimatter

This unit might equally well be headed *charge-symmetric cosmologies*, but *matter-antimatter* sounds less formal and perhaps less forbidding. By whatever name, the question to be considered is the existence of large amounts of antimatter in the universe.

ANTIPARTICLES AND ANTIMATTER

In the late 1920s Paul Dirac published some theoretical work which suggested that particles might exist with all the properties of electrons except that they would have positive charges. Such particles were discovered in 1932, and they were named positrons. The electron and the positron were the first particle-antiparticle pair found. They are identical in every way except that one has positive charge and the other has negative. Positrons are created when energy, usually in the form of radiation, is changed into matter, but when a positron is created an electron is always created also. If a gamma ray having more than enough energy to make both particles ($E > 2m_0c^2$, where m_0 is the mass of an electron or of a positron) interacts with the nucleus of an atom, a positron-electron pair may be created. Nature seems to guard jealously the rule that charge cannot be created, and so when a particle with positive charge is formed, one with negative charge must be formed at the same time.

The life of a positron is short, for as soon as it slows down from whatever high speed it may have had at creation, it is likely to collide with an electron, and when that happens, both disappear and two photons of gamma radiation are sent off in opposite directions, bearing the energy of the masses of the two particles. That process is known as annihilation—for obvious reasons.

Positrons are relatively easy to produce because the energies are not high by the standards used by nuclear physicists, and pair production occurs commonly in cosmic-ray events as well as in many laboratory conditions. The same kind of general argument which might lead one to anticipate the existence of electrons with positive charge also suggests that protons with negative charge should be possible, but the production of negative protons requires such large amounts of energy that it was not observed until many years after positrons had become part of the accepted picture of the world. Negative protons, known as antiprotons, are produced in collisions of particles and targets when the particles have been given tremendously high energies in accelerators. Again, pairs are produced: protons and antiprotons. The total charge in the universe does not change. When antiprotons combine with protons to disappear by annihilation, the immediate product is not gamma rays but particles of lower mass than protons called *pions*. Those in turn decay into *muons*, and those produce electrons and positrons. Eventually all the energy does appear as gamma radiation.

The third particle present in ordinary matter, the neutron, also has its antiparticle, called the antineutron. Since the neutron has no net charge, the difference between it and the antinuetron is less obvious than for the other particles, but differences there are, and they appear to be related to the distribution of charge within the particle. A neutron has both positive and negative charge within it, and the sum of the charges is zero. An antineutron appears to have the same amounts of charge, but their arrangement within the particle is the reverse of the arrangement within the neutron.

Once we know that antiprotons, antineutrons, and antielectrons exist, a natural question is: Could they form antiatoms? Could atoms be formed with negative nuclei made up of antiprotons and antineutrons and with positive electrons (positrons) outside the nucleus? The answer is "yes." So far as we know, such atoms could form with the same ease as the atoms we are more familiar with. It is difficult to form such atoms in the laboratory, for both positrons and antiprotons are likely to disappear by annihilation before they can be brought together, but we have every reason to believe that if large numbers of antiparticles were brought into proximity with one another they would form antiatoms, and the antiatoms would constitute antimatter.

We are confident that antimatter could exist; whether it does exist is an open question. Because contact between matter and antimatter produces instant annihilation, with the release of copious amounts of gamma radiation, we can be sure that the earth contains no large pieces of antimatter, and because the solar wind flowing from the sun bathes the earth and the planets without producing annihilation radiation, we can be sure that neither the sun nor any of the planets is antimatter. But we cannot decide by examining the light they emit whether distant stars or galaxies are normal matter or antimatter. Perhaps some of the stars we observe every night when we look up at the sky are made of antimatter; that fact cannot be deduced from their appearance. Antihydrogen atoms, with a positive electron orbiting a negative proton, should produce a spectrum exactly like the Balmer spectrum of normal hydrogen. Antihelium should have lines placed at the same wavelengths as normal helium. If we are to decide whether the stars are antimatter, we must find some test other than their appearance.

CHARGE SYMMETRY IN THE UNIVERSE

But why worry at all about whether distant stars or galaxies are matter or antimatter? The experimental fact seems to be that the part of the world near us is constructed with negative electrons and positive protons. Why don't we accept that as the way things are and assume that the same thing is true of other stars and other galaxies?

The reason is that this violates a principle of symmetry which seems to be deeply embedded in nature. In probing the structure and behavior of matter at the most fundamental level yet achieved, physicists have found that principles of symmetry seem to express the most profound and most general laws which have been found. Indeed, this is true to such an extent that one's esthetic sense at times seems a better guide to the direction theory should go than preliminary experimental data. And one of the symmetries which nature seems to respect is the symmetry between particle and antiparticle. Every particle has its antiparticle (except for a few neutral particles which are their own antiparticles), and when a particle is created, its antiparticle is also created. Always—except when the universe began? But if that was an exception, then some physical law we do not know must have created the asymmetry, and that law is not observed to be operating now. Stated simply, a world made up of only matter, with negative electrons and positive protons, is asymmetric, and everything else we know about the laws which govern elementary particles would lead us to expect symmetry between matter and antimatter.

The matter can be argued in another way. If the universe began as a hot ball of radiation, if the initial state was so hot that no particles were present and radiant energy was the source of the entire universe, then all the particles now in the universe should have been formed by pair production. When we carry out pair production in the laboratory, a photon must interact with some particle so that it can convert itself into two particles; the reason the third particle is necessary is that both momentum and energy must be conserved, and they cannot be conserved if pair production occurs in a vacuum. (Momentum, like energy, is possessed by photons and moving particles, and it obeys a conservation law similar to that for energy.) In the early state of the universe, however, density and temperature were so high that two photons could have interacted with each other so that one could form a pair of particles. Surely, however, that would have produced as many antiparticles as particles, and the universe now should be half antimatter.

If we have convinced ourselves on such general principles that it might be reasonable to expect the universe to contain large amounts of antimatter, two problems remain. First, can we devise a cosmology to explain the behavior of such a universe? Second, can we experimentally detect any evidence for the existence of that antimatter? We turn to these problems now.

SYMMETRIC COSMOLOGIES

Although many people have taken an interest in symmetric cosmologies and have written about the matter-antimatter problem, three names stand out so prominently that they should be mentioned: Omnes, Alfvén, and Klein. Omnes has worked particularly on the model which begins with radiation and then pair productions; Alfvén and Klein have developed an entirely different kind of model.

Beginning with Radiation

If all the particles in the universe were formed by pair-production processes, they faced an immediate problem. The particles and antiparticles would have been intimately mixed at high density, and annihilation should have destroyed most of them immediately. Since they were not all destroyed, some process must have acted to separate them, and finding a plausible process has been the chief problem of cosmologists beginning from this starting point. Professor Omnes has proposed a process similar to the formation of bubbles in liquid, and he calculates that the mass of each bubble would have been of the order of magnitude of the mass of a galaxy, so that one might suppose that each galaxy is either matter or antimatter, but that neighboring galaxies may not be alike. The initial separation would have had to become effective within perhaps 1 ms after the formation of particles and antiparticles in order to prevent the formation of great amounts of annihilation radiation—radiation which we should see if it had formed.

Once bubbles have formed, one can explain the failure of large amounts of matter and antimatter to mix and annihilate by means of the *Leidenfrost phenomenon*. If a drop of water is dropped onto a very hot skillet or sheet of iron, the drop does

not evaporate immediately, even though the surface is hot
enough to vaporize it in an instant. Often the droplet will
bounce and skitter about the surface for some time. It can do
that because a layer of steam forms beneath the water, and the
steam insulates the water from the hot surface below. That
layer of steam is called the Leidenfrost layer.

In a similar way, if two cells of matter and antimatter meet,
annihilation will occur where the gases of each material touch
at the boundaries, but the heat and pressure produced by that
annihilation will serve to push most of the material from the
boundary and will insulate matter from antimatter. Thus cells,
once separated, can be expected to have long lives, for the annihilation at their boundaries may be a slow process, affecting a
very small amount of the mass of each cell.

Beginning with Thin Gas

Because of the severity of the problems of explaining the separation of matter from antimatter if both formed in a dense, hot
initial phase of the universe, Alfvén and Klein have proposed
another way the early history of the universe could have gone.
They suggest that the initial state was that of a very thin cloud
of equal amounts of matter and antimatter, intermixed. The
temperature was low. Gravitational attraction caused the cloud
to contract, but as it contracted and the density rose, annihilation increased, until finally the pressure of annihilation radiation equaled the gravitational attraction; the contraction
slowed and stopped, and it was replaced by an expansion. We
now live in that expanding phase.

During the long period of contraction and then expansion a
process similar to electrolysis has proceeded, with charges
moving through weak magnetic fields. The combined effects of
magnetic fields, gravitational fields, and electric fields have
separated matter and antimatter into regions, somewhat
smaller than galaxies. Again it is supposed that each region is
pushed away from its neighbor by a Leidenfrost layer at the
boundary surface. Approximately half of each galaxy is expected to be matter and half antimatter. One of the problems
with this model of the universe is that it does not explain the
2.7-K blackbody radiation.

Steady State

The steady-state theory could avoid some of the problems of
evolutionary theories simply by saying that matter and antimatter have always been separated, and that the material which
must be created to maintain constant density of the universe

includes both matter and antimatter. But unless we wish to contend that when a particle-antiparticle pair is created, one is created in the matter region and the other in an antimatter region—a solution of the problem which seems to violate the laws governing pair production as they are known from the laboratory—we must assume that the creation of mass is accompanied by a considerable amount of annihilation radiation. We do not see the gamma-ray flux which would be that annihilation radiation, and therefore the creation of pairs at the rate predicted by the steady-state theory seems inconsistent with observations.

OBSERVATIONAL TESTS

Finally, let us look at the observational data which might help decide whether large quantities of antimatter exist somewhere in the universe. The data are scant, and there is disagreement about the significance of the few facts which are known.

The only observation which can be used to detect the presence of antimatter is the annihilation radiation which results from its contact with normal matter. Annihilation produces gamma rays in a particular part of the spectrum—rays having energies around 70 MeV (1.1×10^{-12} J). If Leidenfrost layers exist within the galaxy or between galaxies or between clusters of galaxies, they should be sources of gamma radiation. Actually a continuous background of gamma radiation is observed, and equipment to study it well is so new that we may assume that the final word has not been spoken on the shape of the spectrum, but the spectrum does have a hump between 1 and 100 MeV. Several mechanisms could produce the extra gamma radiation observed in that energy range, and one of the mechanisms is annihilation of matter and antimatter. Thus cosmologists can take this excess of gamma radiation as indication that annihilation is occurring (or, more accurately, that it occurred in the past, for it is necessary to assume that the gamma rays we are seeing have been redshifted somewhat).

On the other hand, computations of the amount of gamma radiation to be expected if annihilation is occurring now within any reasonable distance of us seem to indicate that we should see far more radiation than we do. This sort of computation is complicated and its results are made tentative by uncertainty about the values of the densities of intergalactic gas which should be assumed.

It has been suggested that the huge energy outpourings from QSOs and from galaxies which appear to be undergoing explosions are the result of annihilation. In annihilation all the mass involved is converted into energy, and the process can be rapid. But if that were the source of the energies of quasars, they should be gamma-ray sources, and they are not.

The strongest argument for the charge-symmetric universe is simply that—its symmetry. But despite the fact that physicists have come to place great emphasis upon symmetry as a guiding principle in the development of models of nature, few cosmologists have been convinced that the universe is in fact charge-symmetric. The reasons are two: the difficulty of explaining the early separation of matter and antimatter, and the lack of convincing experimental or observational evidence. If gamma radiation should be found which could be explained by no other hypothesis, the situation would change quickly. And if a mechanism could be found to explain the separation of the two components that was convincing to most cosmologists the situation would change, but perhaps not quickly. At the present time, however, most cosmologists prefer to work with a universe which appears to have only one kind of matter in it, even though they cannot explain why that should be the case.

QUESTIONS

1 If 0.5 kg of matter were mixed with 0.5 kg of antimatter, both would disappear and be replaced by radiant energy. How much energy, measured in joules, would be produced?

$$Ans \quad 9 \times 10^8 \, J$$

2 Suppose that a sphere of antimatter a mile in diameter entered the solar system and went into orbit around the earth at the distance of the moon. How might we be able to tell that it is antimatter?

3 The MeV (million electron volts) is an energy unit equal to 1.6×10^{-13} J. Verify that 70 MeV is 1.1×10^{-12} J.

23
gravitational radiation

The existence of gravitational radiation is predicted by the general theory of relativity, but not by classical gravitational theory. Einstein recognized this fact when he was developing the general theory, but he did not suggest it as a method to compare the two theories because the technology to detect gravitational waves did not exist. Gravitational waves have been sought for many years, but until recently the search was carried on by one man almost alone; now many men in many countries are working on the problem.

GRAVITATIONAL WAVES

A wave is a disturbance which transports energy without transporting material. A ripple spreading out on the surface of a pond of water can move floating objects far from the source of disturbance even though no water has moved from one position to the other. The water movement is entirely up and down, not outward. Electromagnetic waves travel through empty space, carrying energy. They are produced by movements of electric charges, and they are detected by their ability to make electric charges move again, but nothing material flows from the source charges to the charges which feel the effect of the wave. Instead, the electromagnetic wave is a traveling, fluctuating electromagnetic field. Similarly, gravitational waves are traveling, changing gravitational fields; they should travel through space with the same speed as electromagnetic waves.

Possible Sources

Gravitational radiation should be produced by massive, rotating objects which are not spherically symmetric, by double stars, by matter falling into black holes, and by the collapse and explosion of such objects as supernovae. A star such as the sun, even if it is not spherically symmetric, radiates negligible energy as gravitational radiation; but a neutron star, oblate because of its rapid spin, might radiate considerable amounts of energy during the hours or days after its formation before it has slowed down enough to become virtually spherically symmetric. Ordinary double stars radiate negligibly, but two white dwarfs, close enough together that their period is in minutes, might radiate appreciably.

Possible Detectors

How does one detect gravitational waves? A first thought might be that a pendulum isolated from vibration might suddenly swing if a gravitational wave passed, but that is incorrect. Such a detector would be analogous to the detectors used for electromagnetic waves, but gravitational waves are much more complex than the most commonly used electromagnetic waves, and the detector must be more sophisticated. A gravitational wave produces not so much a transient pull on massive objects as a transient tide within them. Consequently one does not look for sudden motion, as with an accelerometer, but rather for relative motion, as the separation between two massive objects. Two masses connected by a spring might be a reasonable detector, for the passage of a gravitational wave should cause the spring to be stretched or compressed, and, in fact, a wave might cause the mass-spring system to begin to vibrate.

WEBER'S WORK

In 1958 Joseph Weber, at the University of Maryland, began a search for gravitational waves, and in 1969 he reported that he was detecting what might be such waves. Weber's detectors are aluminum cylinders approximately 3 ft in diameter and 5 ft long, each weighing $3\frac{1}{2}$ tons. A cylinder is suspended by wires, attached to a frame resting on vibration isolating pads, so that it is unaffected by disturbances to the tank encasing it or to the earth. The cylinder is sealed into a tank from which much of the air is pumped. Around the center of the cylinder is cemented a group of piezoelectric crystals, crystals which generate an electrical signal when they are stretched or compressed. The crystal detecting system is so sensitive that it can detect deformations in the cylinder of 1 part in 10^{16}; it can detect changes in length of the cylinder equal to $\frac{1}{100}$ the diameter of the nucleus of an atom.

When a gravitational wave passes over the cylinder, it exerts more force on one end than on the other, and the cylinder is stretched or compressed slightly. Once it has been disturbed, however, it continues to vibrate, so that the effect is something like the ringing which would persist for a time and then die out if the cylinder were tapped on the end with a hammer.

Because the detectors depend upon the "ringing" of the cylinders after they have been disturbed, they are particularly sen-

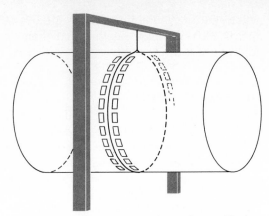

FIGURE 23.1 Weber's gravitational wave detector. The detector is a cylinder of aluminum approximately 3 ft in diameter and 5 ft long, suspended by wires from a frame resting on pads to isolate the system from vibrations. The blocks around the center of the cylinder are piezoelectric crystals, which produce electrical signals when they are stretched or compressed. A pulse of gravitational radiation caused the bar to ring, much as if it had been tapped on the end with a hammer, and the crystals are deformed by the resulting changes in length of the cylinder.

sitive to radiation at the resonant frequency of the cylinders, which is 1,661 Hz; that frequency is determined by the geometry of the cylinders. Furthermore, the detectors are directional, responding much more strongly to waves coming toward them in one plane than from any other directions.

The cylinders are at room temperature, and consequently the atoms of the aluminum have thermal motions, and the random thermal vibrations cause a constant noise output in the electrical system which amplifies the signals from the piezoelectric crystals. The things that Weber looks for as evidence that the cylinder has been affected by a passing gravitational wave is a signal much larger than most of the random signals. Occasionally the needle producing the trace on the paper tape swings 2 or 3 times as far as its normal swings, and such large swings may indicate that the system has responded to a gravitational wave.

But random events can also produce large swings; the larger the swing the more infrequent it is likely to be, but quite large variations in amplitude can be produced by thermal, random vibration. To guard against mistaking large random events for gravitational wave pulses, Weber has constructed several detectors, most of them identical. They are isolated from one another, and he considers significant only events which are registered simultaneously by two, three, or four of the detectors. Most of his detectors are operated at College Park,

Maryland, but one has been operated for many months at
Argonne National Laboratory, near Chicago, about 1,000 km
from the Maryland site. Coincidences are found between the
Argonne detector and the detectors at Maryland.

397

GRAVITATIONAL
RADIATION

Another detector, constructed in a different way and tuned
to 1,030 Hz, has also been set up at Argonne, and coincidences
between large pulses detected by it and pulses detected by the
1,661-Hz detectors in College Park have been found.

By December of 1972 Professor Weber was reporting seeing
about four coincidences per day above the rate that should have
been produced by random events (estimated at one per day),
and the coincidences reached a peak when the earth was
turned in such a way that the detectors were most sensitive to
signals which might have come from the direction of the center
of the galaxy.

The Energy Problem

Weber's detectors are most likely to respond to pulses of radiation, such as might be produced by the explosions of supernovae, or the collapse of bodies to form black holes, or the
swallowing of a star by a black hole. If he had seen one event
each month, or even one per year, they might have been accepted quickly as pulses of gravitational radiation, but a rate of
four per day, or even one per day, raised very perplexing
problems concerning their origins.

One possible source of pulses of gravitational waves is the
explosion of a supernova, but within our galaxy such an
explosion is expected approximately once every 100 years.
That estimate may be slightly low, but certainly the rate of formation of supernovae is not four per day.

A more general analysis can be made in this way. Weber's
equipment responds to waves in a very-narrow-wavelength
region; but we assume that waves of many other wavelengths
are present, undetected by this particular equipment. It is unlikely that any type of pulse source would produce a narrow-
wavelength band of radiation, and even more unlikely that if
such a narrow band were produced, it would coincide with the
one chosen by Weber for his detectors. We can estimate the
energy the detectors are absorbing, and from that estimate the
amount of energy passing them in their wavelength band.
(They are rather insensitive detectors and respond to only a
part of the available energy at 1,661 Hz.) Then we can estimate
the amount of energy that must be present in a band of reasonable width of wavelengths. Finally, we take the indication from
Weber's results that the source is likely to be at the center of the
galaxy, and we assume that it is radiating isotropically, so that

the same energy goes in all directions. These assumptions permit a computation of the amount of energy that is being radiated by the center of the galaxy in the form of gravitational waves; it indicates that about 10,000 solar masses are being converted into gravitational waves each year.

But the galaxy appears to have a mass of about 10^{11} solar masses, of which perhaps 10 percent or 10^{10} solar masses is at the center. If that mass is being destroyed, that is, converted to gravitational radiation, at the rate of 10^4 solar masses per year, the center of the galaxy will be entirely destroyed in a time of 1 million to perhaps 100 million years. If the center of the galaxy were fading away at that rate, the orbits of stars farther out would increase in size as the gravitational attraction providing the centripetal force on them decreased. Consequently the motion of stars and gas in the galaxy have been studied to determine if such an outward drift of the outer parts of the galaxy is occurring. Those studies indicate that over the past 10^8 years the decrease in mass at the center of the galaxy cannot be more than about 200 solar masses per year on the average, and if a period of 10^9 years is taken, the average mass loss cannot be more than 70 solar masses per year.

One possible way to explain the discrepancy between energies observed and the computed energy being radiated is as follows. We have assumed that the source radiates equally in all directions, so that we are struck by a very small part of the total energy it is emitting. Perhaps that is not true; perhaps it beams its radiation in the plane of the galaxy for some reason. We are located almost precisely in the galactic equatorial plane (to within 1 part in 10^5), and so if the radiation were produced by some sort of giant synchrotron effect which sent the radiation out only in the plane of the galaxy we should see part of it, but the total being emitted might be much less than if it is emitted isotropically. It is an interesting idea, but all attempts to find a mechanism which should produce such directional gravitational radiation have been unsuccessful, and it appears that this explanation does not resolve the difficulty.

OTHER EXPERIMENTS

The discovery of gravitational radiation is of such great importance both for its bearing on the general theory of relativity and for its information about happenings in the universe that it cannot be neglected by either astronomers or physicists. Weber's work seemed to be very carefully done, and from 1969 on he constantly improved his equipment and his analysis so that one by one the alternative explanations for his observations seemed to be rendered unlikely. At the same time, however, the frequency of pulses which he observed was so high that they could not be easily accepted as gravitational pulses. The natural response of the scientific world to a situation like this is to attempt to duplicate the work, and by the early 1970s at least a dozen research groups in all parts of the world were attempting to detect gravitational waves. Some groups had built or were building equipment as much as possible like Weber's detectors, but others were attempting to produce more sensitive detectors.

One of the first groups to report on its work is in Russia, at Moscow State University. Their detectors are copies of Weber's, and they have reported observing the same kind of coincidences which Weber sees.

One of the most significant attempts to detect the radiation by detectors of different design has been made by T. Tyson of Bell Telephone Laboratories. He has built a detector 12 ft long which responds to a frequency of 700 Hz. By putting in pulses at one end, he has measured the sensitivity of the system and determined the signal/noise ratio. That ratio is 50 to 100 times better than in Weber's equipment, and the sensitivity of the system to gravitational waves should be about 50 times the sensitivity of Weber's detectors. After many months of running, Tyson reported that he had found no pulses which appear to be caused by gravitational waves.

Detectors are being prepared which will be cooled to liquid helium temperatures to reduce the thermal agitation and improve sensitivity. Some of those detectors weigh 5 tons.

With the single exception of the report from Russia, no one except Weber has found pulses identifiable as gravitational waves. No one else is operating several detectors and looking for coincidences, but numerous experimenters have developed detectors several orders of magnitude more sensitive than Weber's detectors, and they are seeing nothing. It is difficult to draw any conclusion other than that the coincidences which Professor Weber sees are caused by something other than gravitational waves, but what that thing might be has not been explained.

Future Research

As doubt about the reality of the pulses reported by Weber has grown, interest in continuing experiments to detect gravitational waves has also grown. The sensitivity of the detectors now being built or planned is such that they have reasonable probability of being able to detect the waves from events which are rather well understood; one need not hope for some unlikely event which releases remarkable amounts of energy in the form of gravitational waves. Papers are beginning to be written on gravitational wave astronomy, and we may hope some day to "see" the universe by means of those waves. We know what the universe looks like by visible light; we have learned much of its appearance by radio waves; and we are in the process of learning its appearance by x-rays and gamma rays. What it may look like by gravitational waves we do not know.

Present detectors are almost sensitive enough to respond to the gravitational waves emitted by a supernova anywhere in our galaxy, and very soon sensitivity adequate for that observation should be attained. Within a few years we may reasonably expect detectors to become sufficiently sensitive that they could detect radiation from a star falling into a black hole at the center of the galaxy, the explosion of a supernova in any of many nearby galaxies, and perhaps even the rotation of very rapid double stars in our galaxy.

The failure of detectors more sensitive than those of Weber to find pulses which look like gravitational waves would be writing an end to this subject in physics if it were not that that same sensitivity makes further research in the area look promising. Even though Professor Weber may not have found gravitational waves (and the explanation for his coincidences is not known), if such waves are ever found it will be in part because of his persistence, and because he made the subject something which must be discussed and considered.

QUESTIONS

1 What is the wavelength of gravitational waves of frequency 1,661 Hz?   *Ans*   180 km

2 If one assumes that Weber is seeing gravitational waves emitted isotropically from the center of the galaxy 30,000 light years away, the rate at which he sees pulses implies that 10,000 solar masses of matter are being converted into gravitational waves each year. Suppose that the source happens to be in line with the center of the galaxy but is actually only 3,000 light years distant. How much mass would then be required each year to produce the energy apparently seen?

Ans   100 solar masses

3 Why should cooling a detector improve its performance?

4 What is meant by the term "gravitational wave astronomy"?

24
cosmology–
a second look

In Unit 16 we discussed the expansion of the universe, the cosmological principle, the difference between evolutionary and steady-state models, and Olbers's paradox. Now we can consider in more detail different evolutionary models and the experimental evidence which is available to help decide among them, and we shall conclude with a possible scenario of the early universe.

CONSTRUCTION OF MODELS

We assume that the universe is isotropic, which means that it is the same in all directions from us, and that it is homogeneous, which means that another observer in another part of the universe would see the same large-scale features we see. Only two kinds of general motion are possible within these restrictions: uniform expansion or uniform contraction. If any other motions were occurring, parts of the universe would look unlike other parts to either us or to other possible observers. With the benefit of Hubble's observations and the many confirming studies since his time, we assume that the universe is undergoing a uniform expansion.

Hubble Law

Discussion of models of the universe is simplified by the introduction of a concept used by cosmologists: the *co-moving observer*. A co-moving observer is an observer who moves only as the matter of the universe moves because of the expansion; he is not affected by the special motions of stars, galaxies, or other deviations from homogeneity. If we could imagine the universe still filled with a smooth, uniform gas, then the co-moving observer would move like one of the particles of that gas, at rest with respect to the material near him. Two co-moving observers separate only because they are carried apart by the expansion of the universe. Because of Hubble's law, their

401

relative velocity at any time is proportional to their separation at that time: $v = H(t)r$, where $H(t)$ is the Hubble constant and r is their separation. The Hubble constant is written $H(t)$ because it may change with time; it is constant in space but perhaps not in time.

Scale Factor

Because of the validity of Hubble's law, we can write the separation between any two co-moving observers at time t in terms of their separation at some initial time as

$$r = R(t)r_0$$

where r = separation at time t
r_0 = separation when $t = 0$
$R(t)$ = function of time called the *scale factor*

Because of the isotropy and homogeneity of the universe, $R(t)$ is the same everywhere, for all pairs of co-moving observers. Thus if we know the function $R(t)$, and if we take the start of time to be now, so that the distance to a distant galaxy is r_0, then we can compute how that distance will increase with time.

Parameters in the Equations

Let us now consider a model of the universe which is a space filled with dust or gas having no gas pressure; it is expanding. We could use newtonian mechanics to compute how the gas should be expected to move, and we would get the same equations as when we use general relativity, but the interpretation of the results is easier in relativity, and we will discuss the results in those terms. The equations which can be derived by the application of a little mathematics give the scale factor $R(t)$ as a function of time. The following other constants enter into the models:

H_0 = the value of the Hubble constant now [probably about 55 km/(s)(Mpc)]

$1/H_0$ = the time since the expansion began if the Hubble constant has always had its present value [$\frac{1}{55}$ km/(s)(Mpc) = 18 billion years]

k = an indication of the curvature of three-dimensional space; it can be zero, greater than zero, or less than zero (some authors use a slightly different k which can have the values 0, +1, or −1)

q_0 = the deceleration parameter; a measure of the rate at which the expansion is slowing

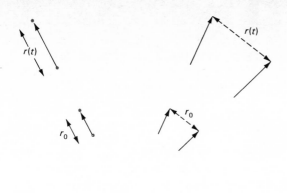

FIGURE 24.1 If the separation of two co-moving observers at time $t = 0$ is r_0, the separation at any later time t is $r(t)$. All co-moving observers appear to us to be receding directly from us.

$T =$ the time since the expansion began
$\rho =$ the average density of matter in space

The equations give relations among these quantities, and at least one relation is so simply visualized that it can be discussed. The expansion is slowing because of the mutual gravitation of the parts of the universe. Thinking of the expansion as it would be described in newtonian theory, we can say that each part of the universe attracts every other part, and the attraction constitutes a pull backward, so that the speed with which everything is flying apart is decreasing. The expansion began with high speed, but now it is slowing, and the rate of slowing is measured by the parameter q_0 (whose exact definition cannot be easily given without the use of more mathematics than we are allowing ourselves). The expansion of the universe with braking of the motion by gravitation can be compared to the motion of a projectile fired upward from the surface of the earth. If the speed of the projectile is below $7\frac{1}{2}$ mi/s, it will slow down, stop, and fall back. That speed, $7\frac{1}{2}$ mi/s, is called the velocity of escape. If the speed is just equal to the velocity of escape, the projectile will always continue going, but its speed will approach zero as it reaches a great distance. But if the speed is above the velocity of escape, the projectile will always continue moving, and its speed will not drop to zero. These three cases correspond to the first three models described below: the first has just the velocity of escape; the second has less than that velocity; and the third has more than that velocity. It is reasonable that if the universe has precisely the velocity of escape, the density of matter, which is producing the braking force, should have a particular value, and that for the other two cases the density should be above or below that single value.

MODELS OF THE UNIVERSE

Four models will be described. These do not exhaust the possibilities, but they serve to illustrate the types of models most often discussed, and they show the general properties of evolutionary models and the steady-state model. The first three are called *Friedmann models*.

Model I

This corresponds to the universe moving at just escape velocity; it is called the *Einstein–de Sitter* model because of the men who did early work on it.

The Einstein–de Sitter model is characterized by $k = 0$. That means that the three-dimensional space is euclidean, so that the interior angles of triangles add to $180°$ and the surface area of a sphere is $4\pi r^2$. The variation of $R(t)$ with t is indicated in Fig. 24.2. The time since the beginning of the expansion is $\frac{2}{3}(1/H)$, or 12 billion years. In the figure the present time is indicated, and the lighter line shows the extrapolation back from the present value of the Hubble constant. If it had remained always at its present value, the expansion would have begun at the time the gray line strikes the time axis. Clearly the actual time since the beginning of the expansion is less than $1/H_0$.

In this model $q_0 = \frac{1}{2}$, and the average density of matter in space is 5.7×10^{-30} g/cm³. The model predicts a precise value for that density through the equation

$$\frac{8\pi G\rho}{3H_0^2} = 1$$

That density, 5.7×10^{-30} g/cm³, is often referred to as the density required to close the universe or as the *critical density*. The reason for that terminology will be apparent from the nature of the next model.

Model II

This is often called the *oscillating model*, although we do not know whether it would actually oscillate even if this is the correct description of the universe. In this model k is greater than

$R(t)$

Now t

FIGURE 24.2 The scale factor $R(t)$ as a function of t for the Einstein–de Sitter model ($k = 0$). $R(t)$ is proportional to $t^{2/3}$.

zero, which means that the space has positive curvature, analogous to the surface of a sphere. Triangles have interior angles adding to more than 180°, and the area of a sphere in this model is less than $4\pi r^2$. The behavior of $R(t)$ is shown in Fig. 24.3; $R(t)$ is the geometric figure known as a cycloid.

If this model describes our universe, we are on the up side of the curve, at about the position shown. Eventually the value of $R(t)$ will reach a maximum value and will begin to decrease. At that time, billions of years in the future, when $R(t)$ is decreasing, distant objects will have a blue shift rather than a redshift, and everything will be moving back toward a state of high density. We do not know what would follow the collapse —whether another cycle would begin or not. If cycles repeat one another, then the universe truly oscillates. Estimates of the time for the expansion and contraction are approximately 70 billion years.

The time since the beginning of the expansion T is less than $\frac{2}{3}(1/H)$; q_0 is greater than $\frac{1}{2}$; and the density of matter ρ is greater than the critical density. The density of matter is great enough to eventually stop the expansion and change it to contraction. Because the curvature of three-dimensional space is positive, the space is *closed*; a light ray sent in one direction could eventually return to its starting point (if the expansion were not always making the space larger). But that closure of space, with positive curvature, is related to the density of matter, which must be greater than that required for the Einstein–de Sitter universe.

Model III

This model corresponds to a universe expanding at greater than escape velocity; it will continue expanding indefinitely, without ever approaching a zero expansion rate.

Three-dimensional space is hyperbolic; k is less than zero. Triangles have interior angles adding to less than 180° and spheres have surface areas greater than $4\pi r^2$. Figure 24.4 shows the variation of $R(t)$ with t. The deceleration parameter lies be-

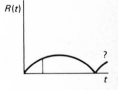

FIGURE 24.3 The scale factor R(t) as a function of t for the oscillating model (k > 0). R(t) is a cycloid.

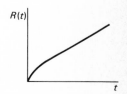

FIGURE 24.4 The scale factor R(t) as a function of t for the ever-expanding model (k < 0). R(t) is proportional to $t^{2/3}$ for small t, but it becomes proportional to t for large t.

tween zero and $\frac{1}{2}$, and the density is below the critical density. The time since the beginning of the expansion is greater than $\frac{2}{3}(1/H_0)$.

Model IV

This is the steady-state model, and it is included for comparison. In it H is a constant, but $R(t)$ is not. In fact, $R(t)$ increases exponentially, so that $R(t)$ is proportional to e^{Ht}. Figure 24.5 shows $R(t)$ as a function of t. The deceleration parameter for the steady-state model is $q = -1$. The model makes no prediction about the value of the density of matter, but it relates the observed density to the required rate of creation of matter.

The properties of the first three models are summarized in Table 24.1.

OBSERVATIONAL TESTS

Now that we have sketched the main features of four types of models, let us consider the observational evidence which can be used to help choose among them. We begin by eliminating the steady state.

Steady State

There is sufficient uncertainty in all the measurements and observations having a bearing upon choice of a model that it is premature to completely eliminate the steady-state model from consideration, but so much evidence weighs against it that at the present time it is not a serious contender for first place. Some of that evidence has already been discussed, but it will be mentioned again.

The fundamental fact about the steady-state theory is that if it is correct, the universe has always looked the same and has always had the same constitution it now has. Two pieces of evidence cast doubt upon the accuracy of that prediction. One is counts of radio sources which suggest that they were more dense in the past than they are now. The other is similar counts of QSOs, which suggests that they also were far more numerous in the distant past than they are now. In both cases, the observa-

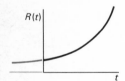

FIGURE 24.5 The scale factor $R(t)$ as a function of t for the steady-state model ($H =$ constant). $R(t)$ increases exponentially with time.

TABLE 24.1 CHARACTERISTICS OF THREE EVOLUTIONARY MODELS OF THE UNIVERSE

Model	k	Three-dimensional space type	Area of sphere	T	ρ, g/cm^3	q_0
I	0	Euclidean	$= 4\pi r^2$	$\frac{2}{3}(1/H_0)$	$= 5.7 \times 10^{-30}$	$= \frac{1}{2}$
II	> 0	Spherical	$< 4\pi r^2$	$< \frac{2}{3}(1/H_0)$	$> 5.7 \times 10^{-30}$	$> \frac{1}{2}$
III	< 0	Hyperbolic	$> 4\pi r^2$	$> \frac{2}{3}(1/H_0)$	$< 5.7 \times 10^{-30}$	$0 < q_0 < \frac{1}{2}$

tion is that as fainter and fainter objects are counted, their numbers increase more rapidly than should be expected if they had always been present in the same numbers. Clearly if things have changed, the universe is not in a steady state.

Probably the most difficult observation to reconcile with the steady-state model is the cosmic blackbody radiation, already discussed at some length. If future work confirms that that radiation is indeed blackbody in nature, and unless a mechanism for producing it can be found which is compatible with the steady-state model, its existence must be considered very strong evidence for an initial hot phase from which the universe has evolved.

Evolutionary Models

At the present time, therefore, the steady state seems the least likely of the types of models we are considering, and we now leave it and try to choose among the three evolutionary types.

The three evolutionary types can be differentiated by knowledge of k, q_0, or ρ. If any one of these three parameters is known, the others can be computed from equations derived from general relativity, and so if any one can be determined, two types of models can be eliminated from consideration.

Measurement of q_0 In principle a direct measurement of q_0 should be possible from the Hubble plot of recessional speed versus distance for galaxies. If the expansion of the universe is slowing, galaxies seen at very great distances and hence very great times in the past should show the higher speed appropriate for that earlier time, and the Hubble plot should not be a straight line. In 1972 Allan Sandage published such a plot using the brightest galaxy in each of 84 clusters of galaxies; it indicated that q_0 is about $+ 1$. Since that time, however, it has been recognized that several factors can affect that result, and each of them has the effect of lowering the value of q_0. Best estimates of q_0 now place it below $\frac{1}{2}$.

The most important factor which changes the measurement of q_0 is the aging of galaxies. Distances are estimated by combining the observed apparent magnitudes of the galaxies with the assumed common absolute magnitude. All the galaxies used in this type study are giant ellipticals, and that type galaxy seems to have ended star formation long ago, unlike spirals which are still forming stars. If all the galaxies to be studied formed their stars at about the same time, the most distant members of the group are seen as they appeared when their stars were young and closer members are seen as they appeared after their stars had aged. Consequently, the assumption that all had the same absolute magnitude when the light we see left them is invalid, and when reasonable estimates of the effect that aging should have are placed into the calculation, q_0 comes out to be very small or even slightly negative. Apparently at the present time no confidence can be placed in direct measurements of q_0.

Measurement of k The measurement of k, the index giving the sign of curvature of three-dimensional space, is the most difficult and uncertain of the three. The only observation which is likely to give information directly about k is one in which sources are counted at great distances and an attempt is made to estimate whether the number seen increases at a rate proportional to the square of the distance, or more or less rapidly than the square of the distance. In principle such a measurement could be made by observing galaxies at great distances with the 200-in telescope, but the effect expected is much too small for useful results to be obtained by that method. If we knew enough about quasars to judge their intrinsic luminosities, they might be used as indicators of the areas of spheres of different sizes, but we know almost nothing about the intrinsic luminosity of those objects.

In the five years ending in 1970 a radio telescope group at Ohio State University surveyed the sources visible at 21.2 cm over a large belt of the sky, cataloging 8,100 radio sources. If all the sources are equally bright intrinsically, the number at great distances falls off more rapidly than it should if space were flat; the observations seem to imply that three-dimensional space has positive curvature and is closed. On the other hand, a survey carried out at the National Radio Astronomy Observatory at a wavelength of 6 cm and reported in 1971 found no evidence that the number of sources detected depends on their age. The only reasonable conclusion seems to be that because of the dubious validity of an assumption that the sources are all equally bright, such observations cannot be convincing about anything.

Measurement of ρ Attempts to choose a model of the universe by measurements of either q_0 or k seem almost hopeless, but when we try to measure the density of matter in the universe, several different methods give consistent results.

We begin by estimating the mass bound into galaxies. The

masses of a few galaxies can be computed either because their rotation is observed so that Kepler's laws can be applied to stars at the periphery or because they are members of binary pairs, rotating about one another. For such galaxies we then calculate a mass-luminosity ratio. That ratio would be 1 if the galaxies were composed of stars like the sun, but for actual galaxies it is more nearly 10. That implies that much of the mass of a galaxy is in stars which are less luminous than the sun. We then measure the luminosity of the galaxies we can see, multiply the luminosity by the mass-luminosity ratio, and compute the masses of the galaxies within view. From this computation we find that if all the mass in galaxies were spread uniformly throughout space, the density would be approximately 3×10^{-32} g/cm^3, about 0.005 the critical density required to close the universe.

We can estimate the masses of galaxies in clusters by use of the virial theorem from mechanics. If the galaxies in a cluster are gravitationally bound and if they have interacted long enough that their energies have been distributed, the theorem can be used to determine the average mass of the members of the cluster. Such calculations have been made for many clusters, and the results always give masses higher than we would have computed from the luminosities of individual galaxies. If we accept the masses estimated from the virial theorem, we compute mass-luminosity ratios of 100 or more. That leads to a computed density of matter in space about 0.05 the critical density. The discrepancy between the masses of galaxies computed from their luminosities and from the virial theorem probably indicates that the clusters still contain gas which lies between the individual galaxies. If mass is present outside the galaxies in the form of black holes, its effect is also included in the virial theorem computation.

The density of matter can be estimated by an indirect method, totally independent of the direct observations just described. We suppose that the universe began in a fireball stage and that the temperature of the relic radiation from that time is now 2.7 K. In the initial seconds of the expansion, while the temperature was very high, helium, deuterium, and small amounts of other light elements were formed. The amounts of those elements formed and persisting to the present depend upon the density of matter at the time of their formation, but because we know the temperatures at which they formed and the temperature now, we know how much expansion has occurred; hence we can compute the density of matter now from analysis of the initial conditions.

Measurements of the amount of helium present support the belief that the expansion began with a fireball, for more helium is seen than would have been produced by stars, but the quantity of helium does not depend strongly on the density of matter. The most sensitive indicator of the density of matter is deuterium. Deuterium is hard to make and easy to destroy. The big bang would have produced it, but it is destroyed in stars, where it is converted into helium, and it would have been destroyed immediately after it was produced if the density of matter had been sufficiently high. If we find large amounts of deuterium in space, therefore, we can set a limit on the density of matter which can have been present when it formed and hence on the density of matter now.

Substantial amounts of deuterium are seen, on the earth, on Jupiter, in the sun, and in interstellar clouds. At least most of it is presumed to have been formed in the big bang, and the quantity we see indicates that the density of matter in the universe now must not be more than 6×10^{-31} g/cm^3, about 0.1 the critical density.

That number is consistent with the computation from the virial theorem. The deuterium calculation gives an upper limit; the mass must be equal to or less than 0.1 the critical density. The virial-theorem calculation gives the mass in clusters of galaxies but neglects mass which may lie between clusters. Arguments can be given that that mass must be low, but its possible presence plus uncertainties in the calculations can easily explain the difference of a factor of 2 in the numbers given.

Summary Until 1973 or perhaps 1974 one could say that the open and closed models of the universe had about equal observational support. Arguments which appeared good could be made for each. That situation has changed now. The strongest argument for the closed model, the direct measurement of q_0, has been shown to be unreliable; in fact, when proper corrections are applied it indicates that the universe is open. And the measurements of the density of matter all agree that the material in the universe is too small by perhaps a factor of 10 to close it. A statement that the decision is final, that the universe will continue expanding forever, would be premature. Enough uncertainties remain that we must admit that further observations and computations may present a different result, but as this is being written, in early 1975, the evidence for the open model seems overwhelming.

The stars of the globular clusters appear to be more than 11 billion and less than 18 billion years old. (The uncertainty in their ages arises because of uncertainty about their helium content.) Those limits on their ages provide an additional constraint on the model of the universe and narrow the range of possible values of H_0 and q_0. A model which satisfies all known conditions has H_0 between 55 and 60 km/(s)(Mpc), q_0 between 0.025 and 0.05, and the density of matter between 0.05 and 0.01 the critical density.

411
COSMOLOGY—A
SECOND LOOK

SCENARIO OF THE EARLY UNIVERSE

If we accept what is coming to be called the standard model of the universe, we can give a description of the development of events from the start down to the present. Many details remain to be filled in, but in broad outline things must have gone somewhat in this way.

Beginning

When the cosmic clock began ticking, the temperature of the universe was somewhat above 10^{10} K, but within a few seconds the temperature had dropped to 10^{10} K. What existed before that initial moment we cannot say; the equations we use to deduce the early state predict a singularity at the starting point, and they cannot be extrapolated further back. During the first second, photons, neutrinos, electrons, positrons, and other particles must have been in existence, in equilibrium with one another. Very quickly, however, as the temperature began to fall, the positrons combined with electrons and disappeared and any other antiparticles also combined and disappeared from the scene. This assumes, of course, that the universe is not charge-symmetric. When the temperature was 10^{10} K, the density was of the order of 10^5 g/cm^3.

During the time from about 2 s until 1,000 s, as the temperature fell from 10^{10} K to about 10^9 K, helium formed. Most of it was helium 4, but traces of helium 3 formed also. Almost no heavier elements formed during that period; everything heavier has formed since that time in stars.

Effects of Coupling of Matter and Radiation

For about the next 100,000 years nothing exciting happened. The universe continued to expand, the density dropped from about 10^{-1} to 10^{-20} g/cm^3, and radiation and matter remained intimately coupled together. If we neglect the neutrinos, which may have been present but had no influence on the other more visible parts of the universe, the universe during those 100,000 years can be said to have comprised two parts—matter, in the form of hydrogen and helium gas, and radiation. If the two had been independent of one another, the expansion would have cooled the gas faster than the radiation so that the temperatures of the two components would have been different. During that initial 10^5 years, however, many of the photons of the radiation had enough energy to ionize the hydrogen and helium, so that in reality the matter was in the form of a plasma (an ionized gas) consisting to a great extent of positive ions and electrons. But the photons interact strongly with electrons, so that as the matter tried to cool, energy was fed to it from the radiation with the result that both cooled together, remaining at the same temperature. The coupling between matter and radiation could have had another effect. If the gas began to form large clumps, such as might have been the ancestors of galaxies, it would have trapped radiation inside. Because of the general expansion of the universe which was proceeding, however, the radiation pressure inside such a condensing region of matter would have became greater than the pressure outside, and the effect would have been to push the matter apart again. In short, such condensations cannot have formed or at least cannot have become very important during the era while radiation and matter were coupled. Thus at about 100,000 years the matter must have been rather uniformly spread throughout the universe.

Decoupling

That time, 100,000 years, is particularly important because at about that point in history the photons of the radiation began to have too little energy to ionize hydrogen, and recombination of the protons and electrons to form neutral atoms became common, so that the gas changed from a plasma with many free electrons to a neutral gas of hydrogen and helium. The temperature must have been 3000 to 4000 K.

From that time matter and radiation have followed their separate courses, cooling at different rates. The cosmic blackbody radiation we see now originated at that time of decoupling, and it brings us information of the state of affairs at that time, some 100,000 years after the big bang.

Galaxy Formation

Perhaps by the time 1 billion years had passed galaxies had begun to form, and from that point on the story is one of successive generations of stars forming, gradually enriching the interstellar medium with heavy elements produced in early-generation stars. And now, some 12 billion years into the expansion, creatures have appeared on earth capable of reconstructing (or inventing!) the history of the universe.

Conclusion

Surely the contemplation of such a story of the universe is the ultimate mind-expanding experience. As we contemplate the sweep of history during our lifetimes, then during the life of the nation, during recorded human history, throughout the life of the earth as deduced from geological evidence, and finally during the life of the universe, gradually increasing the field of view from a few years to hundreds, thousands, millions, and finally billions of years, we must be exhilarated and excited by the experience. One's perspective on life and on the earth must be permanently affected by this vision. A first reaction may be a downgrading of human concerns to a position of total insignificance. For billions of years before the earth formed, stars were born, aged, and died. From the debris of stars which had gone through a life cycle came the dust from which the earth and the human body are made. After the sun has made life on the earth impossible and erased from the surface of the earth every trace that human beings may have left, so that the universe contains no memory of a creature we call *homo sapiens,* stars will be coming into existence and the expansion of the universe will be continuing as if we had never been here. Whether viewed on the scale of size, energy release, or life expectancy, compared to stars and galaxies we seem insignificant. On the other hand, we have something that those stars and galaxies do not have, and, so far as we know, we may be the only part of the universe with this possession: We have self-consciousness and the power to know about the rest of the universe. Stars live with no thought or consciousness; galactic arms form and dissolve without any will; the pirouettes of the galaxies in their cosmic dance reveal no planned choreography. But we know ourselves and our actions, and we can plan and reflect upon our motions. We know about the stars; they do not know about us. Because of this qualitative difference between us and the inanimate universe, we have open before us the possibility of different value systems, systems which place emphasis upon our humanness rather than upon size or longevity. We are both inferior and superior to the stars and galaxies, but above all else, we are different.

QUESTIONS

1 Imagine types of motion other than uniform expansion or contraction (for example, one part rotating relative to the remainder, expansion at greater speed into one hemisphere than into the other, etc.) and convince yourself that any motion other than those two would cause some observer somewhere to see anisotropy as he looks about him.

2 Verify that the critical density, that required for the Einstein–de Sitter universe, is 5.7×10^{-30} g/cm^3 (= 5.7×10^{-27} kg/m^3).

3 Explain why the density of matter in space must be greater than the critical value if the expansion of the universe is to stop and reverse.

4 Suppose that a galaxy is 100 million light years from us now. How will its distance from us change over the next 50 billion years according to each of the four models?

5 Why do we think that galaxies did not begin to form before the decoupling of matter and radiation?

6 Why is helium rather than carbon studied for clues to the early history of the universe?

physical and astronomical constants

PHYSICAL CONSTANTS

Speed of light	$c = 2.998 \times 10^8$ m/s
Universal gravitational constant	$G = 6.668 \times 10^{-11}$ (N) (m²)/kg² $[\text{m}^3/(\text{kg}) \,(\text{s})^2]$
Planck's constant	$h = 6.624 \times 10^{-34}$ J-s
Boltzmann's constant	$k = 1.380 \times 10^{-23}$ J/K
Wien displacement constant	$C = 2.9 \times 10^6$ nm-K
Stefan-Boltzmann constant	$\sigma = 5.67 \times 10^{-8}$ J/(deg⁴) (m²) (s)
Mass of proton	$m_p = 1.673 \times 10^{-27}$ kg
Mass of electron	$m_e = 9.109 \times 10^{-31}$ kg
Mass of neutron	$m_n = 1.675 \times 10^{-27}$ kg

ASTRONOMICAL CONSTANTS

Astronomical unit	$AU = 1.5 \times 10^8$ km
Parsec	$pc = 206{,}264 \, AU = 3.262$ light years $= 3.1 \times 10^{13}$ km
Light year	Light year $= 9.5 \times 10^{12}$ km
Calendar (tropical) year	Tropical year $= 3.16 \times 10^7$ s
Mass of sun	$M_\odot = 2 \times 10^{30}$ kg
Radius of sun	$R_\odot = 7 \times 10^8$ m
Mass of earth	$M_E = 6 \times 10^{24}$ kg
Equatorial radius of earth	$R_E = 6{,}378.24$ km
Luminosity of sun	$L_\odot = 3.86 \times 10^{26}$ J/s
Mean distance center of earth to center of moon	$= 384{,}403$ km
Radius of moon	$R_M = 1{,}738$ km
Mass of moon	$M_M = 7.4 \times 10^{22}$ kg

2
nomenclature of astronomical objects

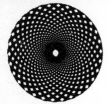

Some bright stars such as Castor and Pollux, have names given them by ancient Hellenic peoples, and many of the names, such as Aldebaran, Algol, and Deneb, are of Arabic origin. A few, such as Barnard's Star and Kapteyn's Star, are named for modern astronomers who discovered or studied them. However, most stars and other objects which can be said to have been "named" are known by their numbers in one or more catalogs. Of course most of the 100 billion stars in the Milky Way are not listed in any catalog, and in fact, many of the faint stars recorded on photographs are not listed. Of the making of catalogs there has been no end, and many special-purpose catalogs exist (catalogs of double stars, of variable stars, etc.), but the most common designations are from three or four particularly important collections.

When a star is designated by a Greek letter and the genitive of the name of the constellation (Alpha Centauri, Epsilon Eridani, Beta Persei) the name goes back to 1603, when J. Bayer, a German astronomer, assigned Greek letters to all the major stars in each of the constellations he depicted in his diagrams of the heavens. Usually, although not always, the brightest star in the constellation was called alpha, the next brightest beta, etc.

In the eighteenth century, John Flamsteed, the first Astronomer Royal of England, prepared another catalog, again using the names of the constellations. But he assigned numbers in the order of the time in which the star passed the north-south line passing directly above him. From his catalog we obtain names such as 61 Cygni and 40 Eridani.

Beginning in 1837, F. W. Argelander of the Bonn Observatory compiled a catalog of stars which eventually included about $\frac{1}{3}$ million stars; it was supplemented by a catalog made in Cordoba, Argentina, that included stars too far south to be observed from Europe. This catalog, the *Bonn Catalog*, is known by its German name, the *Bonner Durchmusterung*, and the stars in it are known by BD numbers, as for example, BD +43°44.

Another catalog which is widely used was prepared by the Harvard College Observatory in the early years of this century. It lists stars by position and gives their spectral types, and it is called the *Henry Draper Catalogue*. Stars listed in it are known by HD numbers, such as HD 120274.

The same star may have many names; for example, the brightest star in the constellation Boötes is Arcturus, Alpha Bootis, 16 Bootis, HD 124897, and BD +19°2777. This does not exhaust its list of names, however, for in other catalogs it is given other numbers; in the *Yale Catalogue of Bright Stars* the same star is BS 5340.

Objects which are not stars are listed in other catalogs. The first important catalog of nonstellar objects was prepared in 1781 by a French astronomer, Charles Messier. Messier was a comet hunter, and he was bothered by objects which might easily be mistaken for comets, and so he published a list of

slightly over 100 objects which appear as fuzzy objects in the telescope. They
include star clusters, nebulae, and galaxies, and some of the objects are still
commonly designated by their Messier numbers. For example, the Andromeda
galaxy is M 31; M 13 is a globular cluster; M 51 is the Whirlpool galaxy; M 57 is
the Ring nebula in Lyra; and M 42 is the Great Nebula in Orion.

The most extensive catalogs of nebulae are those based on the work begun
by William Herschel. Between 1786 and 1802 he published three catalogs with
a total of 2,500 nebulae. His son, John, expanded the number to 5,079 objects in
the *General Catalogue of Nebulae* which he published in 1864. In 1888 that
catalog was revised and expanded by J. L. E. Dreyer and called the *New General
Catalogue*; objects listed in it are known by their NGC numbers, such as
NGC 224, which is M 31, the Andromeda galaxy. Between 1888 and 1908 supplements to the *New General Catalogue* were issued under the name *Index Catalogue* (IC), and by 1908 almost 15,000 nebulae were listed in these two
catalogs, each with its own number.

Variable stars are named by a particular system, and because the names of
variables occur frequently, an explanation of the system may be interesting, although it will not appear simple.

If a star which is found to be variable already has a Greek letter assigned to
it, it keeps that name. Thus, the first Cepheid found was Delta Cephei, and it is
still known by that name. If the star does not have such a name already assigned, it is given a designation of Roman letters with the genitive of the name
of the constellation. The first unnamed variable in a constellation is designated
R, as R Andromedae; the next is S, then T, and on to Z. The next found in that
constellation is RR, as RR Lyrae; then RS, RT, . . . , RZ. After that come SS, ST,
and so on to ZZ. If still more variables are found in that constellation, double letters beginning with AA, AB, . . . , AZ, BB, BC, etc., are used with J always omitted. That system of single and double Roman letters can designate 334 stars,
and if still more variables are found in the constellation, the others are numbered, with a V preceding the number, beginning with V 335. For example,
V 335 Cygni is the 335th variable discovered in the constellation Cygnus.

3
the constellations

Latin name	Genitive	Abbreviation	Translation
Andromeda	Andromedae	And	Andromeda*
Antlia	Antliae	Ant	Pump
Apus	Apodis	Aps	Bird of Paradise
Aquarius	Aquarii	Aqr	Water Bearer
Aquila	Aquilae	Aql	Eagle
Ara	Arae	Ara	Altar
Aries	Arietis	Ari	Ram
Auriga	Aurigae	Aur	Charioteer
Boötes	Bootis	Boo	Herdsman
Caelum	Caeli	Cae	Chisel
Camelopardalis	Camelopardalis	Cam	Giraffe
Cancer	Cancri	Cnc	Crab
Canes Venatici	Canum Venaticorum	CVn	Hunting Dogs
Canis Major	Canis Majoris	CMa	Big Dog
Canis Minor	Canis Minoris	CMi	Little Dog
Capricornus	Capricorni	Cap	Goat
Carina	Carinae	Car	Ship's Keel
Cassiopeia	Cassiopeiae	Cas	Cassiopeia*
Centaurus	Centauri	Cen	Centaur*
Cepheus	Cephei	Cep	Cepheus*
Cetus	Ceti	Cet	Whale
Chamaeleon	Chamaeleonis	Cha	Chameleon
Circinus	Circini	Cir	Compass
Columba	Columbae	Col	Dove
Coma Berenices	Comae Berenices	Com	Berenice's Hair*
Corona Austrina	Coronae Austrinae	CrA	Southern Crown
Corona Borealis	Coronae Borlelais	CrB	Northern Crown
Corvus	Corvi	Crv	Crow
Crater	Crateris	Crt	Cup
Crux	Crucis	Cru	Southern Cross
Delphinus	Delphini	Del	Dolphin
Dorado	Doradus	Dor	Swordfish
Draco	Draconis	Dra	Dragon
Equuleus	Equulei	Equ	Little Horse
Eridanus	Eridani	Eri	River Eridanus
Fornax	Fornacis	For	Furnace
Gemini	Geminorum	Gem	Twins
Grus	Gruis	Gru	Crane
Hercules	Herculis	Her	Hercules*
Horologium	Horologii	Hor	Clock

418   *Proper names of mythological persons or animals.

Latin name	Genitive	Abbreviation	Translation
Hydra	Hydrae	Hya	Hydra*
Hydrus	Hydri	Hyi	Water Snake
Indus	Indi	Ind	Indian
Lacerta	Lacertae	Lac	Lizard
Leo	Leonis	Leo	Lion
Leo Minor	Leonis Minoris	LMi	Little Lion
Lepus	Leporis	Lep	Hare
Libra	Librae	Lib	Balance
Lupus	Lipi	Lup	Wolf
Lynx	Lyncis	Lyn	Lynx
Lyra	Lyrae	Lyr	Harp
Mensa	Mensae	Men	Table Mountain
Microscopium	Microscopii	Mic	Microscope
Monoceros	Monocerotis	Mon	Unicorn
Musca	Muscae	Mus	Fly
Norma	Normae	Nor	Level
Octans	Octantis	Oct	Octant
Ophiochus	Ophiochi	Oph	Snake Bearer
Orion	Orionis	Ori	Orion*
Pabo	Pavonis	Pab	Peacock
Pegasus	Pegasi	Peg	Pegasus*
Perseus	Persei	Per	Perseus*
Phoenix	Phoenicis	Phe	Phoenix
Pictor	Pictoris	Pic	Easel
Pisces	Piscium	Psc	Fish
Piscis Austrinus	Piscis Austrini	PsA	Southern Fish
Puppis	Puppis	Pup	Ship's Stern
Pyxis	Pyxidis	Pyx	Ship's Compass
Reticulum	Reticuli	Ret	Net
Sagitta	Sagittae	Sge	Arrow
Sagittarius	Sagittarii	Sgr	Archer
Scorpius	Scorpii	Sco	Scorpion
Sculptor	Sculptoris	Scl	Sculptor's Tools
Scutum	Scuti	Sct	Shield
Serpens	Serpentis	Ser	Snake
Sextans	Sextantis	Sex	Sextant
Taurus	Tauri	Tau	Bull
Telescopium	Telescopii	Tel	Telescope
Triangulum	Trianguli	Tri	Triangle
Triangulum Australe	Trianguli Australis	TrA	Southern Triangle
Tucana	Tucanae	Tuc	Toucan
Ursa Major	Ursae Majoris	UMa	Great Bear
Ursa Minor	Ursae Minoris	UMi	Little Bear
Vela	Velorum	Vel	Ship's Sails
Virgo	Virginis	Vir	Virgin
Volans	Volantis	Vol	Flying Fish
Vulpecula	Vulpeculae	Vul	Fox

*Proper names of mythological persons or animals.

4
twenty brightest stars

Common name	Bayer name	Right ascension, h min		Declination, deg min		Distance, pc	Visual magnitude*
Sirius	α CMa	6	43	−16	39	2.7	−1.42
Canopus	α Car	6	23	−52	40	30	−0.72
α Centauri		14	36	−60	38	1.3	−0.01
Arcturus	α Boo	14	13	+19	27	11	−0.06
Vega	α Lyr	18	35	+38	44	8	+0.04
Capella	α Aur	5	13	+45	57	14	+0.05
Rigel	β Ori	5	12	− 8	15	250	+0.14
Procyon	α CMi	7	37	+ 5	21	3.5	+0.38
Betelgeuse	α Ori	5	52	+ 7	24	150	+0.41
Achernar	α Eri	1	36	−57	29	20	+0.51
β Centauri		14	00	−60	08	90	+0.63
Altair	α Aql	19	48	+ 8	44	5.1	+0.77
α Crucis		12	24	−62	49	120	+1.39
Aldebaran	α Tau	4	33	+16	25	16	+0.86
Spica	α Vir	13	23	−10	54	80	+0.91
Antares	α Sco	16	26	−26	19	120	+0.92
Pollux	β Gem	7	42	+28	09	12	+1.16
Fomalhaut	α PsA	22	55	−29	53	7	+1.19
Deneb	α Cyg	20	40	+45	06	430	+1.26
β Crucis		12	45	−59	24	150	+1.28

*The magnitude at median brightness is given for variable stars.

†All secondary and tertiary members of multiple systems are much fainter than the primary star except for α Crucis.

Absolute magnitude	Spectral type	Remarks†
+1.4	A1 main seq	Has white dwarf companion
−3.1	F0 giant	
+4.4	G2 main seq	Triple system
−0.3	K2 giant	
+0.5	A0 main seq	
−0.7	G giant	Triple system
−6.8	B8 supergiant	Binary system
+2.7	F5 main seq	Has white dwarf companion
−5.5	M2 supergiant	Variable in magnitude
−1.0	B5 main seq	
−4.1	B1 giant	Binary system
+2.2	A7 main seq	
−4.0	B1 giant	Binary; secondary +1.9m_v
−0.2	K5 giant	Binary
−3.6	B1 main seq	Variable in magnitude
−4.5	M1 supergiant	Binary, variable
+0.8	K0 giant	
+2.0	A3 main seq	Binary
−6.9	A2 supergiant	
−4.6	B0 giant	Variable

5
stars within 5 parsecs

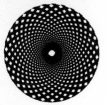

Name	Constellation	Right ascension, h min		Declination, deg min		Distance, pc
α Centauri	Cen	14	36	−60	38	1.31
Barnard's Star	Oph	17	55	+ 4	33	1.83
Wolf 359	Leo	10	54	+ 7	19	2.35
Lalande 21185	UMa	11	00	+36	18	2.49
Sirius	CMa	6	43	−16	39	2.67
Luyten 726-8	Cet	1	36	−18	13	2.67
Ross 154	Sgr	18	47	−23	54	2.94
Ross 248	And	23	39	+43	55	3.16
ε Eridani	Eri	3	31	− 9	38	3.30
Ross 128	Vir	11	45	+ 1	06	3.37
Luyten 789-6	Aqr	22	36	−15	36	3.37
61 Cygni	Cyg	21	05	+38	30	3.40
Procyon	CMi	7	37	+ 5	21	3.47
ε Indi	Ind	21	60	−57	00	3.51
Σ 2398	Dra	18	42	+59	33	3.60
BD +43°44	And	0	16	+43	44	3.60
τ Ceti	Cet	1	42	−16	12	3.64
CD −36°15693	PsA	23	03	−36	08	3.66
BD +5°1668	CMi	7	25	+ 5	23	3.76
CD −39°14192	Mic	21	14	−39	04	3.92
Kruger 60	Cep	22	26	+57	27	3.95
Kapteyn's Star	Pic	5	10	−45	00	3.98
Ross 614	Mon	6	27	− 2	46	4.03
BD −12°4523	Oph	16	28	−12	32	4.10
van Maanen's Star	Psc	0	46	+ 5	09	4.24
Wolf 424	Vir	12	31	+ 9	18	4.45
BD +50°1725	UMa	10	08	+49	42	4.51

*All are main-sequence stars except those indicated to be white dwarfs. An e after the type indicates that the spectrum contains emission lines.

Visual magnitude	Absolute magnitude	Spectral* type
− 0.01	+ 4.4	G2
+ 1.4	+ 5.8	K5
+10.7	+15	M5e
+ 9.5	+13.2	M5
+13.7	+16.8	M6e
+ 7.5	+10.5	M2
− 1.42	+ 1.4	A1
+ 8.7	+11.5	A5, white dwarf
+12.5	+15.4	M5e
+12.9	+15.8	M6e
+10.6	+13.3	M4e
+12.2	+14.7	M6e
+ 3.7	+ 6.1	K2
+11.1	+13.5	M5
+12.58	+14.9	M6e
+ 5.2	+ 7.5	K5
+ 6.0	+ 8.3	K7
+ 0.38	+ 2.7	F5
+10.7	+13.0	F, white dwarf
+ 4.73	+ 7.0	K5
+ 8.9	+11.1	M4
+ 9.69	+11.9	M5
+ 8.1	+10.3	M3e
+11.0	+13.2	M4e
+ 3.50	+ 5.7	G8
+ 7.39	+ 9.6	M2
+ 9.8	+11.9	M4
+ 6.7	+ 8.7	M0
+ 9.8	+11.8	M3
+11.4	+13.4	M5e
+ 8.8	+10.8	M0
+11.1	+13.1	M5e
+14.8	+16.8	?
+10.1	+12.0	M5
+12.4	+14.3	F3, white dwarf
+12.7	+14.4	M6e
+12.7	+14.4	M6e
+ 6.6	+ 8.3	M0

Name	Constellation	Right ascension, h min		Declination, deg min		Distance, pc
CD −37°15492	Scl	0	02	−37	36	4.57
CD −46°11540	Ara	17	25	−46	51	4.70
BD +20°2465	Leo	10	17	+20	07	4.72
CD −44°11909	Sco	17	34	−44	16	4.78
CD −49°13515	Ind	21	30	−49	13	4.78
A0e 17415-6	Dra	17	37	+68	23	4.84
Ross 780	Aqr	22.50		−14	31	4.84
Lalande 25372	Boo	13	43	+15	10	4.87
CC 658	Mus	11	43	−64	33	4.90
40 Eridani	Eri	4	13	− 7	44	5.00

*All are main-sequence stars except those indicated to be white dwarfs. An e after the type indicates that the spectrum contains emission lines.

Visual magnitude	Absolute magnitude	Spectral* type
+ 8.6	+10.3	M3
+ 9.34	+11.3	M4
+ 9.5	+11.1	M4e
+11.2	+12.8	M5
+ 9	+11	M3
+ 9.1	+10.7	M3
+10.2	+11.8	M5
+ 8.6	+10.2	M2
+11	+12.5	A, white dwarf
+ 4.5	+ 6.0	K0
+ 9.2	+10.7	A, white dwarf
+11.0	+12.5	M5e

bibliography

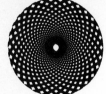

General

This bibliography is not intended to be comprehensive; in particular, it will not list any of the many excellent astronomy texts which are on approximately the same level. References to the literature are primarily to journals which students taking a course at the level of this text are likely to be able to read. Thus *Scientific American* and *Physics Today* are heavily represented; *Science* and *Nature* are cited less often; and the *Astrophysical Journal* is cited only when it seems necessary.

For units which are adequately covered in standard textbooks, no references are given. The references are chosen to help the reader who wants to read further on a topic which is not well treated in most textbooks, and they are intended to give him or her a starting point rather than to list all the literature which may be found by a short search.

The following three texts have more detailed and complete discussions of many of the topics treated in Secs. I and II. The first two assume that the reader knows little mathematics; the third uses somewhat more mathematics.

Abell, George, "Exploration of the Universe," 2d ed., Holt, Rinehart and Winston, New York, 1969.

Wyatt, Stanley P., "Principles of Astronomy," 2d ed., Allyn and Bacon, Inc., Boston, 1971.

Smith, Elske, and Kenneth Jacobs, "Introductory Astronomy and Astrophysics," W. B. Saunders Co., Philadelphia, 1973.

Unit 1

Probably the best source of information about Greek science are the books by George Sarton, particularly

Sarton, George, "A History of Science," Harvard University Press, Cambridge, Mass., 1959.

A description of the Ptolemaic system and of the Copernican revolution is given in

Holton, Gerald, and Duane Roller, "Foundations of Modern Physical Science," Addison-Wesley Publishing Company, Inc., Reading, 1958.

The following provide information about the men whose names are associated with the Copernican revolution:

Armitage, A., "Sun, Stand Thou Still," Henry Schuman, New York, 1947. (A biography of Copernicus.)

Koestler, A., "The Sleepwalkers," The Macmillan Company, New York, 1959. (Includes biographies of Copernicus, Kepler, and Galileo.)

———, "The Watershed," Doubleday & Co., Inc., New York, 1960. (The biography of Kepler taken from "The Sleepwalkers.")

Christianson, J., The Celestial Palace of Tycho Brahe, *Scientific American*, vol. 204, p. 118, February, 1961.

Gingrich, O., Copernicus and Tycho, *Scientific American*, vol. 229, p. 86, December, 1973.

Wilson, C., How Did Kepler Discover His First Two Laws? *Scientific American*, vol. 226, p. 92, March, 1972.

Unit 2

Most astronomy texts, including the three listed at the beginning of this bibliography, include discussions of the motions of the earth and of the appearance of the celestial sphere. For the person who wishes to learn constellations or to view stars and other celestial objects either with the naked eye or with a small telescope, the best aid probably is

"Season Star Charts," Hubbard Press, Northbrook, Ill., 1972.

This is a set of large charts with detailed descriptions of objects which may be viewed by naked eye or with binoculars or a small telescope. A volume with similar coverage that has been a standard reference for many years is

Norton, A. P., "Norton's Star Atlas," Sky Publishing Company, Cambridge, Mass., 1969.

Unit 3

Many of the primary reports on recent research on the moon and on the planets carried out by manned landings or by unmanned space vehicles have appeared in *Science*. Reports on the Apollo program on the moon are found in the issues of *Science* for March 30, 1973; August 17, 1973; and November 16, 1973.

Three other articles analyzing the results of this research are the following:

Anderson, D. L., The Interior of the Moon, *Physics Today*, vol. 27, p. 44, March, 1974.

Eglinton, G., J. R. Maxwell, and C. T. Pillinger, The Carbon Chemistry of the Moon, *Scientific American*, vol. 227, p. 80, October, 1972.

Mason, Brian, The Lunar Rocks, *Scientific American*, vol. 225, p. 48, October, 1971.

Reports on the Mariner 10 examination of Mercury are found in *Science* for July 12, 1974, and April 26, 1974.

Reports on the Mariner 10 encounter with Venus are in *Science* for March 29, 1974.

Some of the latest information available about Mars is in *Science* for January 18, 1974. A longer discussion of Mars is

Murray, Bruce, Mars From Mariner 9, *Scientific American*, vol. 228, p. 48, January, 1973.

A discussion of Jupiter is

Ingersoll, A., Jupiter's Great Red Spot: A Free Atmospheric Vortex? *Science*, vol. 182, p. 1346, December 28, 1973.

The January 25, 1974, issue of *Science* contains 15 papers on the Pioneer 10 flyby of Jupiter which occurred in December, 1973.

Unit 7

Books and articles about the sun are so numerous that no listing will be attempted. The solar neutrino work, however, is not well covered in books, and enough references to reports on it will be given to provide a starting point for anyone wishing to read further.

Wick, Gerald, Neutrino Astronomy: Probing the Sun's Interior, *Science*, vol. 173, p. 1011, September 10, 1971.

Lubkin, G. B., The Case of the Missing Solar Neutrinos, *Physics Today*, vol. 25, p. 17, August, 1972.

Rood, R. T., A Mixed-up Sun and Solar Neutrinos, *Nature Physical Science*, vol. 240, p. 178, December 25, 1972.

Ezer, D., and A. G. W. Cameron, Effects of Sudden Mixing in the Solar Core on Solar Neutrinos and Ice Ages, *Nature Physical Science*, vol. 240, p. 180, December 25, 1972.

Unit 14

Two paperback books that give more information about galaxies are

Shapley, Harlow, "Galaxies," Atheneum, New York, 1967.

Hodge, Paul W., "Galaxies and Cosmology," McGraw-Hill Book Co., New York, 1966.

Recent research on galaxies is described in

Rees, Martin J., and Joseph Silk, The Origin of Galaxies, *Scientific American*, vol. 222, p. 26, June, 1970.

Toomre, Alar, and Juri Toomre, Violent Tides Between Galaxies, *Scientific American*, vol. 229, p. 38, December, 1973.

Rubin, Vera, The Dynamics of the Andromeda Nebula, *Scientific American*, vol. 228, p. 30, June, 1973.

Unit 15

The two books mentioned within the text are

Dole, S. H., Habitable Planets for Man, 2d ed., American Elsevier Publishing Company, Inc., New York, 1970.

Sagan, Carl, and I. S. Shklovsky, "Intelligent Life in the Universe," Dell Publishing Co., Inc., New York, 1968.

Unit 16

The following books provide essential information for a study of cosmology; the book by Peebles assumes more knowledge of mathematics, physics, and astronomy than the other three.

Bondi, H., "Cosmology," Cambridge University Press, Cambridge, 1968.

Kaufmann, W. J. III, "Relativity and Cosmology," Harper and Row, New York, 1973.

Peebles, P. J. E., "Physical Cosmology," Princeton University Press, Princeton, N. J., 1971.

Sciama, D. W., "Modern Cosmology," Cambridge University Press, Cambridge, 1971.

The paper on Olbers's paradox is

Harrison, E. R., Why the Sky is Dark at Night, *Physics Today*, vol. 27, p. 30, February, 1974.

Unit 17

QSOs are discussed in

Kaufmann, W. J. III, "Relativity and Cosmology," Harper and Row, New York, 1973.

Sciama, D. W., "Modern Cosmology," Cambridge University Press, Cambridge, 1971.

Although it is somewhat out of date, the following book gives a good summary of information available at the time of its publication:

Burbidge, G., and M. Burbidge, "Quasi-stellar Objects," W. H. Freeman and Company, San Francisco, 1967.

The following journal articles, listed in order of date, will introduce the reader to current controversies in this field.

Schmidt, Maarten, and Francis Bello, The Evolution of Quasars, *Scientific American*, vol 224, p. 54, May, 1971.

Arp, Halton, Observational Paradoxes in Extragalactic Astronomy, *Science*, vol. 174, p. 1189, December 17, 1971.

Morrison, Philip, Resolving the Mystery of the Quasars? *Physics Today*, vol. 26, p. 23, March, 1973.

Metz, W. D., Quasars: Are They Near or Far, Young Galaxies or Not? *Science*, vol. 181, p. 1154, September 21, 1973.

Burbidge, G. R., Problem of the Redshifts, *Nature Physical Science*, vol. 246, p. 17, November 12, 1973.

Stockton, Alan, Galaxy Associated with QSO 4C37.43, *Nature Physical Science*, vol. 246, p. 25, November 12, 1973.

Bahcall, J., and L. Woltjer, Close Pairs of QSOs, *Nature*, vol. 247, p. 23, January 4, 1974.

A discussion of the redshift controversy is contained in

Hodge, Paul W., Some Current Studies of Galaxies, *Sky and Telescope*, vol. 44, p. 23, July, 1972.

Unit 18

A good discussion of the early work on pulsars is

Ostriker, Jeremiah, The Nature of Pulsars, *Scientific American*, vol. 224, p. 48, January, 1971.

Unit 19

Discussions of the cosmic blackbody radiation are included in

Peebles, P. J. E., "Physical Cosmology," Princeton University Press, Princeton, N. J., 1971.

Sciama, D. W., "Modern Cosmology," Cambridge University Press, Cambridge, 1971.

Two more recent journal references are

Collins, C., and S. Hawking, Why is the Universe Isotropic? *Astrophysical Journal*, vol. 180, p. 317, March 1, 1973.

Webster, A., The Cosmic Background Radiation, *Scientific American*, vol. 231, p. 26, August, 1974.

Unit 20

Perhaps a useful division of books which might serve as additional reading on the subjects of geometry and general relativity can be made on the basis of the mathematical preparation they expect of their readers. The first group are written for persons with little knowledge of advanced mathematics or physics.

Clotfelter, B., "Reference Systems and Inertia," Iowa State University Press, Ames, Iowa, 1970.

Kaufmann, W. J. III, "Relativity and Cosmology," Harper and Row, New York, 1973.

McVittie, G. C., "Fact and Theory in Cosmology," The Macmillan Company, New York, 1961.

A readable source of information about noneuclidean geometry is

Kline, Morris, "Mathematics in Western Culture," Oxford University Press, New York, 1953.

The notion of space-time is introduced along with a discussion of special relativity in

Taylor, E., and J. Wheeler, "Spacetime Physics," W. H. Freeman and Company, San Francisco, 1966.

More difficult books, written for readers with considerable knowledge of mathematics and physics, include

Lanczos, C., "Space through the Ages," Academic Press, London, 1970.

Zeldovich, Ya. B., and I. D. Novikov, "Relativistic Astrophysics," vol. 1, The University of Chicago Press, Chicago, 1971.

A book which includes "easy" reading and very difficult material, but which has the best discussions available now of tests of general relativity is

Misner, C., K. Thorne, and J. Wheeler, "Gravitation," W. H. Freeman and Company, San Francisco, 1973.

An article which makes moderate demands upon the reader is

Will, C. M., Gravitation Theory, *Scientific American*, vol. 231, p. 24, November, 1974.

Unit 21

A brief discussion of black holes will be found in the book already cited by Kaufmann. A more extended but relatively high-level discussion is in the book cited immediately above by Misner, Thorne, and Wheeler.

The following journal articles, arranged in order of date, provide good introductions to the ideas of black holes.

Thorne, Kip S., Gravitational Collapse and the Death of a Star, *Science*, vol. 150, p. 1671, December 24, 1965.

——, Gravitational Collapse, *Scientific American*, vol. 217, p. 88, November, 1967.

Ruffini, R., and J. Wheeler, Introducing the Black Hole, *Physics Today*, vol. 24, p. 30, January, 1971.

Hammond, A. L., Stellar Old Age: White Dwarfs, Neutron Stars, and Black Holes, *Science*, vol. 171, p. 994, March 12, 1971; p. 1133, March 19, 1971; p. 1228, March 26, 1971. (A report in three parts in successive issues.)

Penrose, Roger, Black Holes, *Scientific American*, vol. 226, p. 38, May, 1972.

Two brief reports on the arguments that Cygnus X-1 may be a black hole are given in

Black Hole and Blue Star, *Scientific American*, vol. 229, p. 48, November, 1973.

The Case of a Black Hole in Cygnus X-1, *Science News*, vol. 104, p. 341, December 1, 1973.

A more extensive article on the possibility that Cygnus X-1 contains a black hole is

Thorne, K. S., The Search for Black Holes, *Scientific American*, vol. 231, p. 32, December, 1974.

The suggestion that black holes may radiate is presented in

Hawking, S. W., Black Hole Explosions?, *Nature*, vol. 248, p. 30, March 1, 1974.

Unit 22

A useful starting point for one who wishes to read more about matter and antimatter is a little book by Alfvén, which includes a discussion of antiparticles.

Alfvén, H., "Worlds-Antiworlds," W. H. Freeman and Company, San Francisco, 1966.

More recent papers include

Steigman, G., Antimatter and Cosmology, *Nature*, vol. 224, p. 477, November 1, 1969.

Alfvén, H., Symmetric Cosmology, *Nature*, vol. 229, p. 1841, January 15, 1971.

Omnes, R., Antimatter in the Universe, *Nature*, vol. 230, p. 26, March 5, 1971.

Stecker, F. W., Diffuse Cosmic Gamma Rays: Present Status of Theory and Observations, *Nature Physical Science*, vol. 241, p. 74, January 22, 1973.

Unit 23

Reasonable summaries of the work on gravitational waves are found in

Weber, J., The Detection of Gravitational Waves, *Scientific American*, vol. 224, p. 22, May, 1971.

Logan, J., Gravitational Waves—A Progress Report, *Physics Today*, vol. 26, p. 44, March, 1973.

Sejnowski, T. J., Sources of Gravity Waves, *Physics Today*, vol. 27, p. 40, January, 1974.

Unit 24

Bondi, H., "Cosmology," Cambridge University Press, Cambridge, 1968.

Kaufmann, W. J. III, "Relativity and Cosmology," Harper and Row, New York, 1973.

Peebles, P. J. E., "Physical Cosmology," Princeton University Press, Princeton, N. J., 1971.

Sciama, D. W., "Modern Cosmology," Cambridge University Press, Cambridge, 1971.

Two excellent reviews of the state of cosmology today are

McCrea, W. H., Cosmology Today, *American Scientist*, vol. 58, p. 521, September-October, 1970.

Sandage, Allan, Cosmology: A Search for Two Numbers, *Physics Today*, vol. 23, p. 34, February, 1970.

Evidence for a closed universe is found in

Sandage, Allan, The Redshift-Distance Relation, II, The Hubble Diagram and Its Scatter for First-ranked Cluster Galaxies: A Formal Value for q_0, *The Astrophysical Journal*, vol. 178, p. 1, November 15, 1972.

Evidence for an open universe is found in

Gott, J. R. III, and J. E. Gunn, The Coma Cluster as an X-Ray Source: Some Cosmological Implications, *The Astrophysical Journal*, vol. 169, p. L 13, October 1, 1971.

The suggestion that we see an isotropic universe because only in an isotropic universe would conditions develop which permit life, and hence if the universe were not isotropic we would not be here to see it, is made in

Collins, C., and S. Hawking, Why Is the Universe Isotropic? *The Astrophysical Journal*, vol. 180, p. 317, March 1, 1973.

The possibility that measurements of the deuterium content of the universe may tell whether the universe is open or closed is discussed in

Pasachoff, J., and W. Fowler, Deuterium in the Universe, *Scientific American*, vol. 230, p. 108, May, 1974.

Schramm, D. N., and R. V. Wagoner, What Can Deuterium Tell Us? *Physics Today*, vol. 27, p. 40, December, 1974

The arguments for an open universe that are mentioned in this unit are presented in detail in

Gott, J. Richard, III, James E. Gunn, David N. Schramm, and Beatrice M. Tinsley, An Unbound Universe? *The Astrophysical Journal*, vol. 194, p. 543, December 15, 1974.

index

Aberration:
  chromatic, 108
  coma, 108
  spherical, 105, 108
Absolute magnitude, 186–187
Absorption spectrum, 124–128
Acceleration:
  centripetal, 86
  linear, 85
Adams, John, 70–71
Airy, Sir George, 70–71
Albireo, 42
Alcor, 41, 199
Alfvén, Hannes, 390, 391
Algol, 207–209
Almagest, 12
Alpha Centauri, 184
Alpha particle, 142
Amino acids in meteorites, 80, 292
Andromeda Galaxy, 42, 214, 219, 237, 262,
  263
Angstrom unit, 114
Angular momentum, 285–286, 288–289
Annihilation, 138, 325, 387, 388, 392
Annual parallax, 10, 18, 183–185, 198
Antimatter, 139, 387–393
Antineutrino, 139
Antineutron, 139, 388
Antiparticles, 138–139, 387–389
Antiproton, 138, 388
Aperture synthesis, 118
Aphelion, 24
Archimedes, 11
Arcturus, 42, 194
Aristarchus, 10–11, 17
Aristotelian physics, 19–20, 83
Aristotle, 10, 11, 19
Arp, Halton, 306, 322, 327
Associations, stellar, 253–254
Asteroids, 72
Astronomical unit, 25, 46, 49, 50
Atomic mass, 138
Atomic number, 137
Aurora, 153, 163
Autumnal equinox, 32

Baade, Walter, 237
Babylon, 1
Balmer series, 126–127
Barnard's star, 283
Barycenter, 96
Bessel, F., 183, 209
Beta particle, 140
Betelgeuse, 44, 194
Bethe, Hans, 144
Big-bang model, 307

Big dipper, 9, 41, 42, 198
Binary stars, 198–210, 218–219
  with black holes, 382–384
  eclipsing, 199, 205–209
  optical, 41, 199
  spectroscopic, 199, 202–205
  spectrum, 205
  visual, 41, 199–202
BL Lacertae, 329
Black hole, 6
  description, 378–379, 382
  formation, 259, 379–381
  search for, 381–384
Blackbody, 129–130
Blackbody radiation, 130–133, 176–177,
  343–352
Bode's law, 51–53
Bolometric magnitude, 179, 193
Bolyai, J., 371
Bondi, H., 313
Boötes, 42
Brackett series, 126
Brahe, Tycho, 15, 20, 21, 75, 87, 220
Brans, C., 370
Brans-Dicke theory, 370
Bright-line spectrum, 122–124
Burbidge, Geoffrey, 318, 322, 327
Burbidge, Margaret, 318, 322, 327

Calendar, 7–8
Cambridge catalogs, 275–276
Canis Major, 44
Carbon-dating, 141
Cassegrain focus, 107
Cassiopeia, 39, 41, 42, 224
Castor, 198, 199
Cavendish, H., 94
Cavity radiation, 130–133
Celestial equator, 38
Celestial poles, 37
Celestial sphere, 29, 37, 38
Cepheid variables, 41, 42, 211–216, 253,
  263, 304
  period-magnitude relation, 212–215
  Types I and II, 214
Cepheus, 42
Ceres, 52, 73
Chamberlain, T. C., 285
Chandrasekhar limit, 257–259, 377
China, 8
Chromosphere of sun, 156
Clark, Alvan, 209
Clouds of Megellan, 212, 267, 270, 274
Clusters, stellar: galactic (open), 226–228
  globular, 225–227, 253, 255–256, 264,
  274

433

CNO cycle, 144
Color index, 178, 190
Columbus, C., 12, 16
Comets, 75–78
Communication, interstellar,
   293–297
Co-moving observer, 401
Comte, Auguste, 147
Constellations, 40–45
Continuous creation, 314
Continuous spectrum, 128–133
Copernican revolution, 15–27
Copernicus, N., 15–20, 22, 58, 87
Corona of sun, 156–158
Coronagraph, 156, 163
Cosmic rays, 140, 340
Cosmogony, 300
Cosmological constant, 312
Cosmological models, 311–315, 401,
   404–411
Cosmological principle, 313
Cosmological redshift, 305
Coudé focus, 107
Crab Nebula, 220–221, 276, 332–334,
   340
Craters:
   on Mars, 64–65
   on Mercury, 60–61
   on moon, 54–56
   on satellites of Mars, 63
Critical density, 404, 408–411
Curvature:
   Gaussian, 354–359
   of line, 353–354
   of space, 6, 370–373
   of space-time, 361–364
Cygnus, 42, 224
Cygnus A, 275
Cygnus X-1, 383–384

Darwin, George, 57
Davis, Raymond, 167–170
Day:
   lengthening of, 37
   mean solar, 31
   sidereal, 30
   solar, 30
Deceleration parameter, 373, 402
Declination, 37–38
Decoupling of matter and radiation,
   351, 412
Deferent, 14, 17
Degeneracy:
   electron, 250–251, 257, 377
   neutron, 258–259, 336–337,
   377
Deimos, 63
Deneb, 42

Density of matter in universe, 373,
   408–410
Deuterium, 168, 409–411
Deuteron, 143
Dicke, Robert, 343, 363, 370
Differentiation, 57, 60
Diffraction, 115–119, 121, 317
Diffraction grating, 120
Dirac, Paul, 387
Distance:
   and apparent magnitude, 187–189
   to galaxies, 263–265
   to QSOs, 325–327
   within solar system, 46–51
   to stars, 51, 183–184, 191–192,
   213–214, 216
Dole, S. H., 283, 290, 291
Doppler effect, 133–135, 157, 234,
   288, 305
Drake, Frank, 293
Dust, interstellar, 228–232

Earth:
   circumference, 11
   density, 53
   mass, 93–94
   motions, 29
   tilt of axis, 31
Eccentric, 13, 14, 17
Eccentricity of ellipse, 23
Eclipses, 35–36, 156
Ecliptic, 29, 31, 32, 35, 58
Egypt, 1, 8
Eichhorn, Heinrich, 283
Einstein, Albert, 353, 362, 367, 370
Eisley, Loren, 296
Electromagnetic radiation, 82
Electromagnetic spectrum, 111,
   114–115
Electromagnetic waves, 102, 111, 113
Electron, 137, 138
Ellipse, 22–25, 76, 99
Emissivity, 131
Encounter hypothesis, 285–286
Energy, 97
   in galaxies, 276–281
   kinetic, 97
   mass, 100, 142
   potential, 98
   in QSOs, 323–325
   quantized, 123
   radiation, 100
Enke, comet, 76
Eötvös, R. V., 362
Epicycle, 13–15, 17
Equant, 13, 14, 17
Equator, 31, 32, 58
Equinox:
   autumnal, 32
   vernal, 32, 33, 40
Equinoxes, precession of, 33, 40
Eratosthenes, 11

Erg (unit), 98
Eros, 49, 73
Evening star, 61
Exclusion principle, 251
Expanding universe, 6, 304–308

Fabricus, David, 211
Faculae, 162
Field, electric and magnetic, 113
Filaments, 163–164
Fission, 139, 141–142
Flares (on sun), 163
Flash spectrum, 156
Focal point, 103, 105
Force, 87–91
   centripetal, 91–92
Forces, fundamental, 375–379
Fraunhofer lines, 156, 157
Free-free transitions, 349
Frequency, definition, 113
Friedmann models, 404
Fusion, 139, 142–145

Galaxies, 262–281
   active, 275–281
   clusters of, 272–273
   elliptical, 267–268
   evolution, 273–274
   exploding, 276
   irregular, 266
   local group, 265, 269–270
   Maffei, 269
   masses, 270–273
   N-type, 328
   and QSOs, 328–329
   radio, 275–276
   Seyfert, 279–280, 328
   spiral, 268
   types, 265
Galaxy, 224–240
   disk, 226
   dust in, 228–232
   evolution, 239–240
   gas in, 232–235
   halo, 226
   mass, 236
   nucleus, 237
   rotation, 236
   spiral arms, 235–236
Galileo, 15, 20, 26–28, 63, 67, 83
Galle, J. G., 71
Gamma radiation, 114, 119, 138,
   387, 392
Gamow, George, 343
Gas, interstellar, 232–235
Gatewood, George, 283
Gauss, Karl Friedrich, 371
Gauss (magnetic field unit), 159, 338
Geodesic, 365–367
Geometry, 353–373
Glitches, 334, 337–338

Globules, 246
Gold, Thomas, 313
Goodricke, John, 208
Granulation, 155–156
Gravitation, 92–97, 342
Gravitational collapse, 325, 378,
    380–381
Gravitational radiation, 394–400
    detectors, 395–399
    sources, 394
Gravitational radius, 380
Greek astronomy, 8–15
Greenstein, Jesse, 317
Guest stars, 8, 219, 220
Gulliver's Travels, 63
Gunn, James, 322, 327

H II regions, 233, 264
Half-life, 141
Halley, Edmund, 75, 195, 308
Halley's comet, 75–76, 78
Harrison, E. R., 310
Hawking, Stephen, 378, 385
Helium, 122, 128, 143–145, 243, 245,
    247, 409–411
    burning, 144–145, 248–250
    flash, 249
Hercules, 40, 226
Hermes, 74
Herschel, John, 198
Herschel, William, 69, 176, 183,
    198, 224
Hertz (unit), 114
Hertzsprung, Ejnar, 190
Hertzsprung-Russell diagram,
    190–191, 246
Hipparchus, 33, 173
Horizontal branch, 256
Hoyle, Fred, 287, 313
Hubble, Edwin, 214, 263, 273, 304
Hubble constant, 306
Hubble law, 304–308, 401–402
Hubble time, 308
Huggins, Sir William, 262
Hven, 20
Hyades, 228
Hydrogen:
    spectrum of, 123–127
    21-cm line, 233–234
Hyperbola, 76, 99

Image:
    real, 103
    virtual, 103
Inertia, 83, 87
Infrared, 114, 115, 237
Interference, 115–119, 321
Interferometer:
    radio, 118, 369
    stellar, 193–194
    stellar intensity, 195
Interstellar absorption, 225

Interstellar dust, 228–232
Interstellar gas, 232–235
Interstellar grains, 231–232
Interstellar lines, 233
Interstellar matter, 228–235
Interstellar molecules, 234–235,
    292–293, 347–349
Interstellar reddening, 229–230
Interval in space-time, 360–361, 365
Inverse-square law:
    for gravitation, 92–93
    for light, 185–186
Isotopes, 138
Isotropy:
    of primordial radiation, 349–351
    of universe, 353, 401, 402

Jansky, Karl G., 109
Jeans, Sir James, 285
Joule (unit), 98
Juno, 73
Jupiter, 14, 26, 67–78

Kant, Immanuel, 262, 263, 286, 287,
    371
Kapteyn's star, 195
Kepler, Johann, 15, 21, 22, 63, 87,
    220
Kepler's laws, 22–25, 82, 199–200
Kinetic energy, 97, 157, 276
Kirkwood gaps, 73
Klein, O., 390, 391
Kohoutek, comet, 78
Kuiper, G., 287

Lagrange points, 73–74
Laplace, P. S., 286–287
Laws of motion, Newton's, 87–92
Leavitt, Henrietta, 212
Leidenfrost phenomenon, 390–391
Leverrier, U. J., 70–71
Life:
    outside solar system, 282–297
    probability of, 291–297
    in solar system, 68, 282–283
Light:
    deflection by sun, 368–369
    diffraction, 115
    interference, 115
    scattering, 231–232
    waves, 111
Light curves:
    of binaries, 206–207
    of variables, 211–212
Light year, 171
Lippershey, H., 26
Little dipper, 41
Lobachevsky, N. I., 371
Local group. 269–270, 305
Lowell, Percival, 71, 72
Lyman series, 126–127
Lyot, Bernard, 156

Maffei galaxies, 269
Magellanic Clouds, 212, 267, 270,
    274
Magnetic field:
    in early solar system, 287
    in the galaxy, 231
    in pulsars, 338–339
    in the sun, 159–162
Magnitude of stars:
    absolute, 185–189
    apparent, 173–176
    bolometric, 179, 193
    and distance, 187–189
Main sequence stars, 191, 246–247,
    252
Mariner 9, 63–65
Mariner 10, 60–62
Mars, 22, 62–67
Mass, 89–91
    of earth, 93–94
    of planets, 95
    of stars, 95–97
    of sun, 94
Mass-energy, 100, 142
Mass-luminosity relation, 209–210
Matthews, T. A., 317
Mean solar day, 31
Mercury, 15, 58–61
    perihelion precession, 369–370
Mesopotamia, 8
Meteorites, 55, 57, 78–80
Meteoroids, 78–80
Meteors, 78–80
Metric coefficients, 364
MeV, 167–168
Michelson, Albert, 193
Micrometeorites, 79
Microwave region, 114
Milky Way, 26, 42, 109, 172, 214, 224,
    226, 262–263, 304
Minor Planet Ephemeris, 72
Minor planets, 49, 72–75
Mira, 193, 211, 217
Mirror, 104, 105
Mizar, 41, 198, 199
Models of the universe, 6, 401–413
    Einstein-de Sitter, 404
    hyperbolic, 405–406
    oscillating, 404–405
    steady-state, 406
    tests of, 406–411
Molecules, interstellar, 234–235,
    292–293, 347–349
Momentum, 389
Month:
    sidereal, 34
    synodic, 34
Montonari, G., 207

436

INDEX

Moon, 34–36, 50, 53–58
  age, 55–56
  origin, 57–58
Morning star, 61
Moulton, F. R., 285
Music of the spheres, 21

Nebulae:
  dark, 229
  emission, 232
  nature of, 262
  reflection, 229
Neptune, 70–71
Neutrino, 139, 167–170
Neutron, 137, 138
Neutron star, 335–338, 377–378
Newton, Issac, 20, 28, 84, 87
Newton (unit), 90
Newtonian focus, 107
Newton's laws of motion, 87–92
North star, 33, 41, 215
Novae, 217–219, 264
Nuclear reactions, 139–142
Nucleosynthesis, 259–260
Nutation, 33

Objective lens, 104
Olbers, Heinrich, 52, 308–309
Olbers's paradox, 308–311
Omnes, R., 390
Oort, J. H., 78
Orion, 43, 44, 196, 224
Orion arm, 235
Orion nebula, 44, 221
Osculating circle, 353, 354
Ozma, project, 293

Pallas, 52, 73
Parabola, 76, 99
Parabolic mirror, 106, 110
Parallax, 10, 18, 183–185
  spectroscopic, 191–192
Parmenides, 8
Parsec, 171, 184
Paschen series, 126
Peebles, P. J. E., 343
Pegasus, great square of, 42
Penrose, Roger, 385
Penrose's theorem, 385–386
Penumbra, 35, 159
Penzias, A. A., 343, 345
Perfect cosmological principle, 314
Perihelion, 24, 369–370
Period-luminosity relation, 212–215
Perseus arm, 235
Philolaos of Croton, 8
Phobos, 63
Photodisintegration, 258

Photographic emulsion, 176–177
Photons, 119
Photosphere, 127, 154–156
Piazzi, Giuseppi, 52, 72
Pickering, W. H., 71
Pioneer 10, 67, 69
Pisces, 39
Planck radiation law, 132, 155
Planck's constant, 119
Planetary nebulae, 222–223, 257–258
Planetoids, 49
Planets, habitable, 283–291
Pleiades, 40, 227–229, 254–255
Pluto, 71–72
Polaris, 33, 41, 215
Polarization:
  of light, 80, 221, 230–231, 280,
    329
  of starlight, 230–231
Pope, Alexander, 87
Populations I and II, 214, 227,
    237–239, 260–261
Positron, 138, 387
Potential energy, 98–100
Pound, R. V., 368
Precession of equinoxes, 33
Prime focus of telescope, 107
Principia, 87
Proper motion, 195
Proton, 137, 138
Proton-proton cycle, 143–144
Protoplanets, 287
Protostar, 245–246
Protosun, 287
Ptolemaic system, 12–15, 26
Ptolemy, Claudius, 12, 15
Pulsars, 330–342
Pythagoras of Samos, 8
Pythagoreans, 9

Quantization, 123
Quasars, 7, 316–329, 369

Radar, 50, 59, 62, 69
Radial velocity of stars, 196
Radiation, energy transfer by, 100
Radiation laws, 131–132
Radio doubling, 276
Radio waves, 163
Radioactive decay, 139–141
Radioactivity, 57
Raisin-bread model, 307
Reber, Grote, 109, 275
Rebka, G. A., 368
Red giant, 193, 209, 216, 219, 242,
    248, 255, 256, 379
Redshift:
  controversy, 326–328
  cosmological, 305
  from galaxies, 265
  gravitational, 367–368
  multiple, 320–321

Reflecting telescope, 104–109
Refraction, 102
Relativity:
  general, 7, 312, 353, 364–370
  special, 295, 319, 321–322, 353
  tests of, 367–370
Resolving power, 116–118
Retrograde motion of planets, 9, 14,
    19
Riccioli, J. B., 198
Riemann, B., 371
Rigel, 44
Right ascension, 37, 40
Rigollet, comet, 76
Roll, P. G., 343
Rosse, Lord, 262
RR Lyrae stars, 214, 216, 225,
    274
Russell, H. N., 190
Ryle, Sir Martin, 118

Sagan, Carl, 283
Sagittarius, 225, 226
Sagittarius A, 237
Sagittarius arm, 235
Sandage, Allan, 306, 318, 407
Sarton, George, 8, 12
Saturn, 26, 68–69
Scale factor (of cosmological models),
    402
Scattering of light, 229–232
Schiaparelli, Giovanni, 62
Schmidt, Maarten, 317, 328
Schmidt camera, 108
Schwarzschild solution, 364
Schwarzschild surface, 378–380
Scintillation of stars and radio sources,
    331
Semimajor axis, 24, 46
Seyfert, Carl, 279
Shapley, Harlow, 213, 225, 229, 273
Shklovsky, I. S., 283
Sidereal day, 30
Sidereal month, 34
Sidereal period, 24
Sidereal year, 33
Singularity, 381, 385
Sirius, 1, 44, 176, 209
61 Cygni, 183
Sizi, Francesco, 27
Skylab, 78
Snider, J. L., 368
Socrates, 8
Solar constant, 170
Solar system:
  definition, 46
  origin, 284–288
Solar wind, 77, 289
Solstice:
  summer, 11, 32
  winter, 32
Space-time, 359–364

Spectra, 119
  absorption (dark-line), 124–128
  bright-line, 122–124
  continuous, 128–133
Spectral lines, 122
Spectrograph, 120–122
Spectroscopic parallax, 191–192
Spica, 191
Spicules, 158
Stars, 37–45
  absolute magnitudes, 185–189
  classes, 179–182
  color-indices, 178
  colors, 176–179
  diameters, 192–195
  evolution, 242–261
  intrinsic luminosities, 172–173,
    185–189
  life expectancies, 242–244
  luminosities, 172–173, 185–189
  magnitudes, 173–176
  models, 244–245
  proper motions, 195
  radial velocities, 196
  rotation rates, 288–290
  space velocities, 196
  tangential velocities, 195
  temperatures, 176–179
Steady state theory, 313–315,
    406–407
Stefan-Boltzmann law, 132, 155,
    192–193
Stonehenge, 1
Struve, F., 183
Sun, 153–170
  age, 166
  chromosphere, 156
  corona, 156–158
  density, 153
  faculae, 162
  filaments, 163–164
  flares, 163
  granulation, 155
  interior, 165–170
  mass, 153
  models, 165–170
  motion, 196
  neutrinos, 167–170
  photosphere, 154–156
Sun:
  prominences, 163
  rotation rate, 153, 159
  spicules, 158
  temperature, 154
Sunspots, 158–162
  cycles, 161–162
  magnetic fields, 159–162
Supernovae, 217, 219–221, 258–259,
    333
Swift, Jonathan, 63
Symmetry, charge, 389–390
Synchrotron radiation:
  electromagnetic, 220–221, 280,
    329, 340
  gravitational, 398
Synodic month, 34

T Tauri stars, 221, 238, 254
Tangential (transverse) velocity, 195
Taurus, 228
Telescopes, 102–111
  radio, 109–111
  reflector, 104–107
  refractor, 102–104
  Schmidt, 108
Thorne, Kip, 384
Tidal friction, 37
Tidal locking, 34, 59, 62
Tides, 36–37
"Tired light" hypothesis, 305
Titius, J. D., 52
Titius-Bode law, 52
Tombaugh, Clyde, 71, 72
Travel, interstellar, 294–296
Triangulation, 46–49
Trigonometric parallax, 184–185
Triple-alpha process, 249
21-cm line, 233–234
Tychonian system, 21
Tyson, T., 399

Ultraviolet, 114, 115, 163, 318
Umbra, 35, 159
Uraniborg, 20
Uranus, 62, 69–70
Ursa Major, 40, 41
Ursa Minor, 41

van de Kamp, Peter, 283
Variable stars, 211–223
  eruptive, 217–221
  long-period, 216–217
  pulsating, 211–217
  T Tauri, 221
Vectors, 85
Vega, 33, 183, 196
Vela pulsar, 332, 333
Velocity, 84–85
  of escape, 403
Velocity curve of binaries, 203
Venera spacecraft, 61, 62
Venus, 15, 27, 50, 61–62
Vernal equinox, 32, 33, 38–40
Vesta, 73
Virial theorem, 271–273, 409–410
Vogel, H., 208

Watt (unit), 98
Wave, definition, 112
Wavelength, definition, 113
Wavelength shift, 135
Weber, Joseph, 395–398
Weight, 90–91
White dwarf, 44, 172, 191, 209,
    250–251, 257, 335, 377
White hole, 6, 325, 385
Wien displacement law, 132, 155
Wilkinson, D. T., 343
Wilson, R. W., 343, 345
World line, 365
Wright, Thomas, 224, 262

X-ray sources, 341, 383–384
X-rays, 114, 115, 119, 163

Year, 29
  sidereal, 33
  tropical, 33

Zeeman effect, 159
Ziggurats, 1
Zone of avoidance, 229